Non-Linear Time Series

Kamil Feridun Turkman
Manuel González Scotto
Patrícia de Zea Bermudez

Non-Linear Time Series

Extreme Events and Integer Value Problems

Springer

Kamil Feridun Turkman
Departmento de Estatística e Investigação Operacional
Faculdade de Ciências
Universidade de Lisboa
Lisboa, Portugal

Manuel González Scotto
Departamento de Matemática
Universidade de Aveiro
Aveiro, Portugal

Patrícia de Zea Bermudez
Departmento de Estatística e Investigação Operacional
Faculdade de Ciências
Universidade de Lisboa
Lisboa, Portugal

ISBN 978-3-319-34871-1 ISBN 978-3-319-07028-5 (eBook)
DOI 10.1007/978-3-319-07028-5
Springer Cham Heidelberg New York Dordrecht London

Softcover reprint of the hardcover 1st edition 2014

Printed on acid-free paper

Springer is part of Springer Science+Business Media (www.springer.com)

Preface

Linear processes have been one of the most fundamental tools in modeling serially dependent data. These models and methods heavily depend on Gaussian processes and their properties and therefore second-order moments have played the central role in this toolbox. Linear Gaussian models do not allow for large fluctuations and the stationary distribution when exists has fairly fast decaying tails, thus are not particularly adequate for modeling high variability. We are more and more aware that data coming from many scientific fields as diverse as telecommunications, hydrology or economics show heavy-tailed phenomena which are not compatible with Gaussian assumption. Therefore, such serially dependent data need richer parametrization based on nonlinear relations; consequently there is need to specify and model adequately observations coming from the tails. For such models, inferential methods based on second-order properties are no longer adequate and likelihood-based methods frequently cannot be used, since analytical expressions of likelihood often are not available. Least squares method, which is equivalent to likelihood-based inference for Gaussian processes, looses its nice properties as models deviate from Gaussian structure. Quasi-likelihood methods such as estimating functions and composite likelihood methods seem to work for some cases, whereas recent advances in sequential Markov Chain Monte Carlo (MCMC) methods and particle filters, as well as likelihood free methods such as approximate Bayesian computation (ABC), are quite promising alternatives in dealing with inference for nonlinear models. The objective of this book is to give an overall view of all these problems, including the consequences of nonlinearity on tails, some nonlinear models and adequate inferential methods.

There are many excellent books on nonlinear time series, responding adequately to these issues. Our approach is to introduce diverse topics that appear in different sources under one title which can serve as a reference source, without entering into details. We believe that the book may be particularly useful to graduate students and other scientists who want to have a starting source for further detailed study of the subject.

Chapter 1 gives many examples of nonlinear time series and different sources of nonlinearity. Chapter 2 introduces the basic notions of nonlinearity, and a collection of nonlinear models. There are many ways a process can be nonlinear, and correspondingly there are many different classes of models to cover such diverse sources. We do not claim that the models we introduce are exhaustive, but we hope this selection covers most sources of nonlinearity. Chapter 3 is on the tail behavior of nonlinear processes. The objective of this chapter is to show how nonlinear relationships generate heavy tails and to quantify the effect of nonlinearity on tails. Much is known on the extremal properties of linear and nonlinear time series and the objective of this chapter is to give a quick reference to these results. Chapter 4 gives inferential methods available for nonlinear processes, including several tests for nonlinearity. We hope that this chapter will be particularly useful as an integrated introductory text on inferential techniques, including Bayesian and simulation-based methods. Finally, in Chap. 5 we give linear models based on thinning operators for integer-valued time series. Although these models are linear in structure, technically they are nonlinear models. Many would favor observation-driven or parameter-driven state space models for dealing with such data sets. However, recently there has been a lot of interest for linear models based on thinning operators and there is an accumulation of a rich and diverse information on this subject. To our knowledge, this chapter is the first attempt to put together in a coherent manner these advances spread out in many journals and other sources.

Several institutions provided support for this work. We would like to thank FCT (Fundação para a Ciência e a Tecnologia) of Portugal, which supported our work through the projects PEst-OE/MAT/UI0006/2011, PEst-C/MAT/UI4106/2011, PEst-OE/MAT/UI4106/2014, and PTDC/MAT/118335/2010. The third author was also partially supported by a FCT Sabbatical Grant (ref FRH/BSAB/1138/2011) from 01 April 2011 to 29 July 2011. We also thank CEAUL – Center of Statistics and its Applications of the University of Lisbon – and CIDMA – Center for Research & Development in Mathematics and Applications – for their role in this initiative. Finally, we would like to thank Profs. John McDermott and Maria Antónia Amaral Turkman and Dr. Sónia Gouveia for reading the manuscript carefully and suggesting many useful comments which improved the content.

Lisbon and Aveiro, Portugal
May 2014

K. F. Turkman
M. G. Scotto
P. de Zea Bermudez

Contents

List of Abbreviations

ABC	Approximate bayesian computation
ACF	Autocorrelation function
A-FIGARCH	Adaptative FIGARCH
APARCH	Asymmetric power ARCH
AR	Autoregressive
ARCH	Autoregressive conditionally heteroskedastic
ARMA	Autoregressive moving average
BL	Bilinear model
cdf	Cumulative distribution function
CINAR	Combined INAR
CLS	Conditional least squares
CMLE	Conditional maximum likelihood estimator
DSD	Discrete self-decomposable
DSINAR	Doubly stochastic integer-valued autoregressive
EAR	Exponential autoregressive
EGARCH	Exponential GARCH
EM	Expectation-maximization algorithm
EXPAR	Exponential autoregressive
FIAPARCH	Fractionally integrated asymmetric power ARCH
FIGARCH	Fractionally IGARCH
GARCH	Generalized ARCH
GGS	Griddy-Gibbs sampler
GINARS	Integer-valued autoregressive model with signed generalized power series thinning operator
GJR-GARCH	Special case of the APARCH model
GPD	Generalized Pareto distribution
i.i.d.	independent and identically distributed
IGARCH	Integrated GARCH
INAR	Integer-valued autoregressive
INARMA	Integer-valued ARMA

INARS	Integer-valued autoregressive based on the signed binomial thinning
INMA	Integer-valued moving average
INARMA	Integer-valued ARMA
INE	Instituto Nacional de Estatística
INGARCH	Integer-valued GARCH
LADE	Least absolute deviations estimator
LGARCH	Logarithmic GARCH
log-ARCH	Special case of the APARCH model
MA	Moving average
MAR	Markov switching autoregressive
MCMC	Markov chain Monte Carlo
MEF	Mean excess function
M-H	Metropolis-Hastings algorithm
ML	Maximum likelihood
NARCH	Special case of the APARCH model
PACF	Partial autocorrelation function
pdf	Probability density function
PINAR	Periodic integer-valued autoregressive
PMCMC	Particle MCMC
PoINAR	Poisson INAR
POT	Peaks over threshold
PSI20	Portuguese Stock Market Index 20
RCA	Random coefficient autoregressive
r.v's	Random variables
SETAR	Self-exciting threshold autoregressive
S-HY-APARCH	Seasonal hyperbolic APARCH
SMCMC	Sequential MCMC
STAR	Smooth transition autoregressive
TAR	Threshold autoregressive
TARCH	Special case of the APARCH model
US GNP	United States Gross National Product
WN	White noise
YW	Yule-Walker
ZTPINAR	Zero-truncated Poisson INAR
4NLGMACH	Fourth order nonlinear generalized moving average conditional heteroskedasticity

Chapter 1
Introduction

1.1 Why Do We Need Nonlinear Models?

The *Wold decomposition theorem* says that under fairly general conditions, a stationary time series has a unique linear causal representation

$$X_t = \sum_{j=0}^{\infty} \psi_j Z_{t-j},\ t \in \mathbb{Z}, \tag{1.1}$$

where $\sum_{j=0}^{\infty} \psi_j^2 < \infty$ and (Z_t) are uncorrelated random variables (r.v's). Expression (1.1) is a representation, but not a model for X_t, in the sense that we can only recover uniquely the moments of X_t up to the second-order, unless, (Z_t) is a Gaussian sequence. If we look for models for X_t, then a theorem by Nisio (1960) states that we should look for such models within the class of convergent Volterra series expansions

$$X_t = \sum_{p=1}^{\infty} \sum_{i_1=-\infty}^{\infty} \sum_{i_2=-\infty}^{\infty} \cdots \sum_{i_p=-\infty}^{\infty} g_{i_1 i_2 \cdots i_p} \prod_{v=1}^{p} Z_{t-i_v},\ t \in \mathbb{Z}, \tag{1.2}$$

where (Z_t) is an independent and identically distributed (i.i.d.) Gaussian sequence (although the assumption of Gaussianity is not essential) and $(g_{i1}), (g_{i1,i_2}), \ldots$ are such that (1.2) converge to a well defined random variable. It is clear that if we want to go beyond the second-order properties, then the class of linear models given in (1.1) with i.i.d. Gaussian input (Z_t) is a small fraction of possible models for stationary time series, corresponding to the first term of the infinite order expansion in (1.2). Finite-order Volterra series expansions are not particularly useful as a possible class of models, because the conditions of stationarity and invertibility are hard, if not impossible, to check. Therefore they have very limited use as models for time series, unless the input series (Z_t) is observable.

K.F. Turkman et al., *Non-Linear Time Series*, DOI 10.1007/978-3-319-07028-5_1,

From a prediction point of view, the projection theorem for Hilbert spaces tells us how to obtain the best linear predictor for X_{t+k} within the linear span of $(X_t, X_{t-1}, \ldots)$. However, when linear predictors are not sufficiently good, it is not straightforward to find, if possible at all, the best predictor within richer subspaces constructed over $(X_t, X_{t-1}, \ldots)$. Therefore, in order to improve upon the linear predictor, it is important to look for classes of nonlinear models which are sufficiently general, but at the same time are sufficiently flexible to work with.

1.2 Content of the Book

A time series can be nonlinear in many different ways. As a consequence, there are many classes of nonlinear models to explain such nonlinearities. On one hand, we have mathematically tractable piecewise linear models that switch from one model to another at random moments of time; on the other hand, we have the class of bilinear processes, which are dense within the class of Volterra expansions, but whose probabilistic and statistical properties are difficult to study. Within such a vast amount of literature on nonlinear processes, the choice of the material, as well as the level of mathematical language that appears in a book, will evidently depend on the personal choice. This book is designed for postgraduate students of mathematics or probability who are interested in having a general knowledge of nonlinear time series and some insight into the process of model building for nonlinear processes. Knowledge of linear time series would be highly useful. We tried to keep a balance between the probabilistic arguments and the practical statistical inference. However, for nonlinear processes, due to inherent difficulties in inferential methods, this balance is very difficult to achieve and overall the manuscript is tilted towards probabilistic arguments.

One of the main consequences of nonlinearity is heavy tails, in the sense that nonlinear processes tend to produce more extreme values, as compared to linear models, even with Gaussian inputs. Extreme value theory for stationary time series is well known, and in this book we place special emphasis on the tail behavior of certain classes of nonlinear processes.

Integer-valued linear models based on thinning operators are also included in this book as a separate chapter. One may find questionable to dedicate a chapter to such a class of linear models in a book on nonlinearity. However, in the strict sense, these models do not have the representation (1.1), hence are not linear processes. Generalized state space models are arguably much more flexible tools for modeling integer time series, particularly with recent advances in hierarchical modeling strategies and simulation-based inferential techniques. However, because there has been extensive accumulated literature on the subject, and because there is no manuscript which covers the material in a coordinated manner, it makes sense to reserve a chapter for this class of models.

In the rest of Chap. 1, real and simulated data sets are presented to highlight different aspects of nonlinear time series.

In Chap. 2, a brief introduction to the different sources of nonlinearity and consequently, different classes of nonlinear models are given. The choice of these classes is by no means exhaustive and to some extent represents the individual choice of the authors.

Chapter 3 provides an introduction to extremal properties of nonlinear processes. The main emphasis is on the connection between nonlinearity and heavy tails. In this chapter, we also show how nonlinear input-output relationships often magnify the error propagation and quantify the effect of this propagation on the tails of some nonlinear representations.

Chapter 4 focuses on some of the inferential methods available for nonlinear processes. Some readers may find this chapter disappointing on the grounds that it does not offer solid, one-fits-all inferential method in a manner that allows the reader to deal with all nonlinear models effectively. It is difficult to speak of satisfactory inferential methods when one cannot check the invertibility condition(s) or write down explicitly the likelihood for some of the nonlinear models at hand. Inferential methods based on pseudo-likelihoods, such as estimating functions and in particular, composite likelihood, seem to be possible solutions for certain classes of nonlinear models. Recent advances in sequential and particle filter Monte Carlo methods are opening new doors to successful bayesian inference for nonlinear time series. Thus, we give a brief introduction to these alternative inferential methods. Excellent detailed accounts of these approaches can be found in many other books. For example, simulation-based inferential methods for time series can be found in Prado and West (2010), whereas inference based on pseudo-likelihoods can be found in Heyde (1997).

Chapter 5 addresses the integer-valued time series models based on thinning operators. One may question the importance and relevance of these models by devoting an entire chapter to them, since parameter-driven, as well as observation-driven state space models for integer-valued time series are given in Chap. 4. The method of maximum likelihood seems to work very well for observation-driven state space models, but conditions for the existence of stationary solutions, except for some simple cases, seem to be very difficult to obtain. Parameter-driven generalized state space models with hierarchical specifications, such as generalized linear mixed models with a latent process in the linear predictor as initially suggested by Diggle et al. (1998) (see also Diggle and Ribeiro 2007), and the consequent simulation-based inference, is probably the best way to model integer time series. However, there has been increasing interest in ARMA type models based on thinning operators for integer time series and consequent publication of many articles in the field, due to their many desirable probabilistic properties. Therefore, a whole chapter is dedicated to this class of models with the objective of giving an integrated treatment of these models.

We do hope that this book will generate enough interest among the postgraduate students to learn more about nonlinearity and a course on nonlinear time series eventually finds its way into some of the MSc courses in Probability and Statistics.

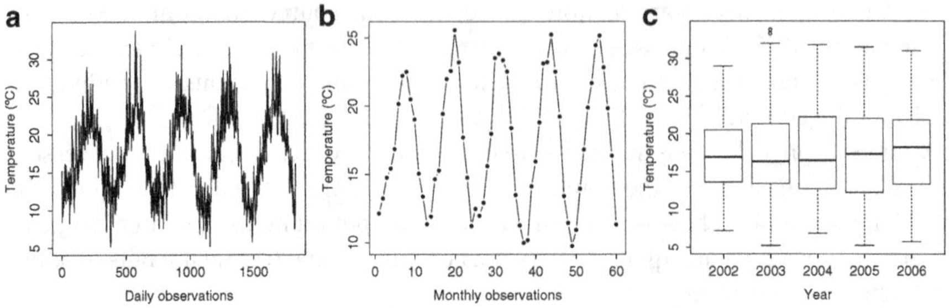

Fig. 1.1 (**a**) Mean daily temperatures observed in Lisbon, Portugal from 2002 to 2006; (**b**) mean monthly temperatures; and (**c**) box-plot of the mean daily observations per year (Data provided by IPMA, Portugal)

1.3 Some Examples of Time Series

The daily and monthly mean temperatures and the box-plot of the daily observations per year, in degrees Celcius, observed in Lisbon, Portugal from the 1st January 2002 to 31th December 2006, are presented in Fig. 1.1.

There seems to be no significant trend or increase in variability throughout the years. However, there is a clear seasonal component. A slight (median) trend in the annual observations also seems to be present, as indicated by the box-plots in Fig. 1.1c. The AutoCorrelation Function (ACF) and the Partial AutoCorrelation Function (PACF) of the monthly data, represented in Fig. 1.2, after differentiating the series at lag 12, resemble a white noise process. There is only a very minor (and possibly neglectable) spike in both ACF and PACF. No sign of nonlinearity can be detected in the ACFs of both squared and absolute values of the series. The fitting of a linear model seems to be the most adequate choice.

The histogram and the normal QQ-plot (plot of the coordinates $(\Phi^{-1}(p_{i:n}), x_{i:n})$, $i = 1, 2, \ldots, n$) strongly support the Gaussian assumption (see Fig. 1.3). This is confirmed by the Shapiro test ($p - value = 0.5205$) and also by the Lilliefors test ($p - value = 0.7258$). Moreover, there is no evidence to reject the null hypotheses of linearity upon applying Keenan's test ($p - value = 0.4433$).

Figure 1.4 shows the monthly average river Tagus flow, observed in Almourol, Portugal from October 1974 to September 2004, together with the plots of the respective ACF and PACF. According to Macedo (2006), Almourol and Tramagal stations are very important for predicting the flow and the hydrometrical levels in the event of a flood. The plot of the log transformed series, as well the corresponding ACF and PACF, are presented in Fig. 1.5. The histogram and the normal QQ-plot are presented in Fig. 1.6. They all reflect an underlying normal distribution for the log transformed data, suggesting that the original distribution for the Tagus river level is lognormal. This result is not surprising. In fact, lognormal distributions are commonly used in hydrology to model river levels.

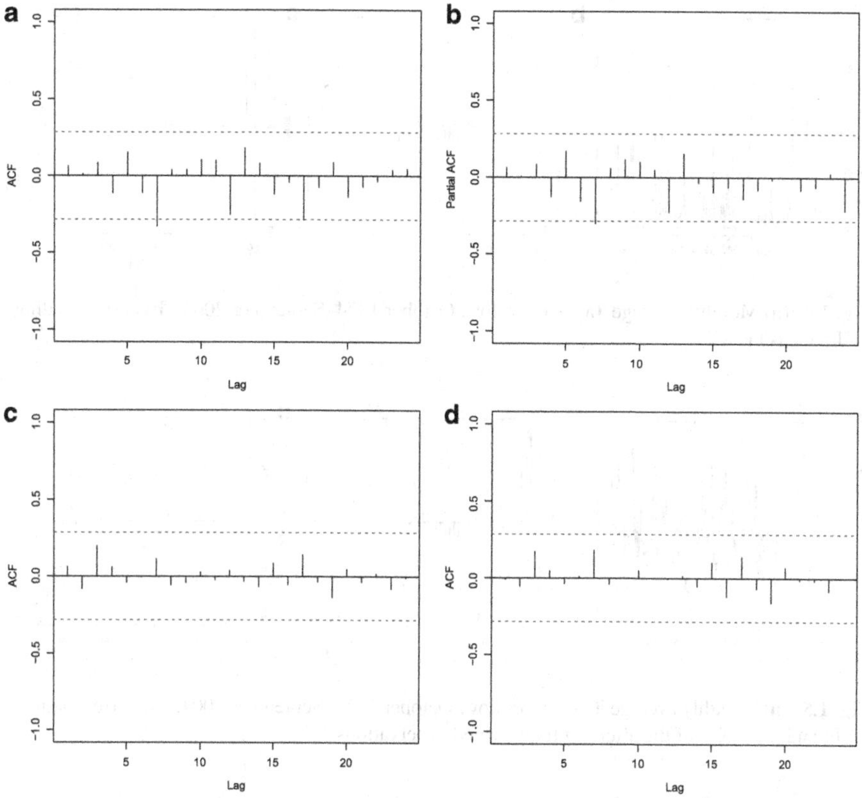

Fig. 1.2 (**a**) ACF; (**b**) PACF of the mean daily temperatures observed in Lisbon, Portugal, after removing the monthly pattern; (**c**) ACF of the squared observations; and (**d**) ACF of the absolute values

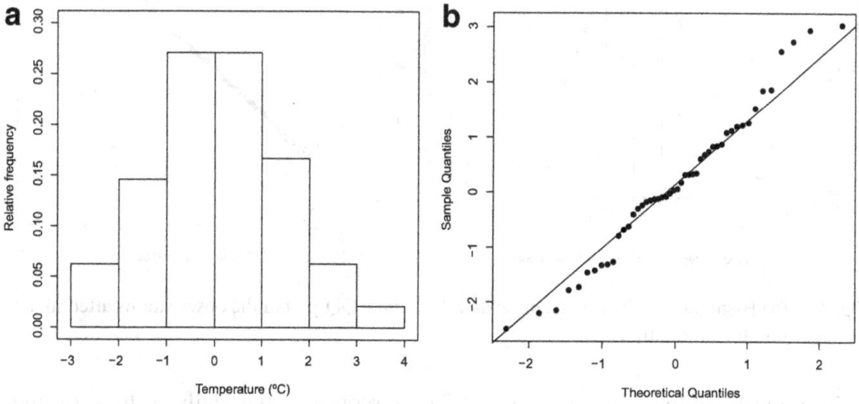

Fig. 1.3 (**a**) Histogram; and (**b**) normal QQ-plot of the mean daily temperatures observed in Lisbon, Portugal, after removing the monthly pattern

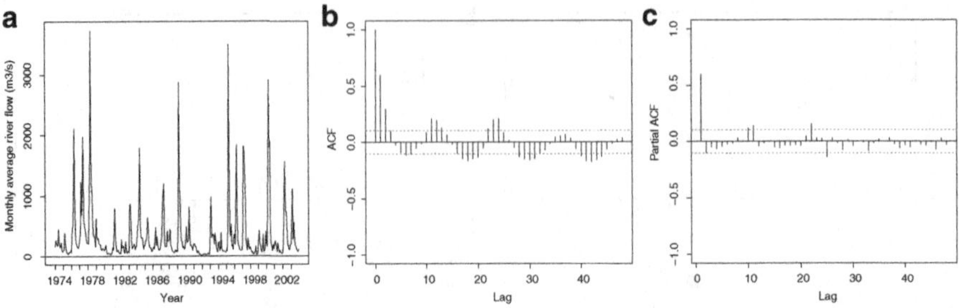

Fig. 1.4 (**a**) Monthly average Tagus river flow, October 1974–September 2004; (**b**) corresponding ACF; and (**c**) PACF

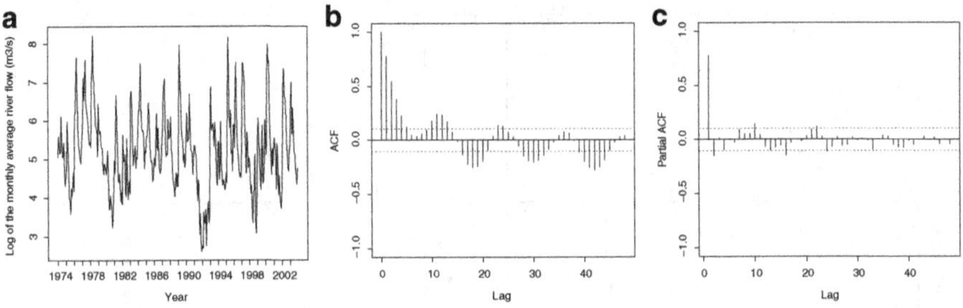

Fig. 1.5 (**a**) Monthly average Tagus river flow, October 1974–September 2004; (**b**) corresponding ACF; and (**c**) PACF of the after log transformed observations

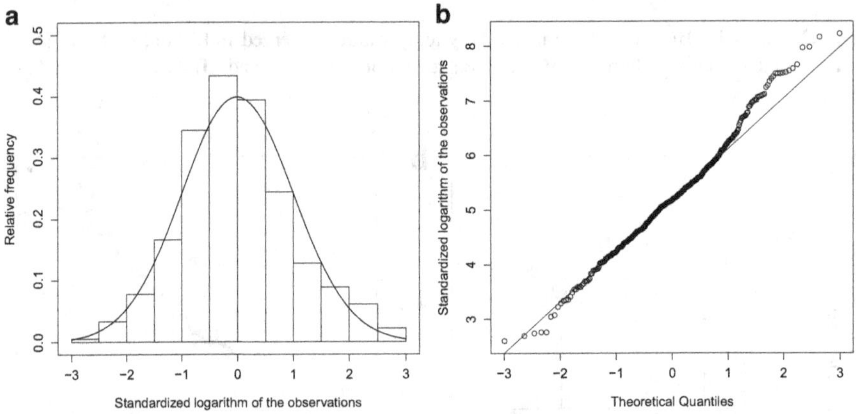

Fig. 1.6 (**a**) Histogram with normal curve; and (**b**) normal QQ-plot of the observations after taking the logarithm transformation

The observations plotted in Fig. 1.7a, b represent the daily values of the Portuguese Stock Market Index (PSI20), at closing time and the corresponding daily log-returns, respectively. The data plotted in Fig. 1.7 dates from 24 January 2000 to 4 January 2012. The plot of the log-returns, defined as

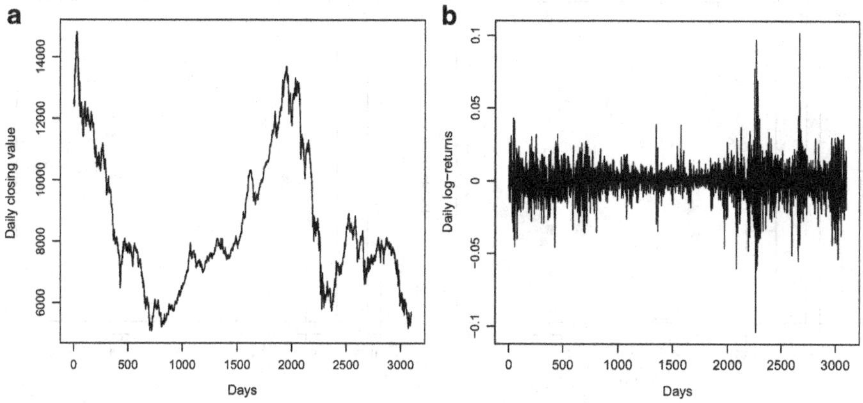

Fig. 1.7 (**a**) Daily values of the PSI20 at closing time; and (**b**) corresponding log-returns (data freely available at http://www.bolsapt.com/historico/)

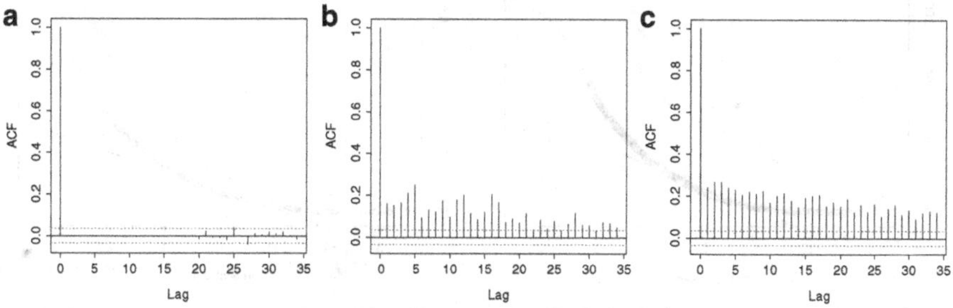

Fig. 1.8 ACFs of the (**a**) residuals; (**b**) squared residuals; and (**c**) of the absolute values of the residuals (**c**)

$$R_t = \log\left(\frac{X_{t+1}}{X_t}\right),$$

where X_t represents the value of the index at day t, is given in Fig. 1.7b. Upon fitting a proper linear model to the series, plots of the ACF of the residuals, squared residuals and absolute values of residuals are given in Fig. 1.8. These plots clearly indicate that although the residuals are uncorrelated, they are not independent. The ACF of squared residuals is often used for testing the presence of nonlinearity in the data, but it does not provide any guidance for the choice of a proper nonlinear model. Typically, to log-returns data, a linear model for the mean and a Generalized AutoRegressive Conditional Heteroskedastic (GARCH) model for the residuals are fitted to explain the nonlinear feature of the variance. The histograms, the normal QQ-plots and the box-plots of the absolute values of the negative and positive

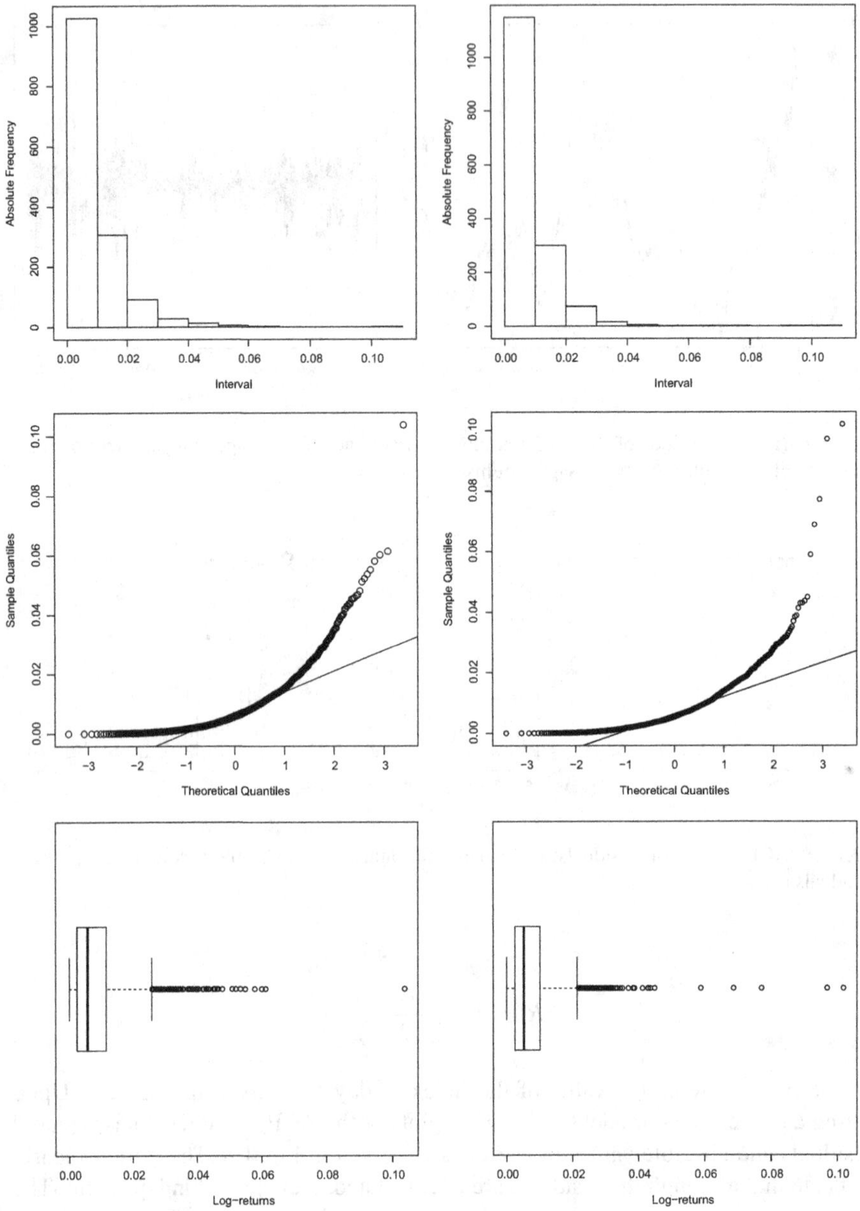

Fig. 1.9 Histograms, QQ-plots and box-plots of the absolute values of the negative (*left column*) and the positive log-returns (*right column*)

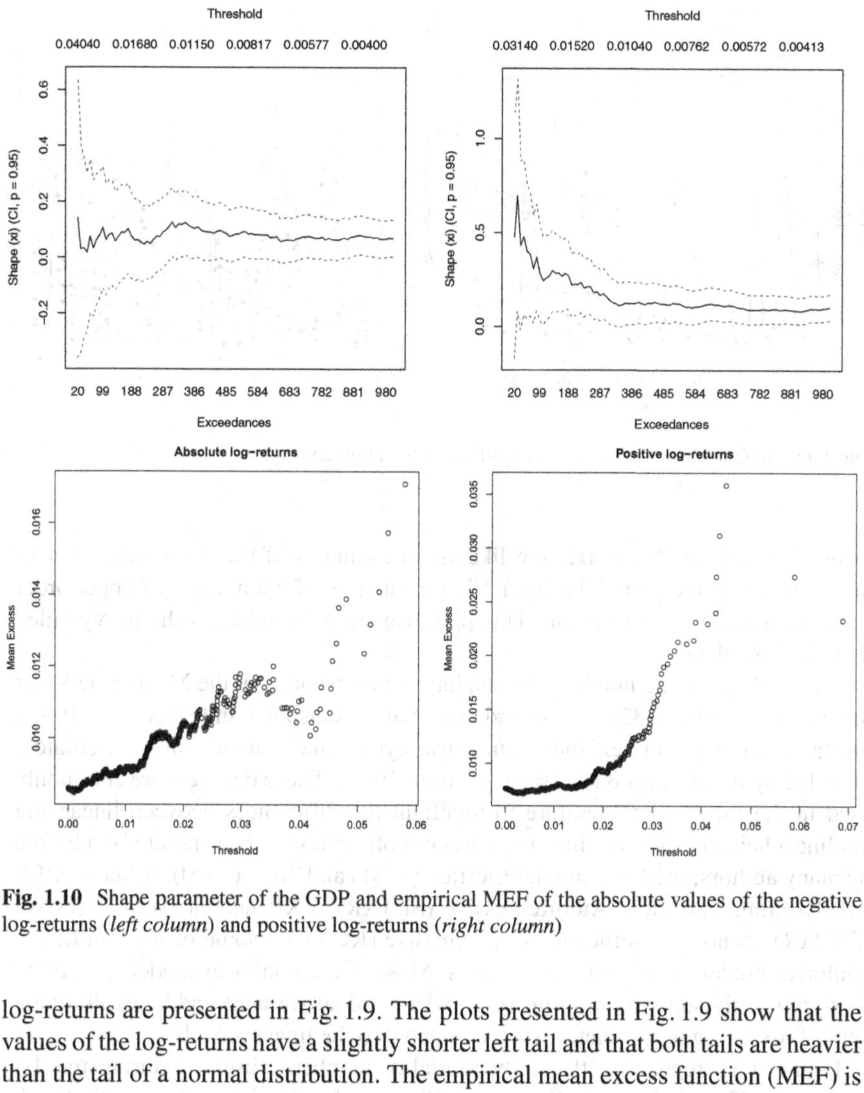

Fig. 1.10 Shape parameter of the GDP and empirical MEF of the absolute values of the negative log-returns (*left column*) and positive log-returns (*right column*)

log-returns are presented in Fig. 1.9. The plots presented in Fig. 1.9 show that the values of the log-returns have a slightly shorter left tail and that both tails are heavier than the tail of a normal distribution. The empirical mean excess function (MEF) is given by

$$e_n(u) := \frac{\sum_{i=1}^{n}(X_i - u)I(X_i > u)}{\sum_{i=1}^{n} I(X_i > u)},$$

where $I(\cdot)$ is the indicator function and u is the threshold. An upwards empirical $e_n(x)$, for all $x > u$, reveals an underlying heavy-tailed distribution. The empirical MEF of the positive log-returns clearly shows a right heavy-tailed distribution (see Fig. 1.10, right). A generalized Pareto distribution (GPD) is fitted to the right tail

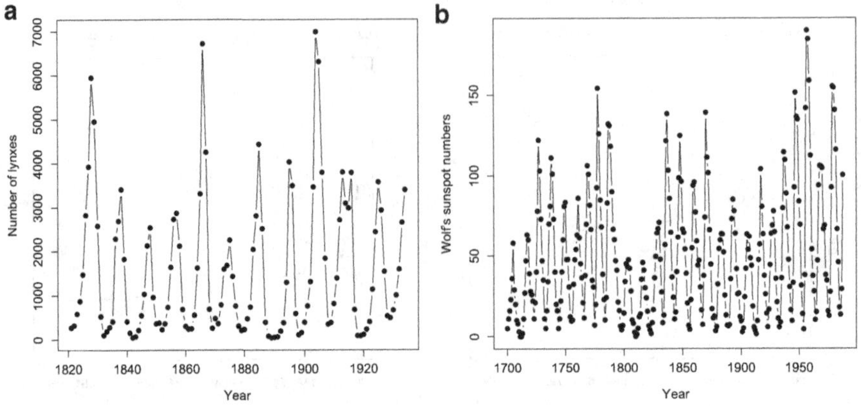

Fig. 1.11 (**a**) Canadian lynxes; and (**b**) Wolf's sunspot numbers

of the distribution. The maximum likelihood estimates of the shape parameter of this distribution are plotted in Fig. 1.10 as a function of the number of upper order statistics used in the estimation. This plot also strongly indicates the heavy-tailed feature of the data.

The analysis of the number of Canadian lynxes trapped in the Mackenzie River district in Canada (1821–1934) and the Wolf's sunspot numbers (1700–1988), displayed in Fig. 1.11, exhibit asymmetric cycles and thus are more accurately modeled by nonlinear models (see e.g., Tong 1990). These data sets are commonly used in the time series literature to highlight the differences between linear and nonlinear behavior. The nonlinear features of both series has been pointed out before by many authors, see for example Priestley (1981) and Tong (1990). Bilinear (BL), Self-Exciting Threshold AutoRegressive (SETAR), EXPonential AutoRegressive (EXPAR), Random Coefficient Autoregressive (RCA) are some of the families of nonlinear models fitted to the two series. Most of the nonlinear models proved to be significantly superior, in terms of a lower residual variance and by exhibiting a better forecasting performance when compared to the linear models.

Figure 1.12 represents the daily mortality numbers observed during the 1st January 1983 and the 31th December 1991 in the district of Évora, Portugal. This data set was provided by the National Statistics Institute (INE) Portugal, Demographic Statistics. Évora is part of a region called the Alentejo. Frequently, in summer, high temperatures are observed in this part of the country. Time periods 9–17 February 1983 and 12–18 July 1991 produced cold and hot events. In order to assess the impact of these two events, the data was plotted (see Fig. 1.13) in the months before and after the two events.

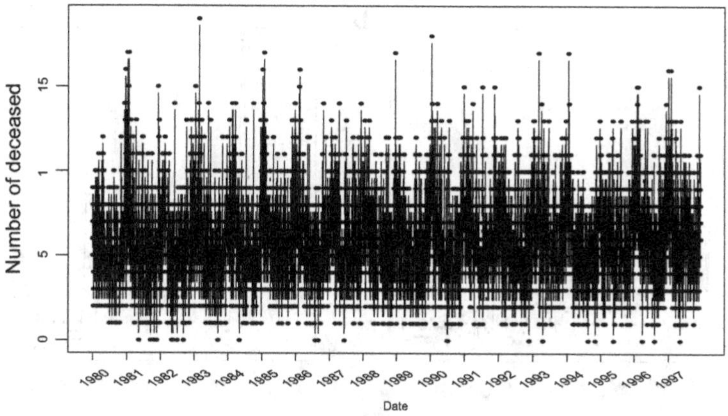

Fig. 1.12 Number of deceased observed in the district of Évora (Portugal) from 1st January 1983 to 31th December 1997 (data provided by INE Portugal, Demographic Statistics)

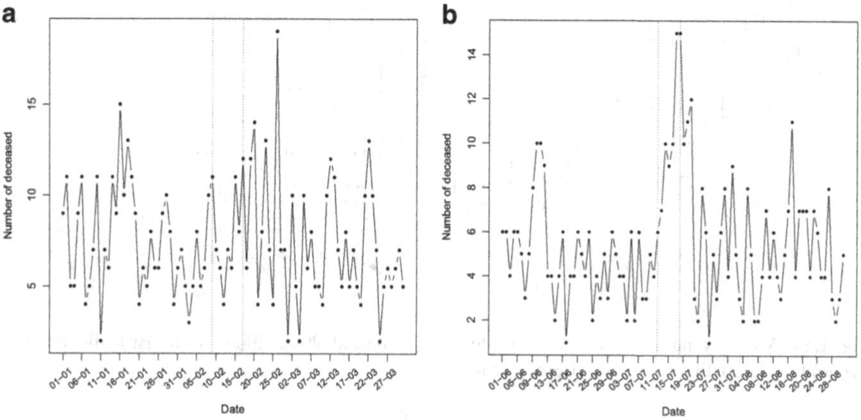

Fig. 1.13 (**a**) Number of deceased observed before and after the cold wave of February 1983; and (**b**) before and after the heat wave that occurred in from in July 1991

The time series plotted in Fig. 1.14 corresponds to the monthly number of cases of Brucellosis, Typhoid, Hepatitis C and Leptospirosis recorded in Portugal from January 2000 to December 2008. It is well known that there is a lag between the notification time and the initial date of appearance of the symptoms. The graphs clearly show a decrease in the number of cases through the years (decreasing trend). They also exhibit over dispersion, which is not compatible the Poisson assumption.

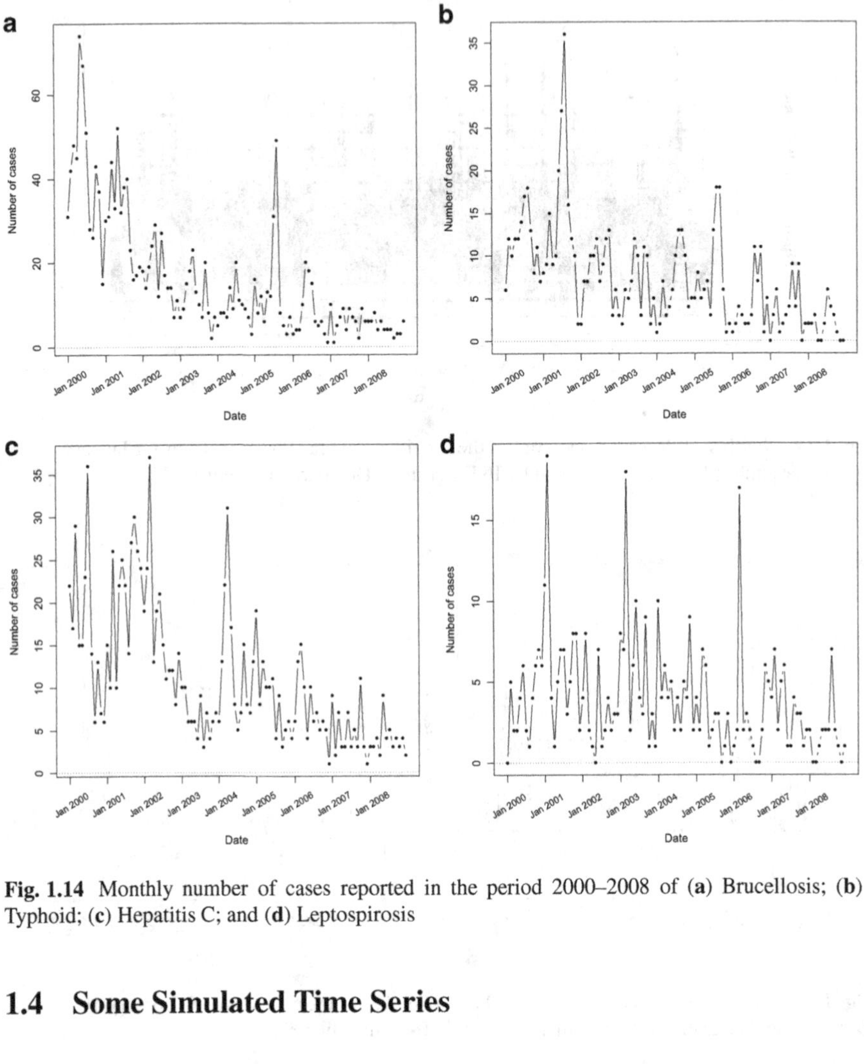

Fig. 1.14 Monthly number of cases reported in the period 2000–2008 of (**a**) Brucellosis; (**b**) Typhoid; (**c**) Hepatitis C; and (**d**) Leptospirosis

1.4 Some Simulated Time Series

In Fig. 1.15, we start by giving two sample paths from linear AR(1) processes

$$X_t = 0.5X_{t-1} + Z_t,$$

where Z_t are i.i.d. r.v's with unit exponential and logistic distributions (with location parameter $\mu = 9$ and scale parameter $\sigma = 4$), respectively. The logistic distribution has a heavier right tail and this is reflected in the sample path. Figure 1.16 shows a sample of $n = 500$ observations simulated from the following Threshold AutoRegressive (TAR(2, 1, 1)) model:

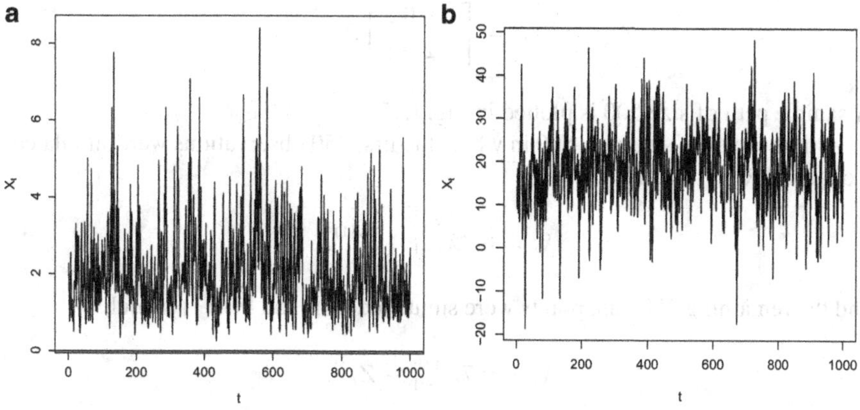

Fig. 1.15 AR(1) model sample paths $n = 1{,}000$ – exponential(1) errors (**a**); and (**b**) logistic errors

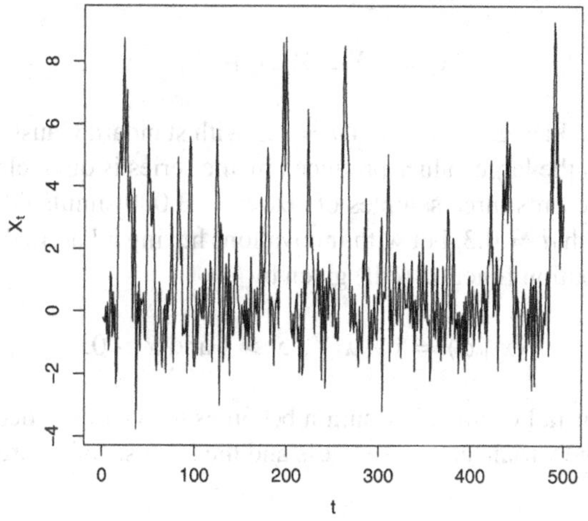

Fig. 1.16 Simulated sample path from a TAR(2,1,1) model

$$X_t = \begin{cases} -0.1X_{t-1} + Z_t, & X_{t-1} \leq 1 \\ 0.8X_{t-1} + 2Z_t, & X_{t-1} > 1 \end{cases}.$$

The innovations are $N(0,1)$. These models are often used for modeling data that exhibit limit cycles. Consider two linear time series models $X_t = 0.75X_{t-1} + Z_{1,t}$ (model 1) and $X_t = -0.75X_{t-1} + Z_{2,t}$ (model 2), both having standard $N(0,1)$ innovations. Suppose that the process switches between these two linear models according to a latent two state Markov chain with transition probability matrix

$$P = \begin{bmatrix} 0.8 & 0.2 \\ 0.2 & 0.8 \end{bmatrix}.$$

A sample path of size 500 is plotted in Fig. 1.17.

A time series of size $n = 500$ in which the first 250 observations were simulated from the AR(1) model

$$X_t = 0.3X_{t-1} + Z_t^{(1)},$$

and the remaining 250 data points were simulated from the MA(1) model,

$$X_t = 0.7Z_{t-1}^{(2)} + Z_t^{(2)},$$

where $Z_t^{(1)}$ and $Z_t^{(2)}$ are two independent i.i.d. samples from the standard Normal distribution, is given in Fig. 1.18. Figure 1.19 shows the plot of two samples of size $n = 500$ simulated from the bilinear model,

$$X_t = aX_{t-1}Z_{t-1} + Z_t, \tag{1.3}$$

for two different values $a = -0.7$ and $a = 0.3$ with standard Gaussian innovations. The effect of a on the large values produced by the series is quite clear.

Figure 1.20 presents three samples of size $n = 3{,}000$ simulated from the same bilinear model with $a = 0.3$, but with innovations having a Pareto distribution with cumulative distribution function (cdf) given by

$$F(x \mid \alpha) = 1 - x^{-\alpha}, \;\; x > 1 \text{ and } \alpha > 0.$$

Note that the right-tail of this distribution becomes heavier as α decreases. In fact, the mean value, the variance, the skewness and the kurtosis of a Pareto(α) r.v X are given by

$$E(X) = \frac{\alpha}{\alpha - 1}, \alpha > 1;$$

$$V(X) = \frac{\alpha}{(\alpha - 1)^2(\alpha - 2)}, \alpha > 2;$$

$$skewness = \frac{2(\alpha + 1)}{(\alpha - 3)}\sqrt{\frac{\alpha - 2}{\alpha}}, \alpha > 3;$$

and

$$kurtosis = \frac{6(\alpha^3 + \alpha^2 - 6\alpha - 2)}{\alpha(\alpha - 3)(\alpha - 4)}, \alpha > 4,$$

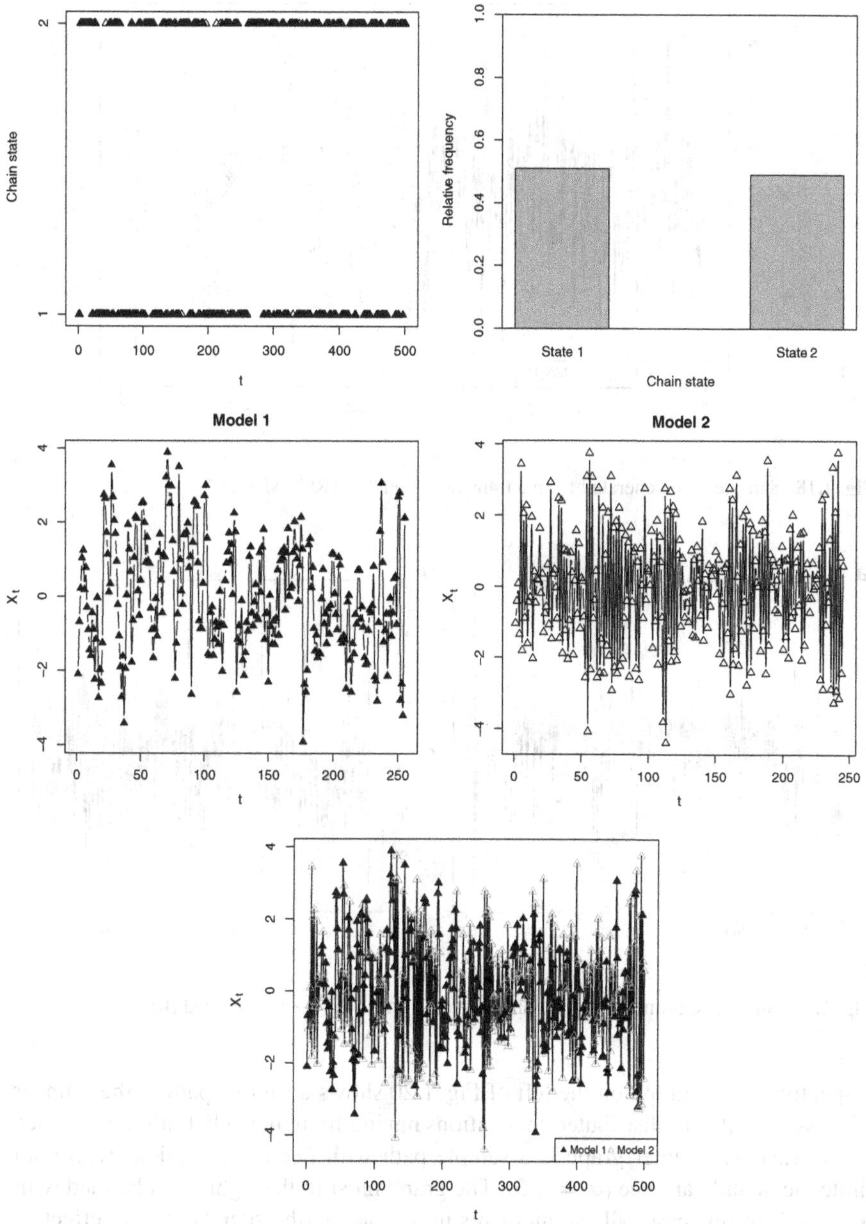

Fig. 1.17 RCA – states of the Markov (*top row*), the data generated for the two models (*middle row*) and the entire time series, $n = 500$ observations (*bottom row*)

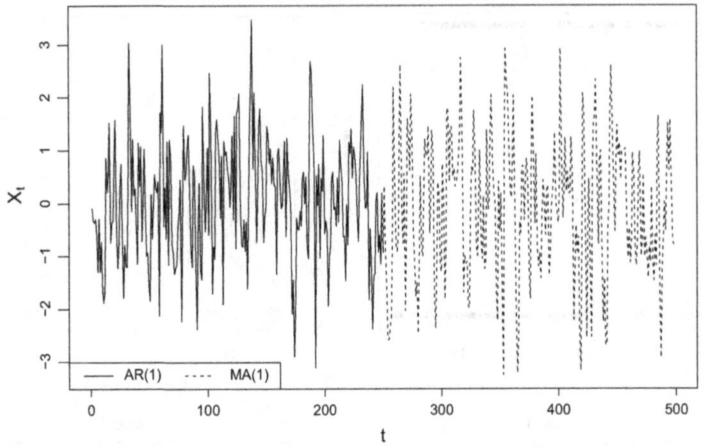

Fig. 1.18 Sample path generated from a transition model – AR(1)/MA(1)

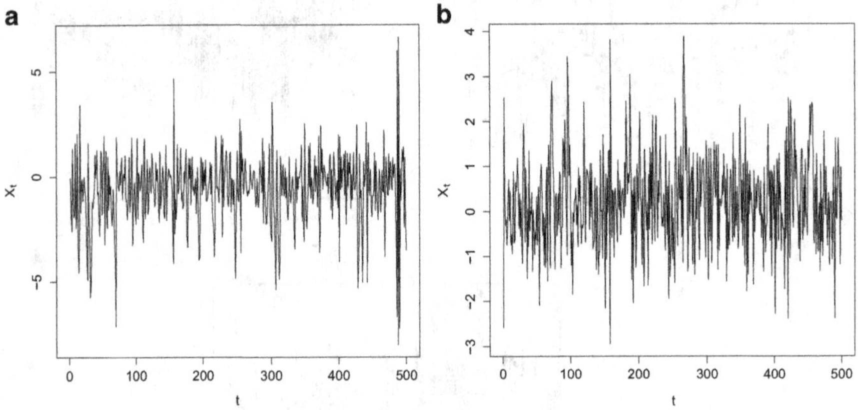

Fig. 1.19 Simulated bilinear model with Gaussian errors – $a = -0.7$ (a); and (b) $a = 0.3$

respectively. The graph on the left of Fig. 1.20 shows a sample path of the bilinear process with Pareto distributed innovations having finite mean but infinite variance ($\alpha = 1.5$). The center graph is a sample path with Pareto innovations with both finite mean and variance ($\alpha = 2.5$). The graph most to the right was obtained with $\alpha = 4.5$. In this case, all the moments up to the fourth order exist. The effect of parameter α on the sample paths is quite clear.

Figure 1.21 shows three sample paths of AR(1) processes

$$X_t = 0.3X_{t-1} + Z_t$$

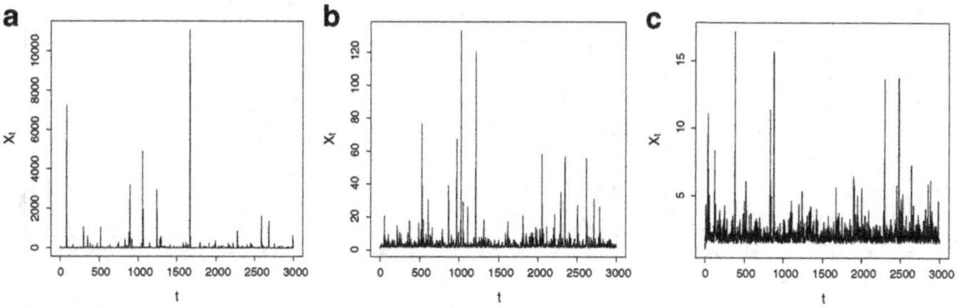

Fig. 1.20 Simulated bilinear model with heavy-tailed errors – Pareto(1.5) (**a**); (**b**) Pareto(2.5); and (**c**) Pareto(4.5)

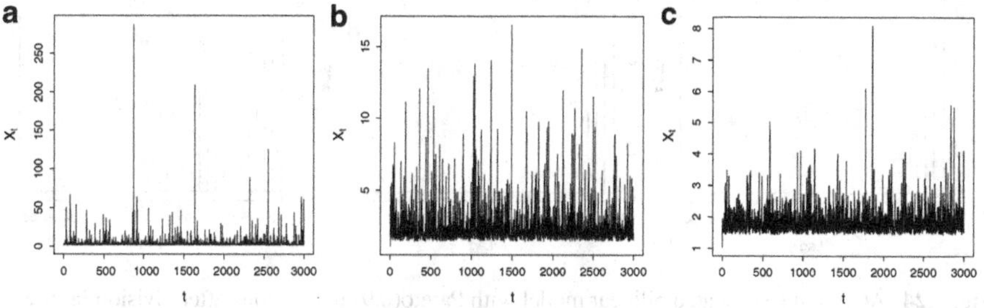

Fig. 1.21 Simulated AR(1) models with heavy-tailed errors – Pareto(1.5) (**a**); (**b**) Pareto(2.5); and (**c**) Pareto(4.5)

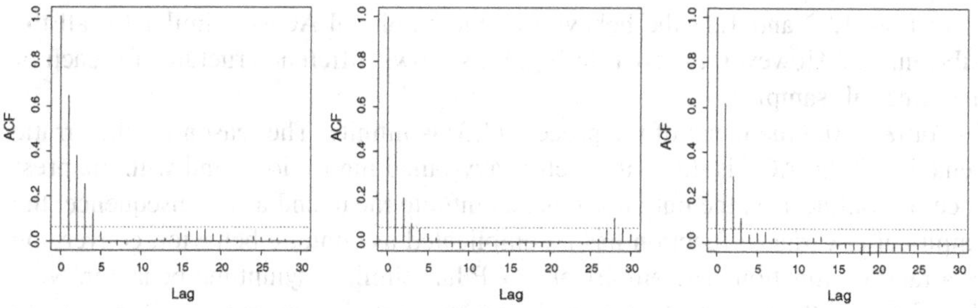

Fig. 1.22 ACF of the simulated bilinear model with heavy-tailed Pareto(2.5) innovations after division in three sets of $n = 1{,}000$ observations

with innovations having the same three different Pareto distributions. The comparison of plots in Figs. 1.20 and 1.21 clearly indicates that not only the value of α, but also the nonlinear relation between the input and output series play an important role in generating large values.

Figures 1.22 and 1.23 represent, respectively, the empirical ACF of three consecutive sub-samples of size 1,000 from 3,000 simulated observations for the bilinear model (1.3), considering $a = 0.3$, with Pareto error distributions $F(x|\alpha = 2.5)$

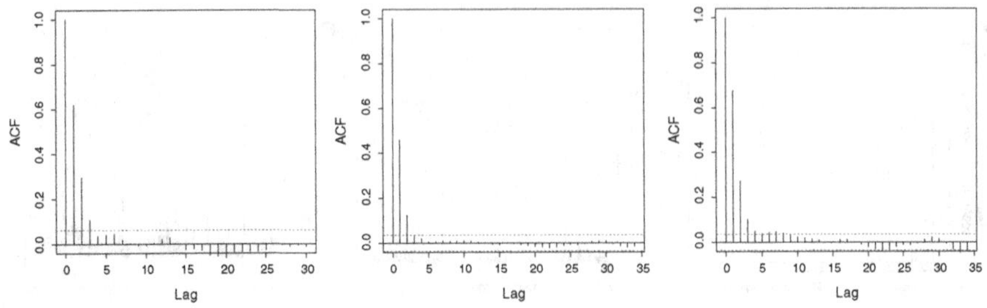

Fig. 1.23 ACF of the $n = 3{,}000$ simulated observations from the bilinear model with Pareto(4.5) innovations

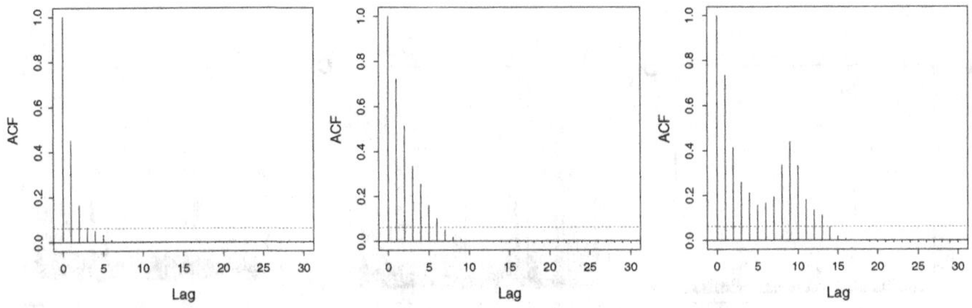

Fig. 1.24 ACF of the simulated bilinear model with Pareto(0.9) innovations after division in three sets of $n = 1{,}000$ observations

and $F(x|\alpha = 4.5)$, whereas Fig. 1.24 represents the ACF of the same process with $F(x|\alpha = 0.9)$.

In Figs. 1.22 and 1.23 the behavior of the empirical ACF is similar for all the sub-samples. However, the ACF in Fig. 1.24 shows different structures for each of the three sub-samples.

For $\alpha = 0.9$ the mean of the process (1.3) is infinite. The reason for the erratic behavior of the ACF is that with such heavy-tailed innovations, and with the presence of nonlinearity, the output series has infinite mean and as a consequence, the empirical covariance function has a complicated asymptotic behavior, converging to a random function. The empirical PACF has similar asymptotic behavior. Note that if we use the empirical autocorrelation and partial autocorrelation functions to identify a linear model for the data, then we not only end up with an inadequate linear model, but also choose a wrong linear model for the data due to the erratic asymptotic behavior of the empirical second order moments.

In general, the tail behavior of the output series is dependent on the tail behavior of the input series, as well as on the type of nonlinear relationship existing between the input and the output series. This relationship between the tails of input and output series will be studied in detail in Chap. 2.

Figure 1.25 represents a sample path of size 300 from the GARCH model

$$X_t = h_t Z_t,$$

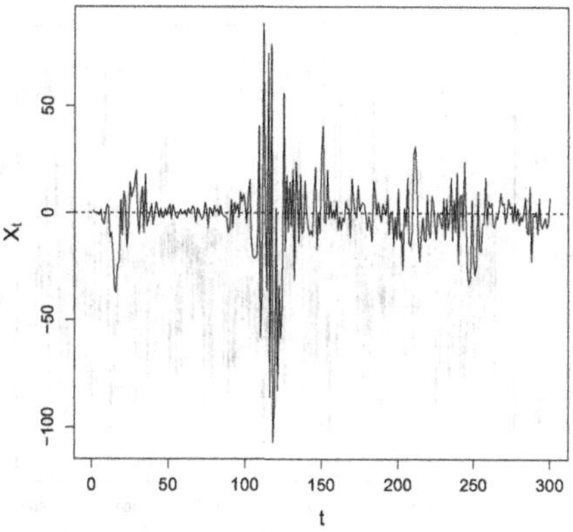

Fig. 1.25 A sample path from the GARCH(1, 1) process with $\alpha_0 = 1.0$, $\alpha_1 = 0.4$ and $\beta = 0.5$

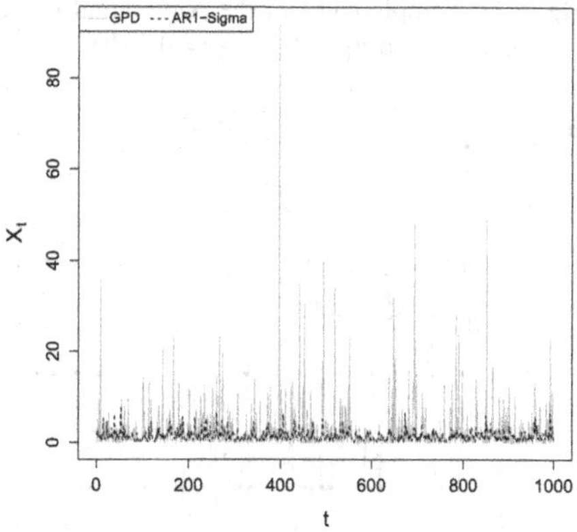

Fig. 1.26 GPD(k,σ_t) parameter driven model – $k = 0.5$; σ_t follows an AR(1) model

where Z_t are i.i.d. $N(0, \sigma^2)$, $\sigma = 1.25$, r.v's and

$$h_t^2 = \alpha_0 + \alpha_1 X_{t-1}^2 + \beta h_{t-1}^2,$$

with $\alpha_0 = 1.0$, $\alpha_1 = 0.4$ and $\beta = 0.5$. GARCH models represent the nonlinear dynamics in the conditional variance, as compared to bilinear models which may represent the nonlinear dynamics in the conditional mean, as well as in the conditional variance. Although the conditional moment structures of these two classes of models are completely different, unconditional moments are quite similar.

Consider the following generalized state space model $X_t \sim \text{GPD}(k, \sigma_t)$, where GPD stands for the generalized Pareto distribution, k is the shape parameter and σ_t is scale parameter which follows an AR(1) process

$$\sigma_t = \omega \sigma_{t-1} + Z_t.$$

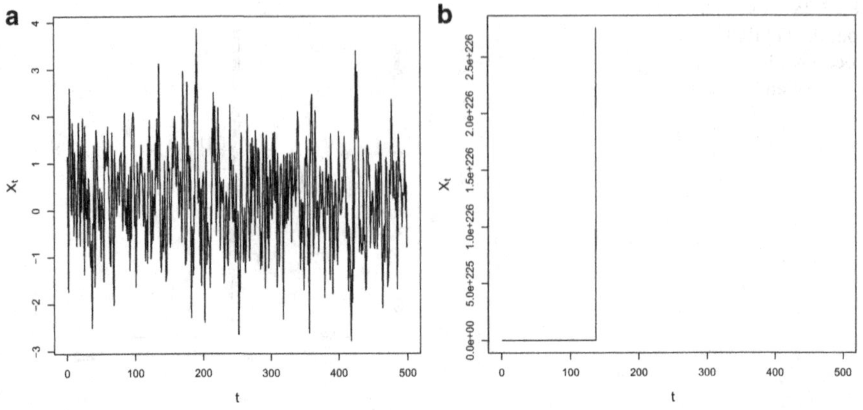

Fig. 1.27 Sample paths of the polynomial model $X_t = a_1 X_{t-1} + a_2 X_{t-1}^2 + Z_t$ for $a_1 = 0.2$, $a_2 = 0.1$ (**a**); and $a_1 = 0.2$, $a_2 = 0.2$ (**b**)

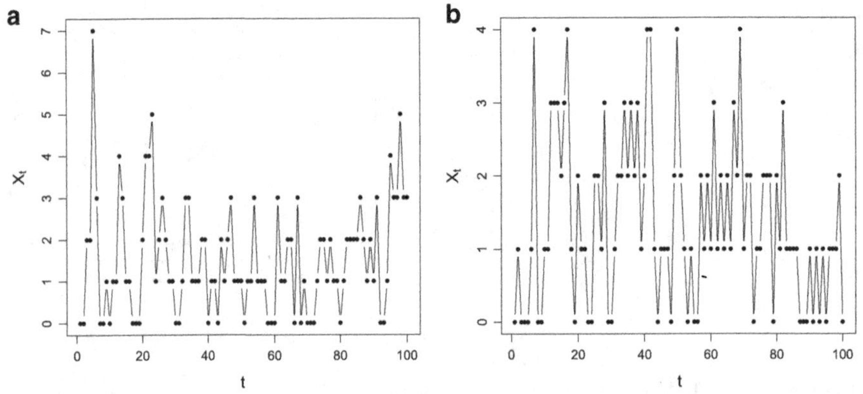

Fig. 1.28 (**a**) Observation-driven integer-valued time series from model (1.4); and (**b**) a sample path from a INAR(1) model given in (1.5) with $a = 0.4$

Here, (Z_t) is an i.i.d. sequence with exponential(1) distribution. We assume that X_t, conditional on σ_t, are independent. Figure 1.26 displays a sample path of $n = 1{,}000$ observations generated from this model with $\omega = 0.4$ and $k = 0.5$ (heavy-tailed). The black line corresponds to the sample path of the latent σ_t process. Two sample paths of size $n = 500$ of the polynomial model

$$X_t = a_1 X_{t-1} + a_2 X_{t-1}^2 + Z_t,$$

where $Z_t \sim N(0, 1)$ are represented in Fig. 1.27a, b with $a_1 = 0.2$, $a_2 = 0.1$ and $a_1 = 0.2$, $a_2 = 0.2$, respectively.

Figure 1.28a represents a sample path of size 100 from the observation-driven Poisson model

$$\begin{cases} X_t|\lambda_t \sim \text{Po}(\lambda_t) \\ \lambda_t = 1.0 + 0.4X_{t-1} \end{cases} \tag{1.4}$$

with $X_0 = 0$.

Figure 1.28b represents a sample path of 100 observations from the integer-valued time series INAR(1) given by the recursion

$$X_t = M_t(X_{t-1}, a) + Z_t. \tag{1.5}$$

$M_t(\cdot, \cdot)$ is the binomial thinning operator, defined as

$$M_t(X_{t-1}, a) := a \circ X_{t-1} \equiv \sum_{j=1}^{X_{t-1}} \xi_j(a),$$

where (ξ_j) is a counting sequence of i.i.d. Bernoulli r.v's with mean $a \in [0, 1]$ and (Z_t) is an i.i.d. sequence of non-negative integer-valued r.v's, stochastically independent of X_{t-1} for all points in time, with finite mean $\mu_Z > 0$ and variance $\sigma_Z^2 \geq 0$. In this simulated series, we take $a = 0.4$ and consider that Z_t are i.i.d. Poisson r.v's with mean one.

References

Diggle PJ, Ribeiro PJ (2007) Model-based geostatistics. Springer, New York

Diggle PJ, Moyeed RA, Tawn JA (1998) Model-based geostatistics. Appl Stat 47:299–350. (With discussion)

Heyde CC (1997) Quasi-likelihood and its application: a general approach to optimal parameter estimation. Springer, New York

Macedo ME (2006) Caracterização de Caudais Rio Tejo, Direção de Serviços de Monitorização Ambiental

Nisio M (1960) On polynomial approximation for strictly stationary processes. J Math Soc Jpn 12:207–226

Prado R, West M (2010) Time series: modeling, computation, and inference. Texts in statistical science. Chapman & Hall/CRC, Boca Raton

Priestley MB (1981) Spectral analysis and time series. Academic, London

Tong H (1990) Non-linear time series. Oxford Science Publications, Oxford

Chapter 2
Nonlinear Time Series Models

2.1 Some Probabilistic Aspects of Nonlinear Processes

2.1.1 Linear Representations and Linear Models

Assume that for $t \in \mathbb{Z}$, (Z_t) and (Z_t^*) are respectively uncorrelated and independent sequences of r.v's having identical marginal distribution $F(\cdot)$, with zero mean and variance $\sigma_Z^2 < \infty$. For any t, define the time series

$$Y_t = \sum_{i=0}^{\infty} \psi_i Z_{t-i} \tag{2.1}$$

and

$$X_t = \sum_{i=0}^{\infty} \psi_i Z_{t-i}^*, \tag{2.2}$$

such that $\sum_{i=0}^{\infty} \psi_i^2 < \infty$, so that both Y_t and X_t are mean-square convergent, having finite variances. In the representation (2.2), specification of the marginal distribution for the independent r.v's (Z_t^*) is enough to specify the finite dimensional distributions of the output series X_t. Therefore (2.2) is a fully specified model for X_t. However, the specification of the marginal distribution for uncorrelated r.v's (Z_t) is not enough to specify fully the finite dimensional distributions of the process Y_t given in (2.1), unless Z_t is a Gaussian sequence. In this case, we can merely calculate uniquely the first two moments, namely the mean, the variance and the autocovariance function of the series Y_t. Therefore (2.1) is not a probabilistic model for Y_t, but can be called a representation. Since this representation uniquely specifies the second-order moments, we will call it the second-order representation for the time series Y_t. *Wold decomposition theorem* (see for example Brockwell and Davis 1991) shows that under fairly general conditions any stationary time series will have

K.F. Turkman et al., *Non-Linear Time Series*, DOI 10.1007/978-3-319-07028-5_2,

the causal linear representation (2.1), but these processes are not necessarily linear processes given in (2.2) and in fact, they can be highly nonlinear processes. Hence, Y_t in (2.1) is a representation for infinitely many time series, having the same (finite) second-order moments. On the other hand, if (Z_t) in (2.1) have marginal Normal distribution $N(0, \sigma^2)$, then they must also be independent. In this case, (2.1) and (2.2) are identical Gaussian processes. Uncorrelated versus independent innovations in (2.1) and (2.2) also have significant different effects on predictions. Y_{t+1} can be written as

$$\begin{aligned} Y_{t+1} &= \sum_{i=0}^{\infty} \psi_i Z_{t+1-i} \\ &= \psi_0 Z_{t+1} + \sum_{i=1}^{\infty} \psi_i Z_{t+1-i} \\ &= \psi_0 Z_{t+1} + \sum_{i=0}^{\infty} \psi_{i+1} Z_{t-i}. \end{aligned}$$

Similarly

$$X_{t+1} = \psi_0 Z^*_{t+1} + \sum_{i=0}^{\infty} \psi_{i+1} Z^*_{t-i}.$$

Let $\mathcal{B}_Z(t)$ be the σ-field generated by the r.v's $(Z_s, s \leq t)$. The best mean-square predictor of Y_{t+1} in terms of $(Z_t, Z_{t-1}, \ldots)$ is given by the conditional expectation

$$E(Y_{t+1}|\mathcal{B}_Z(t)) = \psi_0 E(Z_{t+1}|\mathcal{B}_Z(t)) + \sum_{i=0}^{\infty} \psi_{i+1} Z_{t-i}$$

with

$$E(Z_{t+1}|\mathcal{B}_Z(t)) = \int_x x dF_{Z_{t+1}|\mathcal{B}_Z(t)}(x),$$

where $F_{Z_{t+1}|\mathcal{B}_Z(t)}(x)$ is the distribution of Z_{t+1} conditional on $(Z_t, Z_{t-1}, \ldots)$. Note that Z_{t+1} is not independent of $Z_t, Z_{t-1}, \ldots$, hence, in general

$$F_{Z_{t+1}|\mathcal{B}_Z(t)}(x) \neq F_{Z_{t+1}}(x)$$

and

$$E(Z_{t+1}|\mathcal{B}_Z(t)) \neq 0.$$

In fact, this term will typically be complex, nonlinear function of $(Z_t, Z_{t-1}, \dots)$. Hence, the best predictor of Y_{t+1} in terms of $(Z_t, Z_{t-1}, \dots)$ will be a nonlinear function. If the process (2.1) is invertible, then the sigma fields generated respectively by $(Z_s, s \leq t)$ and $(Y_s, s \leq t)$ are identical, hence

$$E(Y_{t+1}|Y_s, s \leq t) = E(Y_{t+1}|Z_s, s \leq t),$$

so that $E(Y_{t+1}|Y_s, s \leq t)$ in general is a nonlinear function of $(Y_s, s \leq t)$. If (Z_t) have Normal distribution, then the best predictor of Y_{t+1} in this case is a linear function of the past observations $(Y(s), s \leq t)$. On the other hand, $F_{Z^*_{t+1}|\mathcal{B}^*_Z(t)}(x) = F_{Z^*_{t+1}}(x)$ and $E(Z^*_{t+1}|\mathcal{B}^*_Z(t)) = 0$, so that

$$E(X_{t+1}|X_s, s \leq t) = E(X_{t+1}|Z^*_s, s \leq t),$$

is a linear function of $(X_s, s \leq t)$, irrespective of the marginal distribution $F(\cdot)$ of Z_t.

In order to understand better the relation between best predictions and nonlinearity, we look at the geometric interpretation of predictions.

2.1.2 *Linear and Nonlinear Optimal Predictions*

Consider a probability space $(\Omega, \mathcal{F}, P)$ and the collection $\mathcal{C}$ of all r.v's defined on this space with zero-mean and finite second-order moments. For any elements X, Y of $\mathcal{C}$, define the inner product $<X, Y> = E(XY)$, so that the norm is given by $||X|| = \sqrt{E(X^2)}$. Thus, two elements X and Y of $\mathcal{C}$ are orthogonal iff $E(XY) = 0$, in which case we write $X \perp Y$. For simplicity in notation, we assume that the elements of $\mathcal{C}$ have zero means. Alternatively, rather than restricting the class to 0 mean r.v's, we can define the inner product on $\mathcal{C}$ as $<X, Y> = E(XY) - E(X)E(Y)$ and the norm $||X|| = \sqrt{E(X - E(X))^2}$, and the properties would still hold. The norm convergence of any sequence X_n is then given by

$$\lim_{n \to \infty} ||X_n - X||^2 = \lim_{n \to \infty} E|X_n - X|^2 = 0,$$

which is the usual mean-square convergence and we denote it by

$$X_n \overset{m.s}{\to} X.$$

Note that $X_n \overset{m.s}{\to} X$, iff

$$E(X_n - X_m)^2 \to 0,$$

as $m, n \to \infty$, in which case we call the sequence a Cauchy sequence.

If all sequences of $\mathcal{C}$ converge in mean-square, then $\mathcal{C}$ is complete and hence is a Hilbert space (e.g., Brockwell and Davis 1991). Let (X_n) be a stationary time series, such that $E(X_n^2) < \infty$. The norm convergence or mean-square convergence implies that if

$$X_n \overset{m.s}{\rightarrow} X$$

and

$$Y_n \overset{m.s}{\rightarrow} Y,$$

then

1. $E(X_n) \rightarrow E(X)$;
2. $E|X_n|^2 \rightarrow E|X|^2$, so that the variance of X_n converges to the variance of X;
3. $E(X_nY_n) \rightarrow E(XY)$, so that the covariance and correlations between X_n and Y_n converge to the covariance and correlation between X and Y.

Now, let $\mathcal{C}_1$ be any closed subspace of $\mathcal{C}$. Then, from the *projection theorem*, for any $Y \in \mathcal{C}$, there is a unique element $\hat{X} = P_{\mathcal{C}_1}X \in \mathcal{C}_1$ such that

$$||Y - P_{\mathcal{C}_1}X||^2 = \inf_{X \in \mathcal{C}_1} ||Y - X||^2 = \inf_{X \in \mathcal{C}_1} E|Y - X|^2.$$

We know that the value of X which minimizes the mean-square error $E|Y - X|^2$ is given by $E(Y|X)$, so that the projection $P_{\mathcal{C}_1}X$ is the conditional expectation of Y given $\mathcal{C}_1$, and we denote it by $E_{\mathcal{C}_1}(Y)$. By the projection theorem $E_{\mathcal{C}_1}(Y)$ is a unique element X of $\mathcal{C}_1$ which satisfies

$$E(XE_{\mathcal{C}_1}(Y)) = E(XY),$$

for every $X \in \mathcal{C}_1$. We now define this conditional expectation in terms of multivariate r.v's in time series setting: let $(X_1, X_2, \ldots, X_n)$ be r.v's defined on $\{\Omega, \mathcal{F}, P\}$ and $Y \in \mathcal{C}$. Define the subspace $\mathcal{C}_1 = \mathcal{C}_1(X_1, X_2, \ldots, X_n)$ as the space of r.v's consisting of $X_1, X_2, \ldots, X_n$ and all other r.v's obtained by measurable transformations $f(X_1, X_2, \ldots, X_n)$. $\mathcal{C}_1$ is a closed subspace of $\mathcal{C}$. For any $Y \in \mathcal{C}$, let $P_{\mathcal{C}_1}Y = P_{\mathcal{C}_1(X_1,\ldots,X_n)}Y$ the projection of Y in $\mathcal{C}_1(X_1, \ldots, X_n)$. We define $P_{\mathcal{C}_1(X_1,\ldots,X_n)}Y = E_{\mathcal{C}_1(X_1,\ldots,X_n)}(Y)$ to be the conditional expectation of Y given $(X_1, \ldots, X_n)$. By the projection theorem, this conditional expectation is unique and can be obtained from the prediction equation

$$E(XE_{\mathcal{C}_1(X_1,X_2,\ldots,X_n)}(Y)) = E(XY), \tag{2.3}$$

for every element $X \in \mathcal{C}_1(X_1, \ldots, X_n)$. However, elements $X \in \mathcal{C}_1$ are in general nonlinear functions $f(X_1, \ldots, X_n)$ of $(X_1, \ldots, X_n)$ and therefore obtaining this unique conditional mean using the prediction equation (2.3) in general is very difficult. However, there is one particular case, when this unique projection $P_{C_1(X_1,X_2,\ldots,X_n)}X = E_{C_1(X_1,X_2,\ldots,X_n)}(Y) = E_{X_1,\ldots,X_n}(Y)$ can be calculated with ease: Restrict $C_1(X_1,\ldots,X_2)$ to be the closed span of $(X_1,\ldots,X_n)$ so that we only consider linear functions $f(X_1,\ldots,X_n) = \sum_{i=1}^{n}\alpha_i X_i$, and any element of $X \in C_1(X_1,\ldots,X_n)$ is given by $X = \sum_{i=1}^{n}\alpha_i X_i$. In this case the optimal projection of any $Y \in C$ into C_1 is a linear function, and $\hat{Y} = P_{C_1(X_1,\ldots,X_n)}(Y) = E(Y|X_1,\ldots,X_n) = \sum_{i=1}^{n}\alpha_i^* X_i$.

We call $\hat{Y}$ the best linear predictor for Y. This unique function can be obtained from the prediction equation by solving the set of equations

$$\sum_{i=1}^{n}\alpha_i^* E(X_i X_j) = E(YX_j), \tag{2.4}$$

for $j = 1, 2, \ldots, n$. However, the best linear predictor need not be the best predictor, since the best linear predictor is chosen within the closed span of $(X_1,\ldots,X_n)$,

$$C_1 := \{X_1, X_2, \ldots, X_n \text{ and all linear functions of } (X_1,\ldots,X_n)\},$$

whereas the best predictor is chosen within the closed subspace

$$C_1^* := \{X_1, X_2, \ldots, X_n \text{ and all measurable functions of } (X_1,\ldots,X_n)\}.$$

Clearly $C_1 \subset C_1^*$. The following definition is immediate.

Definition 2.1.1. A best linear prediction of Y in terms of a countable collection of r.v's $(X_t, t \in T)$ is defined to be the element of the closed span C_1 of $(X_t, t \in T)$ which has the smallest mean-square distance from Y, and by the projection theorem is unique. On the other hand, the best predictor of Y in terms of the collection $(X_t, t \in T)$ is defined to be the element of the closed subspace C_1^* formed by all measurable functions of $(X_t, t \in T)$.

This definition will be extremely useful in discussing linear and nonlinear time series models. In general, $C_1 \subseteqq C_1^*$ and $C_1 = C_1^*$, if $(Y, X_t, t \in T)$ have joint multivariate Normal distribution.

Example 2.1.1 (Brockwell and Davis 1991). Assume that $Y = X^2 + Z$, where X and Z are independent standard Normal r.v's. Let $C^*(X)$ be the closed space formed by X and all measurable functions ϕ of X. By the projection theorem, the best mean-square predictor of Y in $C^*(X)$ is the unique element $E_{C(X)}(Y)$ of $C(X)$, which satisfies

$$E(\phi(X)E_{C(X)}(Y)) = E(\phi(X)Y).$$

$E_{C^*(X)}(Y)$ is an element of $C^*(X)$, so that $E_{C^*(X)}(Y) = \phi^*(X)$ for some measurable function ϕ^* of X so that

$$E(\phi(X)\phi^*(X)) = E(\phi(X)Y)$$
$$= E(\phi(X)X^2) + E(\phi(X)Z),$$

since X and Z are independent, for any measurable function ϕ, $\phi(X)$ and Z are also independent, hence $E(\phi Z) = E(\phi)E(Z) = 0$. Now, the only measurable function ϕ^* of X which satisfies

$$E(\phi(X)\phi^*(X)) = E(\phi(X)X^2),$$

is $\phi^*(X) = X^2$, hence by the projection theorem, the best mean-square predictor $E_{C^*(X)}(Y)$ of Y is indeed the conditional expectation $E(Y|X) = X^2$ ($E(Z|X) = 0$ due to the independence of X and Z).

Now consider the best linear mean-square predictor of Y, that is, the best mean-square predictor of Y residing in $C(X)$, the closed span of X. Then $E_{C(X)}(Y) = aX + b$, satisfying

$$E[(aX + b)\phi(X)] = E[\phi(X)(X^2 + Z)],$$

for any $\phi(X)$ in $C(X)$. In particular, $\phi(X) = 1$ and $\phi(X) = X$ are in the closed span of X. Consequently from the prediction equations

$$< aX + b, 1 >=< y, 1 >= E(Y) = E(X^2) = 1,$$

and

$$< aX + b, X >= E(YX) = 0.$$

Solving for a and b gives $a = 0$ and $b = 1$, and the best linear predictor of Y in terms of X is given by $P_{C(X)}(X) = 1$. The prediction error of the best predictor is

$$||E(Y|X) - Y||^2 = E(Z^2) = 1,$$

whereas the prediction error of the best linear predictor is

$$||X^2 + Z - 1||^2 = E(X^2 + Z - 1)^2 = E(X^4) + E(Z^2) - 1 = 3.$$

Hence, the best linear predictor has three times as much prediction error as the best mean-square predictor, showing its clear inferior performance.

The above arguments can be applied to predict a future value of a time series. Consider a discrete parameter time series (X_t) defined on $(\Omega, \mathcal{F}, P)$, with zero mean and autocovariance function $\gamma(h)$. Consider the problem of best predictor of X_{n+1} in terms of $X_1, X_2, \ldots, X_n$. Clearly X_{n+1} and $X_1, \ldots, X_n$ are all elements of the Hilbert space with inner product $< X_i, X_{i+h} >:= E(X_i X_{i+h}) = \gamma(h)$, and norm $||X_i||^2 = \gamma(0)$. (Note that the mean is assumed to be zero, so that

$E(X_i X_{i+h}) = \gamma(h)$. Otherwise, we either study the series $X_t - E(X_t)$, or equivalently define the inner product to be

$$< X_i, X_{i+h} >= E((X_i - E(X_i))(X_{i+h} - E(X_{i+h}))) = \gamma(h).$$

Therefore, the assumption of zero mean is not restrictive.)

Consider the closed subspace C_1^* which includes the r.v's $X_1, \ldots, X_n$ and all measurable functions of $(X_1, \ldots, X_n)$. Clearly such closed subspace will include the closed span C_1 of $(X_1, \ldots, X_n)$. From the projection theorem, the best predictor of X_{n+1} as a function of $(X_1, \ldots, X_n)$ is a unique element of $Y \in C_1$ which has the smallest mean-square distance from X_{n+1}, that is a function $\hat{Y} = f(X_1, \ldots, X_n)$ such that

$$||X_{n+1} - \hat{Y}||^2 = \inf_{Y \in C_1^*} E|X_{n+1} - Y|^2.$$

The projection theorem also says that $\hat{Y} = E_{C_1^*}(X_{n+1}) = E(X_{n+1}|X_1, X_2, \ldots, X_n)$, can uniquely be obtained by solving the prediction equations

$$E(Y\hat{Y}) = E(YX_{n+1}),$$

for every $Y \in C_1^*$. Since, Y is any (nonlinear) measurable function $f(X_1, \ldots, X_n)$, it is not easy to get the optimal predictor of X_{n+1} using the prediction equation (2.3). However, if we restrict ourselves to the closed span C_1 of $(X_1, \ldots, X_n)$, we can solve the prediction equation to obtain the best projection of X_{n+1} into the closed span C_1, namely the unique best linear mean-square predictor. In this case, all elements of C_1 are of the form $Y = \sum_{i=1}^{n} \alpha_i X_i$, for some real numbers $\alpha_i, i = 1, \ldots, n$ therefore the best linear predictor of X_{n+1} is an element

$$\hat{X}_{n+1} = \sum_{i=1}^{n} \alpha_i^* X_i,$$

where α_i^* are obtained uniquely from the prediction equations given by

$$\sum_{i=1}^{n} \alpha_i^* E(X_i X_j) = E(X_{n+1} X_j), \; j = 1, 2, \ldots, n. \qquad (2.5)$$

Writing $\boldsymbol{\alpha}^* := (\alpha_1^*, \ldots, \alpha_n^*)$, and

$$\boldsymbol{\Gamma}_n := \begin{pmatrix} \sigma^2 & \gamma(1) \cdots \gamma(n-1) \\ \gamma(1) & \sigma^2 \;\; \cdots \gamma(n-2) \\ \vdots & \vdots \;\; \ddots \;\;\;\; \vdots \\ \gamma(n-1) & \cdots \;\; \cdots \;\;\;\; \sigma^2 \end{pmatrix}$$

and $\boldsymbol{\gamma}_n := (\gamma(1), \ldots, \gamma(n))'$, we can write (2.5) as

$$\boldsymbol{\Gamma}_n \boldsymbol{\alpha}_n^* = \boldsymbol{\gamma}_n. \tag{2.6}$$

This system of equations will have a unique solution, provided $\boldsymbol{\Gamma}_n$ is not singular, which is satisfied when the function $\gamma(h)$ positive definite. If $\boldsymbol{\Gamma}_n$ is singular, then the best linear predictor of X_{n+1} will have infinitely many alternative representations in terms of $X_1, \ldots, X_n$.

Although simpler to calculate, best linear predictors often are inferior to best predictors, unless the relationship between X_{n+1} and $X_1, \ldots, X_n$ is linear; see Example 2.1.1. Note that if X_n is a Gaussian time series, then the conditional expectation $E(X_{n+1}|X_1, \ldots, X_n)$ is a linear function of $(X_1, \ldots, X_n)$ and in this case the best mean-square predictor and the best linear mean-square predictors coincide.

Example 2.1.2 (Brockwell and Davis 1991). Consider the stationary discrete time series

$$X_t = A\cos(\omega t) + B\sin(\omega t),\ t \in \mathbb{Z}, \tag{2.7}$$

where $\omega \in (0, \pi)$ is a constant, A and B are uncorrelated r.v's with zero-mean and variance σ^2. The mean and the variance of the series are given respectively by $E(X_t) = 0$ and

$$V(X_t) = \cos^2(\omega t)V(A) + \sin^2(\omega t)V(B) = \sigma^2.$$

For any h

$$\begin{aligned}\gamma(h) &= E(X_t X_{t+h}) \\ &= \sigma^2(\cos(\omega t)\cos\omega(t+h) + \sin(\omega t)\sin\omega(t+h)) \\ &= \sigma^2\cos(\omega h),\end{aligned}$$

so that the time series (2.7) is second-order stationary. Now consider the best linear predictor of X_3 given by

$$\hat{X}_3 = \alpha_1 X_1 + \alpha_2 X_2.$$

From (2.6) it follows that

$$\begin{bmatrix} \sigma^2 & \gamma(1) \\ \gamma(1) & \sigma^2 \end{bmatrix} \begin{bmatrix} \alpha_1 \\ \alpha_2 \end{bmatrix} = \begin{bmatrix} \gamma(1) \\ \gamma(2) \end{bmatrix}. \tag{2.8}$$

Solving (2.8) for (α_1, α_2), we get $\alpha_1 = 2\cos(\omega)$, $\alpha_2 = -1$, so that the best linear predictor is given by $\hat{X}_3 = 2\cos(\omega)X_2 - X_1$. Note that the prediction error is

$$\begin{aligned}
E(X_3 - \hat{X}_3)^2 &= E((X_3 - 2\cos(\omega)X_2 + X_1))^2 \\
&= E((X_3 - X_1)^2 - 4\cos(\omega)X_2(X_3 - X_1) + 4\cos^2(\omega)X_2^2) \\
&= 2\sigma^2 - 2\gamma(2) + 4\cos^2(\omega)\sigma^2 \\
&= 2\sigma^2(1 - \cos(2\omega)) + 4\cos^2(\omega)\sigma^2 \\
&= 0,
\end{aligned}$$

since for any ω, $cos(2\omega) = 2\cos^2(\omega) - 1$. Hence X_3 is predicted from X_2 and X_1 without any error, which means that

$$X_3 \equiv 2\cos(\omega)X_2 - X_1.$$

Similarly, from stationarity

$$\hat{X}_4 = 2\cos(\omega)X_3 - X_2,$$

with a mean-square error 0. The projection theorem guarantees that there is a uniquely defined predictor $\hat{X}_4$. However, $\hat{X}_4$ has infinitely many linear representations in terms of X_1, X_2, X_3, but by the projection theorem they should give the same predictor. This is due to the fact that $(\alpha_1, \alpha_2, \alpha_3)$ in the representation $\hat{X}_4 = \sum_{i=1}^{3} \alpha_i X_i$ satisfies

$$\begin{bmatrix} \sigma^2 & \gamma(1) & \gamma(2) \\ \gamma(1) & \sigma^2 & \gamma(1) \\ \gamma(2) & \gamma(1) & \sigma^2 \end{bmatrix} \begin{bmatrix} \alpha_1 \\ \alpha_2 \\ \alpha_3 \end{bmatrix} = \begin{bmatrix} \gamma(1) \\ \gamma(2) \\ \gamma(3) \end{bmatrix}. \tag{2.9}$$

However, the 3×3 matrix on the right and side of Eq. (2.9) is singular, giving infinitely many solutions for $(\alpha_1, \alpha_2, \alpha_3)$. It is easy to check that the determinant

$$|\Gamma_3| = \begin{vmatrix} 1 & \cos(\omega) & \cos(2\omega) \\ \cos(\omega) & 1 & \cos(\omega) \\ \cos(2\omega) & \cos(\omega) & 1 \end{vmatrix} = 0.$$

In fact, for any $h > 0$, the future values of the time series X_{t+h} given in (2.7) can be predicted with 0 mean-square error in terms of the linear combination of its observed values. Notice also that the time series (2.7), as well as its covariance function, is periodical with period 2π.

Definition 2.1.2. We call a time series deterministic, if for any $h > 0$, the optimal predictor of X_{t+h}, $\hat{X}_{t+h}$ can be predicted in terms of $(X_t, X_{t-1}, \ldots)$ with zero prediction error.

2.1.3 Nonlinear Representations

In the previous section, we saw that if we are interested only in linear predictors due to its simplicity, then from (2.6), we only need to know the second-order moments to calculate the best linear predictor. Due to the Wold decomposition theorem, (2.1) is the most general model we can use for obtaining such linear predictions. However, we also see that unless the process is Gaussian, the best linear predictor is inferior to the best predictor which is a nonlinear function of the observed time series. Suppose that our time series is not Gaussian and we are not content with the best linear predictor. In this case, we will have to look beyond linear processes and second-order covariance structures. This situation is very common particularly in environmental sciences and economy.

The crucial restriction in the Wold decomposition theorem is that the linear representation is given in terms of an uncorrelated white noise process, so that this representation serves as a model only for the second order moments of the stationary process. Under what conditions, can we represent a (strictly) stationary process in terms of an independent and identically distributed input process (Z_t)? If this is possible, then we should be able to model the whole probability structure of the process in terms of this independent and identically distributed input process.

In Sect. 2.1.2, in order to obtain best linear predictor we looked at the Hilbert space generated by the closed span C_1 of $(X_s, s \leq t)$, with the inner product $< X, Y >= Cov(X, Y)$. The members of this Hilbert space are made up of only the linear combinations of $(X_s, s \leq t)$ and their mean-square limits. The projection theorem then gave us the optimal linear predictors for X_{t+h} as projection of X_{t+h} in this closed span. If we want to extend these results to optimal (nonlinear) projections, we need to look for much more general setup. Now, consider again the set $(X_s, s \leq t)$ and consider the set of all r.v's with finite variance which are measurable with respect to this set, that is the set

$$C_2 := \{Y = g(X_s), s \leq t : G \text{ measurable and } V(Y) \leq \infty\}.$$

This subspace is a Hilbert space and clearly contains the closed span of $(X_s, s \leq t)$. If we can find a closed orthogonal basis for this subspace, then any element Y of this subspace can be written as a linear combination of the orthogonal basis functions, and projection theorem will give us the optimal projection of X_{t+h} in terms of the elements of this subspace.

Definition 2.1.3. Hermite polynomials $H_n(x)$ of degree n are defined as

$$\int_{-\infty}^{\infty} H_n(x)H_m(x)\frac{1}{\sqrt{2\pi}}\exp(-x^2/2)dx = n!I_{n,m}, \quad n, m = 0, 1, 2, \ldots \tag{2.10}$$

where

$$I_{n,m} = \begin{cases} 1, \, n = m; \\ 0, \, n \neq m. \end{cases}$$

These polynomials form a closed and complete orthogonal system in the Hilbert space $\mathcal{L}^2(\mathbb{R}, \mathcal{B}, \frac{1}{\sqrt{2\pi}} \exp(-x^2/2)dx)$ where the inner product is defined as

$$< f, g >= \int_{-\infty}^{\infty} f(x)g(x)\frac{1}{\sqrt{2\pi}} \exp(-x^2/2)dx. \tag{2.11}$$

Hence, every Borel measurable function g such that

$$\int_{-\infty}^{\infty} g^2(x)\frac{1}{\sqrt{2\pi}} \exp(-x^2/2)dx < \infty,$$

can be written as a linear combination (or as a limit) of these Hermite polynomials

$$g(x) = \lim_{N\to\infty} \sum_{n=0}^{N} \frac{g_n}{n!} H_n(x), \tag{2.12}$$

where, the coefficients g_n are given by

$$g_n = \int_{-\infty}^{\infty} g(x)H_n(x)\frac{1}{\sqrt{2\pi}} \exp(-x^2/2)dx.$$

The convergence of (2.12) is in the mean-square sense

$$\lim_{N\to\infty} \int_{-\infty}^{\infty} (g(x) - \sum_{n=0}^{N} \frac{g_n}{n!} H_n(x))^2 \frac{1}{\sqrt{2\pi}} \exp(-x^2/2)dx = 0.$$

Hermite polynomials are given by

$$H_n(x) = (-1)^n \frac{1}{\sqrt{2\pi}} \exp(x^2/2)\frac{d^n}{dx^n} \exp(-x^2/2),$$

and they can also be calculated from the recursions

$$H_{n+1}(x) = xH_n(x) - \frac{d}{dx} H_n(x),$$

or

$$H_n(x) = xH_n(x) - nH_{n-1}(x).$$

The first five Hermite polynomials are given by

$$H_0(x) = 1$$
$$H_1(x) = x$$

$$H_2(x) = x^2 - 1$$
$$H_3(x) = x^3 - 3x$$
$$H_4(x) = x^4 - 6x^2 + 3.$$

Note that the inner product (2.11) is an integral with respect to the standard Gaussian density and hence the Hermite polynomials are orthogonal with respect to the standard normal probability distribution. Instead of Hermite polynomials, we can define Hermite functions

$$\psi_n(x) := \frac{1}{\sqrt{n!2^n\sqrt{2\pi}}} \exp(-x^2/2) H_n(x).$$

Hermite functions are normalized versions of the Hermite polynomials, therefore they form a closed and complete orthonormal basis for $\mathcal{L}^2(\mathbb{R}, \mathcal{B}, \frac{1}{\sqrt{2\pi}} \exp(-x^2/2)\, dx)$. The Hermite polynomials are orthogonal with respect to the standard Normal distribution, although it is possible to define Hermite polynomials which are orthogonal with respect to the Normal distribution $N(0, \sigma^2)$. The closed linear span of Hermite polynomials is the space of all polynomials, therefore any element of $\mathcal{L}^2(\mathbb{R}, \mathcal{B}, \frac{1}{\sqrt{2\pi}} \exp(-x^2/2)dx)$ can be written as a polynomial of finite- or infinite-order. Elements of $\mathcal{L}^2(\mathbb{R}, \mathcal{B}, \frac{1}{\sqrt{2\pi}} \exp(-x^2/2)dx)$ are deterministic functions. How can we pass from polynomial representation for deterministic functions to random functions? Consider now the simple case: let X be a standard Gaussian random variable and consider the set of all r.v's Y which are measurable functions of X with finite variances, that is the set

$$\mathcal{C}(X) := \{Y = g(X) : g \text{ measurable and } V(Y) < \infty\}.$$

Define the inner product $< Y_1, Y_2 >= Cov(Y_1, Y_2)$ on this set. This set forms a Hilbert space. The above results on Hermite polynomials immediately suggest the construction of the orthogonal base for this Hilbert space; let $H_n(X), n = 0, 1, 2, \ldots$ be r.v's, where $H_n(x)$ are Hermite polynomials defined in (2.10). Then any measurable function Y (of X) can be written as

$$Y = \sum_{n=0}^{\infty} \frac{g_n}{n!} H_n(X),$$

where

$$g_n = Cov(Y, H_n(X))$$

and

$$\lim_{N\to\infty} V(Y - \sum_{n=0}^{N} \frac{g_n}{n!} H_n(X)) = 0.$$

Note that if we restrict ourselves to the Hilbert space of the closed span generated by X, then any member of this space is a linear function of X. If we extend this space to include all measurable functions with finite variances, then the elements are again represented by a linear function, but this time a linear combination of (random and nonlinear) Hermite polynomials or simply polynomials of finite- or infinite-order.

Now let us introduce more complexity and start with a collection of standard Gaussian r.v's $(X_s, s \le t)$ and consider the space of all measurable functions defined on this collection with the usual inner product defined over it. Any element of this Hilbert space can be written as a linear combination of products of Hermite polynomials. Here we will not enter into details, which can be found in Terdik (1999). As an example, consider standard Gaussian r.v's $(X_1, X_2, \ldots, X_n)$ with covariances $r(i, j)$. The first five (random) Hermite polynomials which form the orthogonal basis for the Hilbert space of all measurable functions defined on $(X_1, X_2, \ldots, X_n)$ are given by

$$\begin{aligned}
H_0 &= 1 \\
H_1(X_1) &= X_1 \\
H_2(X_1, X_2) &= X_1X_2 - r(1,2) \\
H_3(X_1, X_2, X_3) &= X_1X_2X_3 - r(1,2)X_3 - r(1,3)X_2 - r(2,3)X_1 \\
H_4(X_1, X_2, X_3, X_4) &= X_1X_2X_3X_4 - r(1,2)X_3X_4 - r(1,3)X_2X_4 \\
&\quad - r(1,4)X_2X_3 - r(2,3)X_1X_4 - r(2,4)X_1X_3 - r(3,4)X_1X_2 \\
&\quad + r(1,2)r(2,3) + r(1,3)r(2,4) + r(1,4)r(2,3).
\end{aligned}$$

Therefore any element of this Hilbert space can be represented as sums of products of polynomials given in the form

$$\sum_{p=0}^{\infty} \sum_{i_1=1}^{\infty} \cdots \sum_{i_p=1}^{\infty} a_{i_1 i_2 \cdots i_p} \prod_{\nu=1}^{p} X_{i_\nu},$$

with the convention $\prod_{\nu=1}^{0} X_{i_\nu} = 1$.

The following remarkable result due to Nisio (1960) extends this polynomial representation to any strictly stationary time series.

Definition 2.1.4. Let Z_t be independent, standard Gaussian r.v's. The polynomial representation

$$
\begin{aligned}
Y_t^{(m)} &= \sum_{p=1}^{m} \sum_{i_1=-\infty}^{\infty} \sum_{i_2=-\infty}^{\infty} \cdots \sum_{i_m=-\infty}^{\infty} g_{i_1 i_2 \cdots i_m} \prod_{v=1}^{p} Z_{t-i_v} \\
&= \sum_{i_1=-\infty}^{\infty} g_{i_1} Z_{t-i_1} \\
&\quad + \sum_{i_1=-\infty}^{\infty} \sum_{i_2=-\infty}^{\infty} g_{i_1 i_2} Z_{t-i_1} Z_{t-i_2} \\
&\quad + \sum_{i_1=-\infty}^{\infty} \sum_{i_2=-\infty}^{\infty} \sum_{i_3=-\infty}^{\infty} g_{i_1 i_2 i_3} Z_{t-i_1} Z_{t-i_2} Z_{t-i_3} \\
&\quad + \cdots \\
&\quad + \sum_{i_1=-\infty}^{\infty} \sum_{i_2=-\infty}^{\infty} \cdots \sum_{i_m=-\infty}^{\infty} g_{i_1 i_2 \cdots i_m} Z_{t-i_1} Z_{t-i_2} \cdots Z_{t-i_m},
\end{aligned}
$$

is called a Volterra series of order m. We will call

$$
Y_t = \sum_{p=1}^{\infty} \sum_{i_1=-\infty}^{\infty} \sum_{i_2=-\infty}^{\infty} \cdots \sum_{i_p=-\infty}^{\infty} g_{i_1 i_2 \cdots i_p} \prod_{v=1}^{p} Z_{t-i_v}, \tag{2.13}
$$

the (infinite-order) Volterra series expansion.

Theorem 2.1.1 (Nisio 1960). *Let X_t be any strictly stationary time series. Then there exists a sequence of Volterra series $Y_t^{(m)}$ such that*

$$
\lim_{m\to\infty} Y_t^{(m)} \stackrel{d}{=} X_t,
$$

in the sense that for any n and for any $\theta_j, |j| \le n$ as $m \to \infty$,

$$
|E\exp(i\theta_{-n}X_{-n} + \cdots + i\theta_n X_n) - E\exp(i\theta_{-n}Y_{-n}^{(m)} + \cdots + i\theta_n Y_n^{(m)})| \to 0.
$$

If further X_t is Gaussian, then X_t can be represented by

$$
X_t = \sum_{j=-\infty}^{\infty} g_j Z_{t-j}.
$$

The proof of the result above is beyond the scope of this book. However, we only mention that the proof is centered around first finding a polynomial representation for a uniformly bounded time series using Hermite polynomials and then extending the results to any time series using Slutsky type arguments. Although assumption of independence of the innovations Z_t is essential, normality is not essential.

One can define Hermite polynomials orthogonal with respect to any probability distribution and therefore the Volterra representation can be given in terms of any other distribution.

Nisio's theorem essentially says that although most stationary time series will have a linear representation in terms of uncorrelated innovations (Wold theorem), it will have very complicated, nonlinear representations in terms of the independent innovations. Therefore Nisio's theorem can be seen as the extension of the Wold decomposition theorem. While modeling with ARMA classes, we often require that the innovations are Gaussian, hence the modeling is restricted to the Volterra series of order 1. Note that (2.13) is a representation for the whole probability structure of the time series as contrast to the representation (2.1), which is representation for the covariance structure of the series. Let us give some examples to highlight this difference.

Example 2.1.3. Consider the process

$$X_t = Z_t + \alpha Z_{t-1} Z_{t-2},\ t \in \mathbb{Z},$$

where (Z_t) is a zero-mean i.i.d. sequence with finite variance. It is easy to verify that X_t is covariance stationary with zero mean and constant variance and

$$Cov(X_t X_{t+h}) = 0.$$

Hence, X_t is an uncorrelated time series, whose correlation structure is equivalent to that of the independent innovation process Z_t. However, the probability structure of X_t is different from that of Z_t. For example,

$$E(X_t | X_{t-1}, X_{t-2}, \ldots) = \alpha Z_{t-1} Z_{t-2},$$

whereas

$$E(Z_t | Z_{t-1}, \ldots) = 0.$$

Hence, by looking at the second-order properties, we can decide that there is no structure in X_t to model, but certainly X_t has structure which should be studied by its higher-order moments. In fact, if Z_t are also Gaussian, then all cumulants higher then the second-order are zero. However, it is easy to check that the higher-order cumulants of X_t are not identically equal to zero.

Example 2.1.4 (All-pass models). The class of uncorrelated but not independent processes is quite rich. In fact, one can encounter uncorrelated but not independent linear processes. The class of all pass models (Andrews et al., 2006) is one example, which can be constructed within the ARMA class by choosing autoregressive and moving average polynomials in such a manner that the roots of the autoregressive polynomial are reciprocals of the roots of the moving average polynomial or vice-versa. Assume that $\phi_p(z) = 1 - \phi_1 z - \cdots - \phi_p z^p$ is a causal autoregressive polynomial so that $\phi_p(z) \neq 0$ for $|z| \leq 1$. Define the moving average polynomial

$$\begin{aligned}\theta_p(z) &= \frac{z^p \phi_p(z^{-1})}{-\phi_p} \\ &= -(B^p - \phi_1 B^{p-1} - \cdots - \phi_p)/\phi_p,\end{aligned}$$

and consider the time series which satisfies the difference equation

$$\phi_p(B)X_t = \theta_p(B)Z_t,$$

where (Z_t) is an i.i.d. sequence with zero-mean and finite variance σ^2. The time series X_t has some interesting properties.

1. X_t is not invertible, but is causal.
2. The time series satisfies the difference equation

$$\begin{aligned}&X_t - \phi_1 X_{t-1} - \cdots - \phi_p X_{t-p} \\ &= Z_t + \frac{\phi_{p-1}}{-\phi_p} Z_{t-1} + \cdots + \frac{\phi_1}{\phi_p} Z_{t-p+1} - \frac{1}{\phi_p} Z_{t-p},\end{aligned}$$

so that, when $p = 1$, and $|\phi_1| < 1$, first order all-pass model is given by

$$X_t - \phi_1 X_{t-1} = Z_t - \frac{1}{\phi_1} Z_{t-1}$$

and the second-order all-pass model is given by

$$X_t - \phi_1 X_{t-1} - \phi_2 X_{t-2} = Z_t + \phi_1/\phi_2 Z_{t-1} - 1/\phi_2 Z_{t-2}.$$

3. The spectral density of X_t is given by the constant function

$$f(w) = \frac{\sigma^2}{\phi_p^2 2\pi},$$

for every $w \in [-\pi, \pi]$, so that the X_t process is uncorrelated. Further, if Z_t are Gaussian, then X_t is an i.i.d. sequence with distribution $N(0, \phi_p^{-2}\sigma^2)$. However, if Z_t are not Gaussian, then for $p \geq 1$, X_t is not an independent sequence.
4. Since all-pass processes are uncorrelated but not independent, the usual second-order techniques based on autocorrelation and partial autocorrelation functions cannot identify an all-pass model, as these functions will report that the data have no structure. Inferential methods based on Gaussian likelihood or least squares do not give the desired results when fitting all-pass models. Instead, inferential techniques based on cumulants of order greater than two are often used; see Andrews et al. (2006) for details. The need for inferential methods

based on cumulants higher than two or approximate methods based on non-Gaussian likelihoods are quite universal while modeling nonlinear data.

Let us make a summary of the results:

1. $Y_t = \sum_{j=0}^{\infty} \psi_j Z_{t-j}$, Z_t uncorrelated r.v's is called a linear, causal representation.
2. $Y_t = \sum_{j=0}^{\infty} \psi_j Z_{t-j}$, Z_t i.i.d. r.v's is called a linear causal model.
3. If further, Z_t are Gaussian, then any linear representation is also a linear model and $Y_t = \sum_{j=0}^{\infty} \psi_j Z_{t-j}$ is called the Gaussian causal linear model.
4. (Almost) all non-deterministic, second-order stationary time series X_t have a unique linear representation in terms of uncorrelated innovations. In this case, moments of X_t and Y_t up to second-order coincide. However, moments of order higher then two, need not coincide, except when X_t is Gaussian.
5. (Almost) all strictly stationary time series X_t has a (infinite-order) Volterra series expansion

$$Y_t = \sum_{p=1}^{\infty} \sum_{i_1=-\infty}^{\infty} \sum_{i_2=-\infty}^{\infty} \cdots \sum_{i_p=-\infty}^{\infty} g_{i_1 i_2 \cdots i_p} \prod_{v=1}^{p} Z_{t-i_v},$$

for some i.i.d. innovation sequence Z_t.
6. Therefore, X_t has a linear causal model in terms of an i.i.d. innovation sequence Z_t iff it has a first order, one-sided Volterra series expansion, that is, iff X_t is a Gaussian process. Hence, the class of causal, linear models is not dense within the class of stationary time series.
7. If we want only the best linear predictors for future values of the time series, then we can work with linear causal representations, as we do not need information other than the second-order moments to obtain best linear predictors.
8. On the other hand, if we want the best predictor, then we need to look for models within the general class of Volterra series expansions.

Working with linear models, particularly with Gaussian linear model, is relatively simple, whereas working directly with the general, infinite order Volterra series is very difficult, if not impossible. For example, it not possible to give conditions of stationarity on the kernels $g_{i_1 i_2 \cdots i_p}$. Further, time series such as

$$X_t = Z_t + \alpha Z_{t-1} Z_{t-2},$$

or

$$X_t = Z_t + \alpha Z_{t-1}^2,$$

where Z_t is a sequence of independent r.v's, are not invertible (Granger and Andersen 1978). Hence, one would expect that Volterra series expansions have limited use as models for predicting future values, unless the input process (Z_t) is observable. Therefore, to model nonlinear data, we need to look for sub-classes of Volterra series expansions which are easier to study.

There are many ways a process can be nonlinear. Therefore, in order to come up with fairly general and useful classes of nonlinear models, we need to look at certain aspects of the probability structure of the processes to understand and describe the underlying nonlinear behavior. Since linear and nonlinear processes differ on moments higher than order two, particular emphasis has to be given to studying the higher moments and tails of the stationary distributions of the processes. We now look at certain aspects of nonlinear processes which may indicate how we should construct useful nonlinear models.

2.1.4 Sensitive Dependence on Initial Conditions, Lyapunov Exponents

The most striking feature of nonlinear processes is the strong dependence on initial conditions and the noise amplification. Let us start with deterministic difference equations, representing some dynamic system in discrete time. Suppose that $x_n = f(x_{n-1})$ defines a deterministic difference equation, for some function f. Starting from the initial condition x_0, let

$$x_n = f^{(n)}(x_0) = f(f(\cdots(f(x_0))))$$

be the value of the system after n iterations. Now let us disturb the initial starting value x_0 by a small number δ_0 to $x_0 + \delta_0$. We would be interested in the impact of this initial disturbance on the dynamic system after n iterations, namely

$$\delta_n = f^{(n)}(x_0 + \delta_0) - f^{(n)}(x_0),$$

and in particular, we may be interested in the limit as $n \to \infty$ and $\delta_0 \to 0$. If f is a linear function so that

$$x_n = \alpha x_{n-1} + \beta,$$

then it is easy to verify that

$$x_n = \alpha^n x_0 + \beta(\alpha^{n-1} + \alpha^{n-2} + \cdots + 1),$$

so that

$$\delta_n = \alpha^n \delta_0$$

and

$$\frac{f^{(n)}(x_0 + \delta_0) - f^{(n)}(x_0)}{f^{(n)}(x_0)} = O(\delta_0).$$

On the other hand, consider the logistic difference equation

$$x_{n+1} = \alpha x_n(1 - x_n). \tag{2.14}$$

Here, α is called the *driving parameter*. Now start with an initial value $x_0 \in (0, 1)$. This difference equation has a peculiar behavior for different values of α. If $\alpha \in [0, 3)$, then as $n \to \infty$, the difference equation converges to a single number. When $\alpha = 3.0$, then x_n no longer converges but oscillates between two values. As α is increased, in the limit x_n oscillates between increasingly different numbers, and for $\alpha > 3.57$ the sample path behavior of $x_1, \ldots, x_n$ is chaotic, resembling a sample path of a random process. This chaotic behavior is due to the sensitive dependence of the difference equation on the initial value x_0 for increasing values of the parameter α. In fact, when $\alpha = 4$, this difference equation has an analytical solution

$$x_n = \sin^2(2^n \beta\pi),$$

where $\beta \in [0, 1)$ is a function of the initial value x_0. When $x_0 \in [0, 1]$ then β is almost surely an irrational number which will have different dyadic representation for each iteration n causing a chaotic behavior of the sample path $x_1, \ldots, x_n$. This chaotic behavior caused by the dependence on the initial condition is quite common for nonlinear difference equations and measuring this dependence on initial conditions may give a degree of nonlinearity that exists in a difference equation (Fig. 2.1).

Lyapunov exponent λ of a dynamic system is a quantity that characterizes this dependence on the initial conditions through the relationship

$$\delta_n \sim e^{n\lambda}\delta_0. \tag{2.15}$$

One can give an heuristic argument for the definition of the Lyapunov exponent. Assume that $x_n = f(x_{n-1})$ and f is everywhere differentiable. Then using first-order Taylor series approximation,

$$\begin{aligned}\delta_n &= f^{(n)}(x_0) - f^{(n)}(x_0 + \delta_0)\\ &\sim \delta_0 \frac{d}{dx} f^{(n)}(x_0).\end{aligned}$$

Here,

$$\frac{d}{dx} f^{(n)}(x_0) = \frac{d}{dx} f^{(n)}(x + \delta_0),$$

calculated at $x = x_0$. Since

$$f^{(n)}(x) = f(f(\cdots(f(x)))),$$

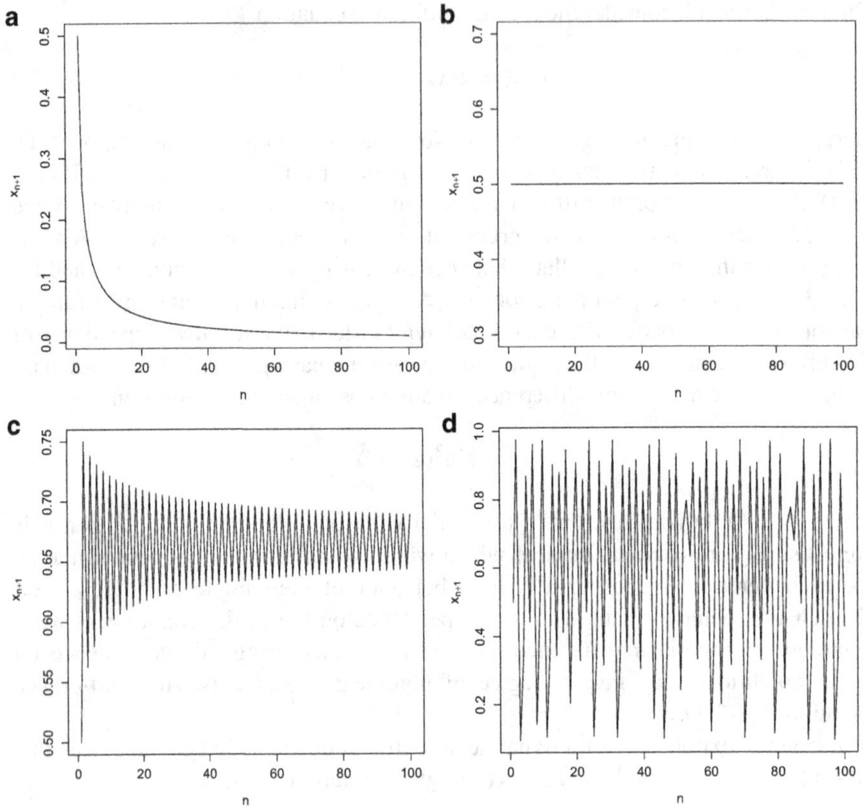

Fig. 2.1 Simulated samples of 100 observations from the logistic difference equation; (**a**) $\alpha = 1.0$; (**b**) $\alpha = 2.0$ with $x_0 = 0.5$; (**c**) $\alpha = 3.0$; and (**d**) $\alpha = 3.9$

by the chain rule

$$\frac{d}{dx} f^{(n)}(x_0) \sim \exp(n \log \frac{d}{dx} f(x_0)),$$

assuming that each of the factors $\frac{d}{dx} f(x_n) \sim \frac{d}{dx} f(x_0)$ have comparable sizes. Therefore it is reasonable to consider

$$\lambda = \lim_{n\to\infty} \frac{1}{n} \ln |\frac{d}{dx} f^{(n)}(x_0)|,$$

as an indicator of the degree of dependence on the initial conditions, or equivalently, as an indicator of the degree of nonlinearity through the relationship (2.15). When

$\lambda < 0$, the dynamic system is called dissipative or non-conservative. Such a dynamic system exhibits asymptotic stability, typically resulting from damped harmonic oscillations. When $\lambda = 0$, the system is called conservative and is said to be Lyapunov stable. The case $\lambda > 0$ corresponds to an unstable system, resulting in chaotic sample paths.

Quantifying the degree of dependence on initial conditions or equivalently quantifying the degree of nonlinearity of stochastic difference equations representing dynamic random systems needs more attention. This is due to the fact that the system in each iteration is perturbed by a random noise. Since each sample path of the dynamic system will have different realizations of random shocks, it makes sense to consider the divergence of expected values (ensemble average) of these sample paths.

Example 2.1.5. Consider the stochastic difference equation

$$X_{n+1} = A_n X_n + B_n, \; n \geq 0. \tag{2.16}$$

Here, for each n, A_n and B_n are dependent scalar r.v's, but the pair (A_n, B_n) is an i.i.d. sequence. The stochastic difference equation in (2.16) and its multivariate versions, where X_n, B_n are random vectors in $\mathbb{R}^d$ and A_n are $d \times d$ matrices, often appear as basis for studying many different forms of nonlinear time series and will be revisited in future chapters. Starting from X_0 and upon n iterations the process will be in the state (Brandt 1986)

$$X_n = \sum_{j=0}^{n-1} \left(\prod_{i=n-j}^{n-1} A_i \right) B_{n-j-1} + \left(\prod_{i=0}^{n-1} A_i \right) X_0.$$

Conditional on the two initial values $X_0 = x_0$, $X_0 = x_0 + \delta_0$, we can quantify the deviation in the sample paths with

$$\begin{aligned} \delta_n &= f^{(n)}(x_0) - f^{(n)}(x_0 + \delta_0) \\ &= \left(\prod_{i=0}^{n-1} A_i \right) \delta_0. \end{aligned}$$

Note that δ_n is a random variable. Lyapunov exponent λ can now be defined as the expected deviation on the sample paths upon n iteration, conditional on two initial realizations $X_0 = x_0$, $X_0 = x_0 + \delta_0$ through the relation

$$E(\delta_n) = e^{n\lambda} \delta_0,$$

in which case

$$\lambda = E[\frac{1}{n}\log(|A_0||A_1|\cdots|A_n|)]$$
$$= E\log(A_0).$$

(Here, $|A_i|$ rather than A_i appear in the expression to insure that $\frac{d}{dx}f(x_n) = A_n$ have comparable sizes and contributions). Brandt (1986) also shows that if the process is dissipative, that is, $\lambda = E\log|A_0| < 0$ and $E(\log|B_0|)^+ < \infty$ then

$$\lim_{n\to\infty} X_n = \sum_{j=0}^{\infty}\left(\prod_{i=n-j}^{n-1} A_i\right) B_{n-j-1},$$

is the unique stationary solution of (2.16). If $\lambda > 0$, one would expect a chaotic behavior, without a stationary limit for the difference equation. Hence, the existence of stationary solutions for the stochastic difference equation given in (2.16) depends on the degree of dependence on the initial conditions.

Example 2.1.6 (Fan and Yao 2003). Consider again the dynamic system defined by the deterministic difference equation $x_t = f(x_{t-1})$, where f is an everywhere differentiable function. But now we disturb the dynamic system at each iteration by a small i.i.d. noise Z_t, resulting in the stochastic difference equation

$$X_t = f(X_{t-1}) + Z_t.$$

The process that satisfies this difference equation is called the first order nonlinear autoregressive model of order 1 (NLA(1)). In order to facilitate arguments, assume further that Z_t are independent of $(X_s, s < t)$. It may be interesting to know how much these additive noises affect the variation in this process after n steps. Again, let us consider two sample paths of this process, starting from $X_0 = x_0$ and $X_0 = x_0 + \delta_0$ and look at how much (on average) these two sample paths diverge after n iterations. Note that, if f is a linear function, then with any uncorrelated noise with finite variance, the divergence between these two sample paths would be of order $O(\delta_0)$. Let $f^{(n)} = f(f(\cdots f(x)))$ be the n fold composition of f. Then by the arguments given in Fan and Yao (2003) which are based on iterative Taylor series expansions,

$$X_n = f^{(n)}(X_0) + \sum_{j=1}^{n-1}\prod_{k=j}^{n-1} f'(X_{n-k})Z_{n-j+1} + Z_n. \tag{2.17}$$

In general the derivatives $f'(X_{n-k})$ are functions of $Z_{n-k}, Z_{n-k-1}, \ldots Z_1$. However, if we assume that the random shocks are of small order, that is $|Z_n| < \eta < 1$ almost surely for every n, then by (2.17) for any fixed n,

$f(X_n) \sim f^{(n)}(X_0) + O(\eta)$, and this can be used as a second-order approximation in the arguments of the derivates to give

$$f(X_n) \sim f^{(n)}(X_0) + \sum_{j=1}^{n-1} \prod_{k=j}^{n-1} f'(f^k(X_0))Z_{n-j} + Z_n + O(\eta). \tag{2.18}$$

Let $\sigma_n^2(x_0) := V(X_n|X_0 = x_0)$ be the variance of the process after n iterations. Then from (2.18),

$$\sigma_n^2(x_0) = (1 + \sum_{j=1}^{n-1} \prod_{k=j}^{n-1} f'(f^{(k)}(x_0)))^2\sigma^2 + o(\eta).$$

Hence, even when the shocks Z_t are almost surely small, the variance of the process after n iterations is amplified by a quantity $(1 + \sum_{j=1}^{n-1} \prod_{k=j}^{n-1} f'(f^{(k)}(x_0)))^2$, which may be quite significant.

2.1.5 Limit Cycles

We have seen that the logistic difference equation given in (2.14) can have very different sample paths, from a constant to total chaotic behavior depending on the value of its parameter α. The region $\alpha \in [3.0, 3.7)$ is interesting, as the sample paths oscillate among a finite number of states. This type of limiting behavior is quite common in deterministic and stochastic dynamic systems, particularly involving population dynamics and is called limit cycle. Typically, dynamics of a population depends on many internal and external factors, the size of the population being one of these factors. As the population increases in size over passing a critical threshold, typically this has a negative influence on the reproductive and survival capacities of the population, lowering its growth rate. Moreover, as the population size goes down, these capacities tend to increase, increasing its growth rate. Hence, under equilibrium conditions the sample paths of a population size will show limit cycles, switching at random epochs. For example, consider the deterministic difference equation

$$N_{t+1} = N_t \frac{b}{(1 + aN_t)^c},$$

often used for modeling annual plant population. Here a, b and c are parameters of the model. The parameter a does not affect the dynamics of the model, whereas the parameter c has a very strong effect on the dynamics. Again, this deterministic difference equation will have very different sample path properties, depending basically on the values of the parameters b and c. For example, when $c = 1$, for

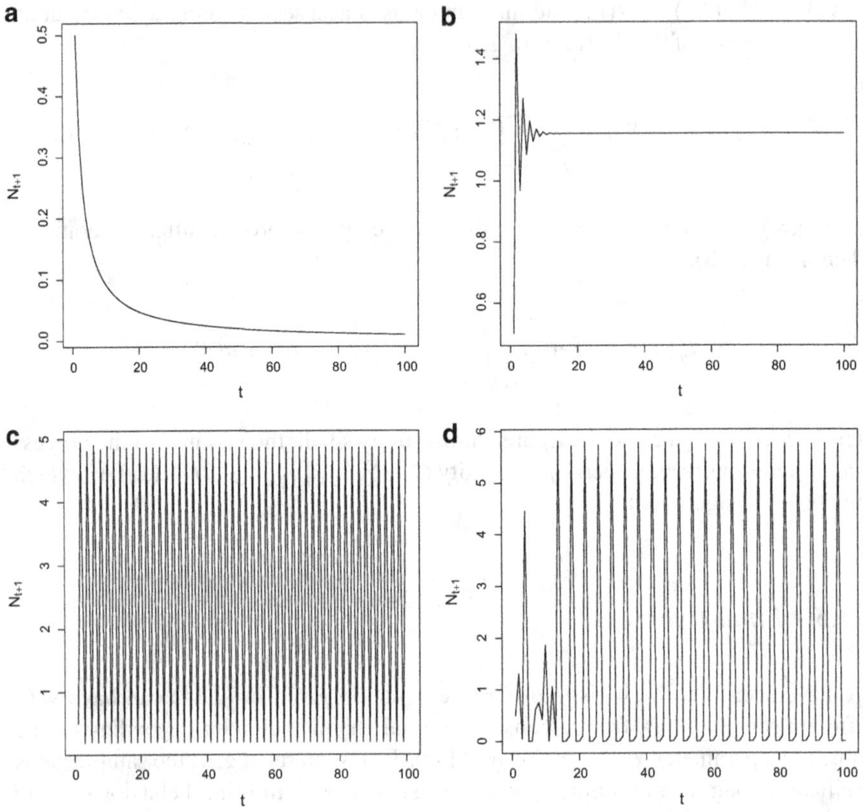

Fig. 2.2 Limit cycles of simulated samples of 100 observations (**a**) $a = 1, b = 1, c = 1$; (**b**) $a = 1, b = 10, c = 3$; (**c**) $a = 1, b = 50, c = 4$; and (**d**) $a = 1, b = 150, c = 10$

any value of b, the sample paths will converge monotonically to a constant, whereas when $c > 2$ and $b = 10$, the sample paths will show damped oscillations, finally converging to a constant. This sample path behavior then starts getting ever more erratic as b and c increase. For values of $b = 50$ and $c > 3.5$, the sample paths oscillate between fixed number of population sizes, and this behavior is called the stable limit cycles. Ultimately, for $b > 100$ and $c > 5$, the sample paths behave in a chaotic way. The limit cycles generated by several samples of size $n = 500$ are presented in Fig. 2.2. In random dynamic systems, stable limit cycle behavior can manifest itself in many different ways. For example, rather than switching between fixed number of values, the process can switch between different linear models at random epochs, depending on internal or external factors, resulting in many different piecewise linear models such as threshold models. These models will be discussed in Sect. 2.2.1.

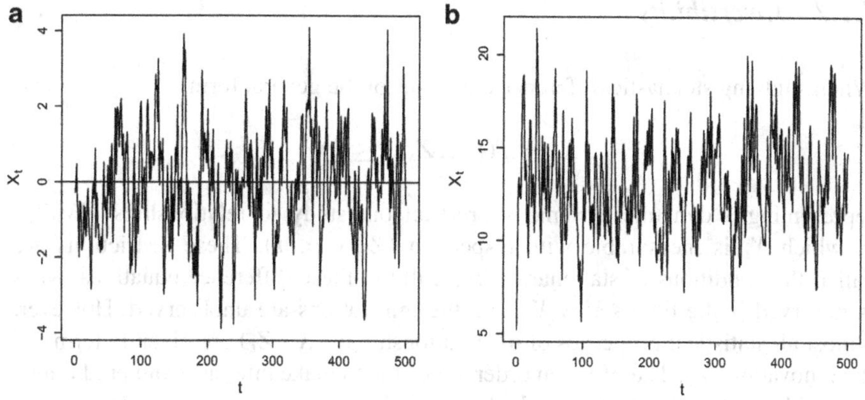

Fig. 2.3 Simulated Gaussian AR(1) model, $n = 500$ and parameter 0.7 (**a**). In (**b**) the same model with Gamma(4,1) residuals

2.1.6 *Time Reversibility*

A process X_t is time reversible if

$$(X_{t_1}, X_{t_2}, \ldots, X_{t_n}) \stackrel{d}{=} (X_{t_n}, X_{t_{n-1}}, \ldots, X_{t_1}),$$

for every n and $t_1, \ldots, t_n$. Gaussian processes are time reversible, and except for few special cases, non-Gaussian processes are time irreversible. In general, if a stationary time series is stationary and time reversible then for every k, kth order cumulants satisfy

$$C(-u_1, -u_2, \ldots, -u_k) = C(u_1, u_2, \ldots, u_k).$$

This a very strong condition and it is very unlikely that there will be many non-Gaussian processes that satisfy this condition. Therefore, time irreversibility must be a rule among nonlinear processes. Note that if X_t is a stationary time series and $Y_t = h(X_t)$ is a one-to-one transformation, then Y_t is time reversible if and only if X_t is time reversible. Therefore, fitting Gaussian time series models to transformed data cannot be a adequate method of dealing with nonlinearity. In other words, in most cases, we cannot get rid of nonlinearity by transformation of the data. The simplest way of checking reversibility is by plotting the data. In general, for a reversible stationary time series, the plots of $x_n, x_{n-1}, \ldots, x_1$ and $x_1, x_2 \ldots, x_n$ should look the same. Similarly, since time irreversibility is a characteristic of Gaussian data rather than linearity, a time series which is non-Gaussian should be treated as irreversible (see Fig. 2.3).

2.1.7 Invertibility

When studying stochastic difference equations of the general form

$$X_t = f(\{X_s, Z_s, s \leq t\}),$$

representing a dynamic system, we restrict our study to relationships (X_t, Z_t) in which X_t is measurable with respect to $(Z_s, s < t)$. These restrictions are called the conditions of stationarity. Typically in these difference equations, what is observed is the time series X_t, and the innovations are unobserved. However, almost all statistical properties of the relationship in (X_t, Z_t) are given in terms of the innovations Z_t. Therefore, in order to be able to make inference and predictions on the dynamic system, the residuals should be recovered from the observations $x_s, s < t$. The set of conditions which guarantee this possibility are called the conditions of invertibility, under which the innovations Z_t are measurable with respect to $(X_s, s < t)$. Note that, conditions of invertibility and stationarity are joint properties of the processes (X_t, Z_t), rather than being a property of X_t alone. Unfortunately these conditions, particularly conditions of invertibility, are not so easy to obtain for general nonlinear difference equations, except for some special cases. We will look at these conditions for specific cases, whenever possible.

2.2 A Selection of Nonlinear Time Series Models

According to Tjøstheim (1994), nonlinear models can be broadly classified into the following categories;

1. Parametric models
   - Parametric models for the conditional mean
   - Parametric models for the conditional variance
   - Mixed parametric models for the conditional mean and variance
   - Generalized state space models
2. Semiparametric and nonparametric models

The above classification is by no means exhaustive and mutually exclusive. In its most general form, a nonlinear model can be written as a stochastic difference equation

$$X_t = f(X_{t-1}, \dots, X_{t-p}, Z_t, Z_{t-1}, \dots, Z_{t-q}, \boldsymbol{\theta}), \tag{2.19}$$

for some integers p and q, model parameters $\boldsymbol{\theta}$ and some measurable function f which renders a stationary causal solution. Such general representation contains both nonlinear conditional mean and conditional variance components for X_t in terms of the past values. Often, it is easier to look at a simpler class of models

$$X_t = f(X_{t-1}, \dots, X_{t-p_1}, Z_{t-1}, \dots, Z_{t-q_1}, \boldsymbol{\theta}_1) + g(X_{t-1}, \dots, X_{t-p_2}, Z_{t-1}, \dots, Z_{t-q_2}, \boldsymbol{\theta}_2) Z_t, \tag{2.20}$$

for measurable functions f and g, separating the nonlinear models for the conditional mean and the conditional variance components. Taking g as a constant, for various combinations of f functions, we get subclasses of nonlinear models for the conditional mean, whereas taking f constant and for various combinations of g, we get subclasses of models for the conditional variance. The general class (2.19) can be classified as the class of mixed models, although Tjøstheim (1994) classifies (2.20) as the class of mixed models.

In this chapter, we give a brief description of some of these models.

2.2.1 *Parametric Models for the Conditional Mean*

These models represent the conditional mean function of the process X_t as a nonlinear function of the past observations, keeping the conditional variance constant. Hence, an appropriate general model is given as

$$X_t = f(\mathcal{F}_{t-1}, \boldsymbol{\theta}) + Z_t.$$

Here, the function f has a known parametric form, $\mathcal{F}_{t-1}$ is the sigma-field generated by X_t up to time $t-1$, Z_t is an i.i.d. sequence and $\boldsymbol{\theta}$ is an unknown parameter vector to be estimated. In some cases, the function f may also depend on other external processes. Several different forms of f give different classes of nonlinear models. One important subclass is the regime models or regime switching models. Models in this class are typically made up of several piecewise linear processes and the generating process switches from one linear model to another, depending on the value of an indicator. This indicator may be a random variable, such as the delayed value of the series itself, or it can be the value of a different, possibly latent process. Depending on the parameter values, such piecewise linear regime models are stationary but nonlinear, in the sense that they cannot be represented in the form (2.1). This class of models include threshold models, first introduced by Tong (1990), and later enriched by other classes of similar nature. The fundamental reason for introducing such classes of models is the need to model random cyclic behavior that exists in many time series; see Sect. 2.1.5 for further details. As we will see, the class of bilinear processes, which is by far the most general class of nonlinear models, in the sense that they form a dense subset of the Volterra expansions, cannot generate limit cycles (e.g., Tong 1990) and therefore the threshold models have gained importance on their own right in modeling time series. For general treatment of regime models see Hamilton (2008), Granger and Teräsvirta (1993), and Franses and Van Dijk (2000). Regime models may also switch at deterministic but unknown times, in which case the process will be linear but not stationary. Such models are called segmented time series (e.g., Davis et al. 2008). We now look at some of the regime models.

Threshold Autoregressive (TAR) and Self-Exciting Threshold (SETAR) Models

The basic idea behind this class is as follows: we start with a linear model for X_t and allow the parameters of the model to vary according to the values of a finite number of past values of X_t, or a finite number of past values of an associated series Y_t. Hence, such regime models in general can be written as

$$X_t = \begin{cases} a_0^{(1)} + \sum_{i=1}^{p_1} a_i^{(1)} X_{t-i} + Z_t, \text{ if } Y_t \leq r \\ \\ a_0^{(2)} + \sum_{i=1}^{p_2} a_i^{(2)} X_{t-i} + Z_t, \text{ if } Y_t > r \end{cases}, \tag{2.21}$$

where r is the threshold and Y_t is a switching process which can be a latent or an observable process, determining which regime describes the process in a certain moment of time. Such processes are called Threshold autoregressive (TAR) models. When the switching process is the time series itself observed at a certain lag, we have the SETAR sub-class. In its simplest form a first-order SETAR is given as

$$X_t = \begin{cases} a_1 X_{t-1} + Z_t, & \text{if } X_{t-1} \in A^{(1)} \\ a_2 X_{t-1} + Z_t, & \text{if } X_{t-1} \in A^{(2)} \end{cases},$$

where $A^{(i)}$ are some regions. Typically, these regions are intervals such as $A^{(1)} = \{X_{t-1} \leq r\}$, and $A^{(2)} = \{X_{t-1} > r\}$, for some threshold r. We can generalize this class of models to

$$X_t = \sum_{j=1}^{p} a_{ij} X_{t-j} + Z_t^{(i)}, \ \ (X_{t-1}, \ldots, X_{t-p}) \in A^{(i)}, \ \ i = 1, 2, \ldots, l. \tag{2.22}$$

having different error structures in each segment. Note that when $l = 1$, the first-order threshold model can be seen as a piecewise linear approximation to the general nonlinear first order model

$$X_t = f(X_{t-1}) + Z_t,$$

whereas the pth order model in (2.22) is a linear piecewise approximation to the general nonlinear equation

$$X_t = f(X_{t-1}, X_{t-2}, \ldots, X_{t-p}) + Z_t.$$

In practice it is not feasible to fit a model of the form (2.22) with a large p, since the identification of the threshold regions would involve search in a p-dimensional space. A sub-class of the form

$$X_t = a_0^{(i)} + \sum_{j=1}^{p} a_{ij} X_{t-j} + Z_t^{(i)}, \quad X_{t-d} \in A^{(i)},$$

where $A^{(i)}$ is in $\mathbb{R}$ can be considered, thus simplifying the identification of such models. These models can still be extended to include cases when switching between sets of parameters is determined by the past values of a different process Y_t, extending the TAR model given in (2.21)

$$X_t = a_{0i} + \sum_{j=1}^{m_i} a_{ij} X_{t-j} + \sum_{j=1}^{l_i} b_{ij} Y_{t-j} + Z_t^{(i)}, \quad Y_{t-d} \in A^{(j)}.$$

Such models are known to be very useful, particularly in modeling data which shows random cyclic movements.

Smooth Threshold Autoregressive (STAR) Models

As mentioned above, TAR/SETAR models should be used when the process to be modeled shifts from one regime to another abruptly. However, if the transition is gradual, then the STAR models are more appropriate. A two-regime STAR(p) model is defined by Chan and Tong (1986) as follows:

$$X_t = c_0 + \sum_{i=1}^{p} a_{0,i} X_{t-i} + G\left(\frac{X_{t-d} - a}{b}\right)\left(c_1 + \sum_{i=1}^{p} a_{1,i} X_{t-i}\right) + Z_t,$$

where d is the delay parameter, a and b represent the location and scale parameters of G, respectively. The transition function G, that enables the transition between one regime to the other, is a smooth, continuous and monotonically increasing function, satisfying the inequality $0 < G(z) < 1$. Two subclasses of STAR models are the exponential and logistic STAR models, when the function G respectively is given by the expressions

$$G(z) = 1 - \exp[-a(z-b)^2], \ a > 0,$$

$$G(z) = \frac{1}{1 + \exp[-a(z-b)]}, \quad a > 0.$$

Chan and Tong (1986) give an alternative STAR model with Gaussian smooth transition function

$$G(z) = \Phi\big[a(z-b)\big],$$

where $\Phi(\cdot)$ the cdf of the standard Normal distribution. The parameter b can be regarded as the threshold and a controls how fast and how abrupt the model shifts from one regime to another (see e.g., Zivot and Wang 2006).

Markov Switching AutoRegressive (MAR) Models

This class of models was developed by Hamilton (1989), based on ideas previously proposed by Goldfeld and Quandt (1973). Let S_t be a discrete first-order homogeneous Markov chain with state space $S = \{0, 1, \ldots, k\}$. Each member of S corresponds to a regime. Let $P(S_t = j \mid S_{t-1} = i) = p_{ij}$ be the transition matrix given by

$$P = \begin{bmatrix} p_{11} & p_{12} & \cdots & p_{1k} \\ p_{21} & p_{22} & \cdots & p_{2k} \\ \vdots & \vdots & \ddots & \vdots \\ p_{k1} & p_{k2} & \cdots & p_{kk} \end{bmatrix}.$$

Each state, at time t, has an associated probability given by $\pi_t := (P_1, P_2, \ldots, P_k)$, where $\pi = P'\pi$. A k-regime MAR model is given as

$$X_t = \mu_{S_t} + \mathbf{X}_{t-1}\boldsymbol{\theta}_{S_t} + Z_t,$$

where $\mathbf{X}_{t-1} := (X_{t-1}, X_{t-2}, \ldots, X_{t-p})$, and μ_{S_t}, $\boldsymbol{\theta}_{S_t}$ are the model parameters that switch between k different values according to the latent Markov chain. Z_t is assumed to be a Gaussian sequence with mean zero and the variance can be taken as constant, or may switch between k different values depending on the realization of S_t. A classical application of a two-state MAR model to the US GNP time series is given in Hamilton (1989).

Random Coefficient Models

Sometimes it may be useful to introduce random regime switch into the model parameters, giving rise to a different class of models. A simple model within this class is the first order AR model

$$X_t = \psi_t X_{t-1} + Z_t,$$

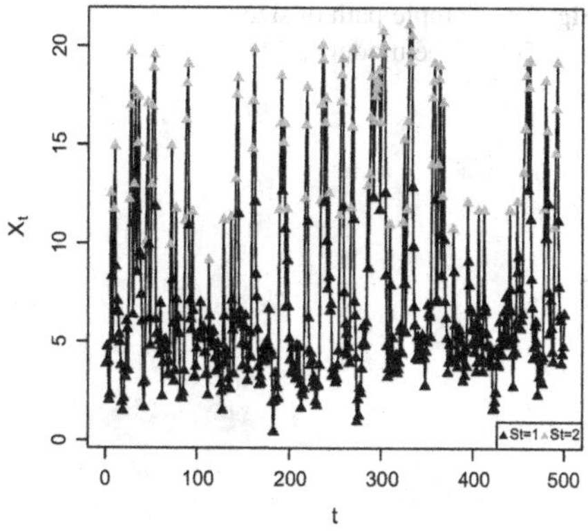

Fig. 2.4 Sample path of size $n = 500$ of the model (2.23)

where ψ_t is a homogeneous Markov chain with a finite space and transition probabilities p_{ij}, for example taking values a_1 and a_2. In this case, the process X_t will alternate between the two processes

$$X_t = a_1 X_{t-1} + Z_t,$$

and

$$X_t = a_2 X_{t-1} + Z_t,$$

according to the transition probabilities (p_{ij}) for $i, j = 1, 2$. Smoother changes in the parameter can be modeled by state space type model

$$X_t = \psi_t X_{t-1} + Z_t,$$

$$\psi_t = a\psi_{t-1} + \nu_t,$$

where Z_t and ν_t are independent i.i.d. sequences. In general, a regime model will take the form (X_t, S_t), where, (S_t) is a latent process, typically a homogeneous Markov chain with a finite state space, such that at any time t, X_t conditional on $S_t = j$ follows a linear model ARMA(p_j, q_j). Hence the process will alternate among various linear models in accordance with the transient behavior of the unobserved process S_t. The estimation, identification and diagnostics for these models are complicated although not impossible, due to the fact that the process S_t is not observed. Note that if the residual process is made to depend on the unobserved Markov chain then the variance of the process also changes from one regime to another. In Fig. 2.4, we have a sample path of the process

$$X_t = C^{(S_t)} + 0.5X_{t-1} + Z_t, \tag{2.23}$$

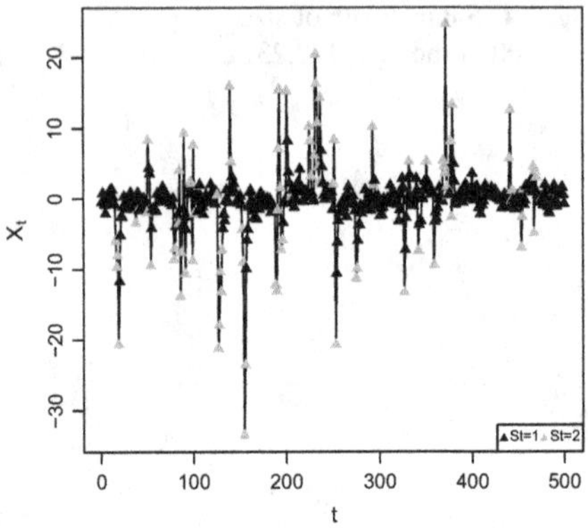

Fig. 2.5 Sample path of size $n = 500$ of the model (2.24)

where Z_t are i.i.d. N(0,1) and

$$C^{(S_t)} = \begin{cases} 2 & \text{if } S_t = 1 \\ 10 & \text{if } S_t = 2 \end{cases},$$

where S_t is a homogeneous Markov chain with transition matrix

$$P = \begin{bmatrix} 0.9 & 0.1 \\ 0.4 & 0.6 \end{bmatrix}.$$

For this matrix, the stationarity limit distribution is given by $\pi = (0.8, 0.2)$. A sample path of the extended process

$$X_t = 0.5X_{t-1} + Z_t^{(S_t)}, \tag{2.24}$$

where

$$Z_t^{(S_t)} = \begin{cases} N(0,1) & \text{if } S_t = 1 \\ N(0,10) & \text{if } S_t = 2 \end{cases},$$

is presented in Fig. 2.5.

Segmented Time Series

Davis et al. (2008) introduced a broad class of nonlinear and non-stationary time series segmented into several pieces. Each segment is assumed to be a stationary time series modeled by a parametric class of time series, whereas the number and the locations of the break points or the segments are treated as unknown model parameters. Thus the observed time series y_t is assumed to be generated by a time series Y_t of the form

$$Y_t = X_{t,j}, \ \tau_{j-1} \le t < \tau_j, \ j = 1, \ldots, m,$$

where $\tau_j, j = 1, \ldots, m$ are the break-points or segments of unknown number m and each segmented $X_{t,j}$ are stationary times series independent of each other. Typically, these time series can be AR(p_j) or GARCH(p_j, q_j) models. In a more general form, each segment may be composed of time series having different state space representations. The model choice, namely the identification of the number of segments and their respective locations, as well as the order of the models in each segment is then performed by using a genetic algorithm. Simulation results indicate that these models perform well and the availability of software to fit these models makes this class of models a good candidate for regime models. The main difference between segmented time series and threshold models is that, whereas in threshold models, the transition between models is triggered by lagged values of the time series, in segmented time series, the changes occur at specified time points. Examples of time series models studied by Davis et al. (2008) include:

- *Segmented AR process:*
  In this case, each segment $X_{t,j}$ is assume to be an AR(p_j) process given by

$$X_{t,j} = a_{1j} X_{t-1,j} + \cdots + a_{p_j,j} X_{t-p_j,j} + Z_{t,j},$$

where $Z_{t,j} \sim \text{WN}(0, \sigma_j^2)$. Here the (unknown) parameters of the model are $\boldsymbol{\theta}_j := (p_j, \boldsymbol{a}_j, \sigma_j^2)$.[1]

- *Segmented GARCH(p_j, q_j) process:*
  In this case, each segment Y_j is modeled by a GARCH(p_j, q_j) given by

$$X_{t,j} = \sigma_{t,j} Z_t,$$

where $Z_t \sim \text{WN}(0, 1)$ and

$$\sigma_{t,j}^2 = a_{j,0} + a_{j,1} X_{t-1,j}^2 + \cdots + a_{j,p_j} X_{t-p_j,j}^2 + b_{j,1} \sigma_{t-1,j}^2 + \cdots + b_{j,q_j} \sigma_{t-q_j,j}^2,$$

$$\tau_{j-1} \le t < \tau_j,$$

where to satisfy stationarity of each segment, the parameters are restricted by $a_{0,j} > 0$, $b_{0,j} \ge 0$ and

$$\sum_{i=1}^{p_j} a_{i,j} + \sum_{i=1}^{q_j} b_{i,j} < 1.$$

Here, the model parameters are given by $\boldsymbol{\theta}_j := (p_j, q_j, \boldsymbol{a}_j, \boldsymbol{b}_j)$.

[1] A discrete counterpart of conventional segmented AR processes, based on the thinning operator in (1.5), was proposed by Kashikar et al. (2013).

- *Segmented state space models:*
  In this case the jth segment $Y_{t,j}$ has a state space representation given by the equations

$$p_j(y_t|\mathbf{x}_{j,t}) = p_j(y_t|x_{t,j}, \mathbf{x}_{t-1,j}, \mathbf{y}_{t-1}), \tau_{j-1} \leq t \leq \tau_j$$

  and the state process $X_{t,j}$ follows a AR(p_j) process.

2.2.2 Exponential Autoregressive Models

Consider for example, the second-order autoregressive model

$$X_t = a_1 X_{t-1} + a_2 X_{t-2} + Z_t,$$

where a_1, a_2 instead of being constants, are functions of X_{t-1}. Specifically, assume that they are exponential functions of X_{t-1}^2 taking the form

$$a_1 = \phi_1 + \pi_1 \exp(-\gamma X_{t-1}^2),$$

$$a_2 = \phi_2 + \pi_2 \exp(-\gamma X_{t-1}^2).$$

Such a model then is called second-order EAR(2) model. Note that for large $|X_{t-1}|$, $a_1 \sim \phi_1, a_2 \sim \phi_2$, whereas for small $|X_{t-1}|$, $a_1 \sim \phi_1 + \pi_1, a_2 \sim \phi_2 + \pi_2$, so that the EAR model behaves like the threshold AR model where the coefficients change smoothly between two extreme parameter values. EAR models are capable of producing amplitude dependent frequency effect, limit cycles and jump phenomena; see Tong (1990) or Priestley (1981). The coefficients a_1, a_2 can be defined as a function of X_{t-1} in different ways to assure smooth transitions. For example, in the case of EAR(1) model, a_1 can be parameterized as

$$a_1 = \theta_1 X_{t-1} + \theta_2 X_{t-1}\{[1 + \exp(\theta_3(X_{t-1} - \theta_4))]^{-1} - 1/2\}, \ \ \theta_3 > 0,$$

in which case the model is called logistic exponential model. These models can be generalized to the form

$$X_t := \boldsymbol{a}'\boldsymbol{X}_{t-1} + \sum_{i=1}^{k} b_i \phi_i(\boldsymbol{\gamma}_i' \boldsymbol{X}_{t-1}) + Z_t,$$

where $\boldsymbol{X}_{t-1} := (X_{t-1}, X_{t-2}, \ldots, X_{t-p})$, and $\boldsymbol{a}' := (a_1, \ldots, a_p)$, $\boldsymbol{\gamma}' := (\gamma_1, \ldots, \gamma_p)$ are p-dimensional parameter vectors and $\phi_i(\cdot)$ are known specific functions. Although such models are used, it is evident that there would be problems of estimation as the parameter space increases.

2.2.3 Polynomial-Type Models

One can also use nonlinear regression type models based on polynomials of the form

$$X_t = \sum_{i=1}^{k} a_i X_{t-1}^i + Z_t.$$

More general polynomial models can be devised by introducing terms depending on $X_{t-2}, X_{t-3}, \ldots, X_{t-p}$ and cross terms. These models are not very much used due to the feedback of X_t into itself, causing explosive behavior.

2.2.4 Bilinear Models

The process X_t is said to be a bilinear process $\mathrm{BL}(p, q, m, l)$ if it satisfies the difference equation

$$X_t = \sum_{j=1}^{p} \phi_j X_{t-j} + \sum_{j=1}^{q} \theta_j Z_{t-j} + \sum_{i=1}^{m} \sum_{j=1}^{l} b_{ij} X_{t-i} Z_{t-j} + Z_t. \tag{2.25}$$

The conditional mean of the process (2.25) is given by

$$E(X_t | \mathcal{F}_{t-1}) = \sum_{j=1}^{p} \phi_j x_{t-j} + \sum_{j=1}^{q} \theta_j z_{t-j} + \sum_{i=1}^{m} \sum_{j=1}^{l} b_{ij} x_{t-i} z_{t-j},$$

whereas the conditional variance is given by $V(X_t | \mathcal{F}_{t-1}) = \sigma_Z^2$. Hence the bilinear model given in (2.25) represents the nonlinear dynamics present in the mean. This class obviously can be extended to include cross terms of $(X_{t-1}, \ldots, X_{t-m})$ with Z_t resulting in models

$$X_t = \sum_{j=1}^{p} \phi_j X_{t-j} + \sum_{j=1}^{q} \theta_j Z_{t-j} + \sum_{i=1}^{m} \sum_{j=0}^{l} b_{ij} X_{t-i} Z_{t-j} + Z_t. \tag{2.26}$$

In this case, $V(X_t | \mathcal{F}_{t-1})$ will also be a function of passed values of the series, therefore bilinear models described in (2.26) fall in the class of mixed models for the conditional mean and variance.

The class of bilinear models plays an important role in modeling nonlinearity for various reasons. The class is an obvious generalization of $\mathrm{ARMA}(p, r)$ models resulting in nonlinear conditional mean. Under fairly general conditions, bilinear processes approximate finite order Volterra series expansions to any desired order of accuracy over finite time intervals (Brockett 1976). Due to Nisio's theorem, Volterra series expansion are a dense class within the class of nonlinear time series, therefore, under fairly general conditions, bilinear processes are also a dense class within nonlinear processes, approximating any nonlinear process to a desired level of accuracy. However, it is well known that bilinear processes cannot capture random cyclic movements, such as limit cycles and jump phenomena. The class is fairly well-studied, and conditions for the existence of unique and stationary solutions are known. Although identification, estimation and diagnostic techniques are available, much of the work on the class remains to be completed. Volterra series expansions and bilinear processes are often used in the control theory and are somewhat different from the context within which they are used in time series. In the control theory, the output X_t, as well as the input process Z_t are observable, making the probabilistic structure simple. For example conditional on the passed values of Z_t, the process X_t is linear, and conditional on the passed values of X_t, the process Z_t is also linear. In the time series context, the input random process Z_t is not observed and unfortunately, the lack of verifiable conditions for invertibility (except for very simple bilinear processes) limits the use of these processes as models. Bilinear processes are capable of producing sudden bursts of large values and hence are suitable for modeling time series showing heavy tailed phenomena.

The bilinear process BL(p, q, m, l) given in (2.26) can be written in the form (Resnick and Van den Berg 2000)

$$\mathbf{X}_t = \mathbf{A}_{t-1}\mathbf{X}_{t-1} + \mathbf{B}_t,$$

where

$$\mathbf{B}_t = \boldsymbol{\Theta}\mathbf{Z}_t,$$

$\boldsymbol{\Theta}$ is a $p \times (1+q)$ matrix given by

$$\boldsymbol{\Theta} = \begin{pmatrix} 1 & \theta_1 & \dots & \theta_{q-1} & \theta_q \\ 0 & 0 & \dots & 0 & 0 \\ 0 & 0 & \dots & 0 & 0 \\ \vdots & \dots & \ddots & \dots & \vdots \\ 0 & 0 & \dots & 0 & 0 \end{pmatrix},$$

$\mathbf{A}_{t-1}$ is a $p \times p$ matrix given by

$$\mathbf{A}_{t-1} = \begin{pmatrix} \phi_1 + \sum_{j=1}^{l} b_{1j} Z_{t-j} & \phi_2 + \sum_{j=1}^{l} b_{2j} Z_{t-j} & \dots & \dots & \phi_p + \sum_{j=1}^{l} b_{pj} Z_{t-j} \\ 1 & 0 & \dots & 0 & 0 \\ \vdots & \vdots & \ddots & \vdots & \vdots \\ 0 & 0 & \dots & 1 & 0 \end{pmatrix}$$

and $\mathbf{X}$ and $\mathbf{Z}_t$ are respectively $(p-1)\times 1$ and $q \times 1$ column vectors

$$\mathbf{X}_t := (X_t, X_{t-1}, \ldots, X_{t-p+1})',$$

$$\mathbf{Z}_t := (Z_t, Z_{t-1}, \ldots, Z_{t-q})'.$$

The general bilinear model (2.25) can also be written in an equivalent state space form with the observation equation

$$X_t = \mathbf{H}'\boldsymbol{W}_{t-1} + Z_t,$$

and the state equation

$$\boldsymbol{W}_t = \mathbf{A}_t\boldsymbol{W}_{t-1} + \mathbf{C}_t.$$

Here, the state vector $\boldsymbol{W}_t$ is a Markov chain, and $\mathbf{A}_t$, $\mathbf{C}_t$ are random matrices, depending on the specific form of the general bilinear process given in (2.25). The general form is quite complicated (e.g., Fan and Yao 2003) but simpler bilinear models can conveniently be written in this form. For example the model BL$(p, 0, p, 1)$ given by

$$X_t = \sum_{j=1}^{p} \phi_j X_{t-j} + \sum_{i=1}^{p} b_{i1} X_{t-i} Z_{t-1} + Z_t,$$

can be written in the vector state space form (e.g., Priestley 1981)

$$\mathbf{X}_t = \mathbf{H}\boldsymbol{W}_t + \mathbf{C}Z_t,$$

$$\boldsymbol{W}_t = (\mathbf{A} + \mathbf{B}W_t)\boldsymbol{W}_{t-1} + (\mathbf{A} + \mathbf{B}Z_t)\mathbf{C}Z_t,$$

where

$$\boldsymbol{W}_t := (X_t, X_{t-1}, \ldots, X_{t-p})',$$

$$\mathbf{H}_{1\times p} := (1, 0, \ldots, 0),$$

$$\mathbf{C}_{p\times 1} := (1, 0, \ldots, 0)',$$

$$\mathbf{A}_{p\times p} := \begin{pmatrix} \phi_1 & \phi_2 & \cdots & a_p \\ 1 & 0 & \cdots & 0 \\ 0 & 1 & \cdots & 0 \\ \vdots & \vdots & \ddots & \vdots \\ 0 & \cdots & 1 & 0 \end{pmatrix},$$

$$\mathbf{B}_{p\times p} := \begin{pmatrix} b_{11} & b_{21} & \dots & b_{p1} \\ 0 & 0 & \dots & 0 \\ \vdots & \vdots & \ddots & \vdots \\ 0 & 0 & \cdots & 0 \end{pmatrix}.$$

Note that $\boldsymbol{W}_{t-1}$ is independent of the coefficient $(\mathbf{A} + \mathbf{B}Z_t)$ and the error $(\mathbf{A} + \mathbf{B}Z_t)\mathbf{C}Z_t$, and is a Markov chain. Note also that the pair $(\mathbf{A}+\mathbf{B}Z_t, (\mathbf{A}+\mathbf{B}Z_t)\mathbf{C}Z_t)$, forms an i.i.d. sequence of random matrices, but the components of this pair are not independent of each other. This state space representation, due to its Markovian nature facilitates the study of the probabilistic properties of the process. For example, if the process $\boldsymbol{W}_t$ is stationary, then so is X_t. Due to its Markovian structure, it is relatively easy to study the conditions under which $\boldsymbol{W}_t$ is stationary; see Meyn and Tweedie (2009) for the study of probabilistic properties of Markov processes.

If we solve the difference equation given by

$$\boldsymbol{W}_t = \mathbf{A}_t \boldsymbol{W}_{t-1} + \mathbf{C}_t,$$

where $(\mathbf{A}_t, \mathbf{C}_t)$ is an i.i.d. sequence of random matrices, iteratively n times, the partial solution for $\boldsymbol{W}_t$ is given by

$$\boldsymbol{W}_t = \prod_{i=0}^{n-1} \mathbf{A}_{t-i} \boldsymbol{W}_{t-n} + \sum_{j=1}^{n-1} \prod_{i=1}^{j-1} \mathbf{A}_{t-i} \mathbf{C}_{t-j},$$

so that the convergence in probability

$$\sum_{j=1}^{n-1} \prod_{i=1}^{j-1} \mathbf{A}_{t-i} \mathbf{C}_{t-j} \to 0,$$

is a sufficient condition for the existence of a stationary solution. For example, consider the simple bilinear process

$$X_t = aX_{t-1} + bX_{t-1}Z_{t-1} + Z_t.$$

Solving iteratively for X_t, upon n iterations we get

$$X_t = \prod_{i=1}^{n} (a + bZ_{t-i}) X_{t-n} + \sum_{j=1}^{n-1} \prod_{i=1}^{j} (a + bZ_{t-i}) Z_{t-j},$$

and, if in probability

$$\prod_{j=1}^{n}(a + bZ_{t-j}) \to 0, \tag{2.27}$$

then

$$X_t = \sum_{j=1}^{\infty}\prod_{i=1}^{j}(a + bZ_{t-i})Z_{t-j}. \tag{2.28}$$

A sufficient condition for (2.27) is given by Pham and Tran (1981). If Z_t are i.i.d. zero-mean r.v's with $E(Z_t^2) = \sigma^2$ and $E(Z_t^4) < \infty$, then (2.27) converges in mean-square if $a^2 + b^2\sigma^2 < 1$, in which case, (2.28) is the unique stationary solution. This is also a sufficient condition for invertibility. However, it is far from being a necessary condition for stationarity and invertibility. Note that (2.28) is a moving average representation

$$X_t = \sum_{j=1}^{\infty}\theta_j Z_{t-j},$$

with random coefficients

$$\theta_j = \prod_{i=1}^{j}(a + bZ_{t-i}).$$

Therefore, one would expect that the second-order properties of this process resembles that of a linear process. Indeed, assuming $\sigma^2 = 1$ simple calculations show that

$$\begin{aligned}\mu = E(X_t) &= \sum_{j=1}^{\infty} a^{j-1}b\sigma^2\\ &= \frac{b}{1-a},\end{aligned}$$

$$E(X_t^2) = \frac{1+2b^2}{1-b^2},$$

$$\gamma(1) = E(X_t X_{t-1}) = 2b^2,$$

and for $k \geq 2$,

$$\gamma(k) = a\gamma(k-1).$$

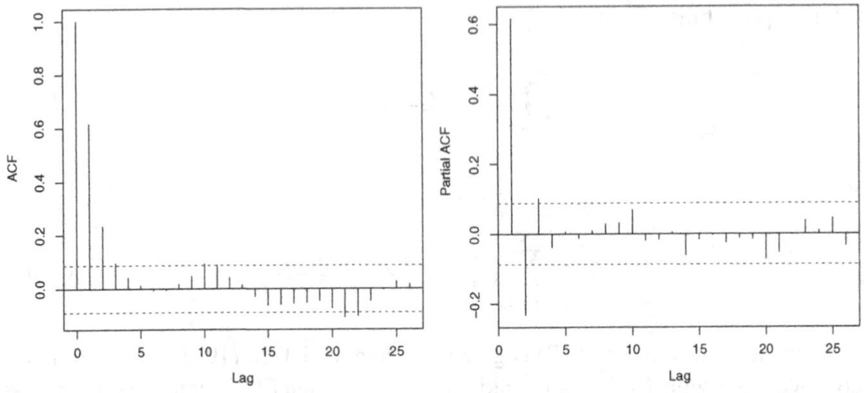

Fig. 2.6 ACF and PACF of the bilinear model (2.29)

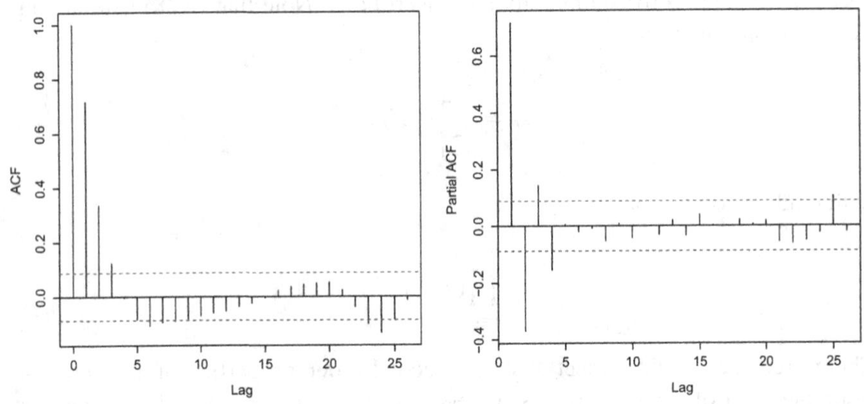

Fig. 2.7 ACF and PACF of the linear model (2.30)

Note that this is exactly the covariance structure of a linear MA(1, 1) process. Similarly, the autocovariance function of a BL(p, q, m, l) process behaves like the autocovariance function of the process MA(p, q_0) where $q_0 := \max(q, l)$; see Fan and Yao (2003) for details. It is clear once again that one cannot differentiate a nonlinear model from a linear model by studying only the second-order properties. In Figs. 2.6 and 2.7 below, we give respectively the autocorrelation and partial autocorrelation functions based on a data of dimension 500, simulated from the models

$$X_t = 0.5X_{t-1} + 0.6X_{t-1}Z_{t-1} + Z_t, \tag{2.29}$$

and

$$X_t = 0.5X_{t-1} + 0.6Z_{t-1} + Z_t \tag{2.30}$$

with i.i.d. standard Normal innovations.

Unfortunately, like for most nonlinear processes, the condition of invertibility which is crucial for estimation and prediction, is not well understood and cannot be checked except for some simple bilinear processes (see Chap. 4, for some empirical ways of checking invertibility). Therefore, although bilinear processes have desired properties as models, their use in practice is quite restricted.

Here we note a fundamental difference between the bilinear and threshold models. Threshold models, as in the case of bilinear models, can be put in the state space representation

$$X_t = \mathbf{H}\boldsymbol{W}_t,$$

$$\boldsymbol{W}_t = \mathbf{A}^{(i)}\boldsymbol{W}_{t-1} + \mathbf{B}^{(i)}Z_t^{(i)},$$

for some properly chosen state vector $\boldsymbol{W}_t$, and constant matrices $\mathbf{A}^{(i)}$,$\mathbf{B}^{(i)}$ and regions $\mathbb{R}^{(i)}$. However, the essential difference between threshold models and bilinear models is that whereas in bilinear processes the nonlinearity is introduced by the cross terms $Z_{t-i}X_{t-j}$, in the threshold models the relation between $\boldsymbol{W}_t$ and $\boldsymbol{W}_{t-1}$ is nonlinear (that is a nonlinear function of $X_s, s < t$), with residuals still entering the model linearly. This difference has strong influence on the type of nonlinear behavior. Bilinear processes, due to this cross terms, in general, are capable of producing extreme observations but cannot produce limit cycle behavior, hence each class has its own use in modeling different nonlinear phenomena.

2.2.5 *Parametric Models for the Conditional Variance*

These models are special case of the representation (2.20) with $f \equiv 0$ and are based on modeling the function g in different forms. A useful conceptual division of these models can be made as

1. Observation-driven models and
2. Parameter-driven models.

Observation-Driven Models

Lets assume that for each t, the time series satisfies

$$X_t|\sigma_t^2 \sim N(0, \sigma_t^2).$$

The observation-driven models are based on representing σ_t^2 as a function of lagged values of X_t taking the general form

$$X_t = g(\mathcal{F}_{t-1}, \boldsymbol{\theta}_2) Z_t,$$

giving rise to the rich classes of ARCH and GARCH models. Since

$$V(X_t | \mathcal{F}_{t-1}) = g^2(\mathcal{F}_{t-1}, \boldsymbol{\theta}_2)\sigma_Z^2,$$

it is customary to represent the function g by σ_t.

Since the seminal paper of Engle (1982) traditional time series tools such as the ARMA models for the mean have been extended to essentially analogous models for the variance. Autoregressive conditional heteroscedasticity (ARCH) models are now widely used to describe and forecast changes in volatility of financial time series. For a survey of ARCH-type models and various extensions, see Bollerslev et al. (1992, 1994), Pagan (1996), Palm (1996), Shephard (1996), Berkes et al. (2003), Bauwens et al. (2006), Silvennoinen and Teräsvirta (2009), and Teräsvirta (2009). According to Engle (2004) the original idea was to find a model to assess the validity of the conjecture of Friedman (1977) that the unpredictability of inflation was a primary cause of business cycles. Uncertainty due to this unpredictability would affect the investor's behavior. Pursuing this idea requires a model which characterizes the time dynamics of this uncertainty.

Financial time series, such as relative returns of stock indices, share prices and foreign exchange rates, often show the following features (usually referred to as *stylized facts*):

- The sample mean of the data is close to zero whereas the sample variance is of the order 10^{-4} or smaller;
- Exceedances of high/low thresholds tend to occur in clusters. This property indicates that there exists dependence in the tails;
- Return data exhibit heavy-tailed marginal distributions;
- The sample autocovariance function of such data is statistically insignificant at all lags (with a possible exception of the first lag), whereas the sample autocovariance function of the absolute values or the squares of the time series are different from zero for a large number of lags and stay almost constant and positive for large lags.
- As one increases the time scale on which returns are calculated, their distribution looks more and more a Gaussian. This means that the *peakedness* around zero and the *heavy-tailedness* of the empirical distribution turn into bell shapedness.

The list above is far from being complete. An exhaustive analysis of stylized facts can be found in Cont (2001).

Most models for financial time series (and in particular for return data) used in practice to accommodate such features are given in the multiplicative form

$$X_t = \sigma_t Z_t, \ t \in \mathbb{Z}, \tag{2.31}$$

where (Z_t) forms an i.i.d. sequence of real-valued innovations or noise variables with zero mean and unit variance, (σ_t) is a stochastic process such that σ_t and

Z_t are independent for fixed t. In general, (σ_t) and (X_t) are assumed to be strictly stationary. Motivation for considering this particular choice of a simple multiplicative model comes from the fact that (a) in practice, the direction of price changes is well modelled by the sign of Z_t, whereas σ_t provides a good description of the order of magnitude of this change; and (b) the volatility σ_t^2 represents the conditional variance of X_t given σ_t.

Engle (1982) suggested the following simple model for the volatility σ_t:

$$\sigma_t^2 = a_0 + a_1 X_{t-1}^2, \; t \in \mathbb{Z}, \tag{2.32}$$

for positive constants a_0 and a_1. Equations (2.31) and (2.32) define an *AutoRegressive Conditionally Heteroscedastic model of order one* (in short ARCH(1)). For example, assume Z_t to be an i.i.d. Gaussian white noise distribution. Then the distribution of tomorrow's return X_{t+1}, conditionally on today's return X_t, has Normal distribution with zero mean and variance $a_0 + a_1 X_t^2$. This allows one to give a *distributional forecast* of X_{t+1} given X_t. The ARCH(1) fit to real-life data can be improved by introducing the ARCH(p) model, with $p \in \mathbb{N}$, where σ_t obeys the recursive equation

$$\sigma_t^2 = a_0 + \sum_{i=1}^{p} a_i X_{t-i}^2, \; t \in \mathbb{Z}, \tag{2.33}$$

with $a_0 > 0, a_1, \ldots, a_{p-1} \geq 0$ and $a_p > 0$. A major improvement upon the expression in (2.33) was achieved by Bollerslev (1986) and Taylor (1986), independently of each other, who introduced the Generalized ARCH (GARCH) models of order p and q. In this model, the conditional variance is also a linear function of its own lags and takes the form

$$\begin{aligned} \sigma_t^2 &= a_0 + \sum_{i=1}^{p} a_i X_{t-i}^2 + \sum_{j=1}^{q} b_j \sigma_{t-j}^2, \\ &:= a_0 + a(B) X_t^2 + b(B) \sigma_t^2, \; t \in \mathbb{Z}, \end{aligned} \tag{2.34}$$

with $a_0 > 0, a_1, \ldots, a_{p-1} \geq 0$, $a_p > 0$, $b_1, \ldots, b_{q-1} \geq 0$ and $b_q > 0$. The requirement that all the coefficients are non-negative ensures that σ_t^2 is also non-negative. The most popular GARCH model in applications has been the GARCH(1, 1) model, with $p = q = 1$ in (2.34).

The family of GARCH models has been generalized and extended in various directions in order to accommodate different features often exhibited by financial time series. One possible generalization of the GARCH models is the so-called ARCH(∞) sequences defined as follows:

Definition 2.2.1. A random sequence (Y_t) is said to satisfy ARCH(∞) equations if there exists a sequence of i.i.d. non-negative r.v's (η_t) such that

$$Y_t = \zeta_t \eta_t, \; t \in \mathbb{Z}, \tag{2.35}$$

and

$$\zeta_t = a_0^* + \sum_{i=1}^{\infty} a_i^* Y_{t-i}, \tag{2.36}$$

with $a_i^* \geq 0$ for $i = 0, 1 \ldots$.

The general framework leading to the model in (2.35) and (2.36) traces back to Robinson (1991). This class of models include, among others, the classical squared ARCH(∞) model, that is the model in (2.31) and (2.33) with $p = \infty$ and $Y_t = X_t^2$, $\zeta_t = \sigma_t^2$, $\eta_t = Z_t^2$ and the coefficients $a_0^* = a_0$ and $a_i^* = a_i$ for $i = 1, \ldots, p$; or the squared GARCH(1, 1) with $Y_t = X_t^2, \zeta_t = \sigma_t^2, \eta_t = Z_t^2, a_0^* = a_0/(1-b_1)$, and $a_i^* = b_1^{i-1} a_1$.

On the other hand, several extensions of the GARCH models aim at accommodating asymmetric response of the volatility for positive and negative shocks. Giving heed to this problem, Ding et al. (1993) proposed the Asymmetric Power ARCH of order (p, q), in short APARCH(p, q), model defined as

$$\sigma_t^\delta = \omega + \sum_{i=1}^{p} a_i (|X_{t-i}| - \gamma_i X_{t-i})^\delta + \sum_{j=1}^{q} b_j \sigma_{t-j}^\delta,$$

where $\omega > 0$, $a_i \geq 0$, $b_j \geq 0$, $\delta \geq 0$ represents the parameter for the power term, and $-1 < \gamma_i < 1$ is the leverage parameter. This model allows detecting asymmetric responses of the volatility for positive or negative shocks. If $\gamma_i > 0$, negative shocks have stronger impact on volatility than positive shocks, as would be expected in the analysis of financial time series. If $\gamma_i < 0$, the reverse happens. The APARCH model includes as special cases the GARCH(p, q) model, the Taylor/Schwert GARCH in standard deviation model (Taylor 1986; Schwert 1989, 1990), the GJR-GARCH model (Glosten et al. 1993), the TARCH model (Rabemananjara and Zakoïan 1993; Zakoïan 1994), the NARCH models (Higgins and Bera 1992) and the log-ARCH model (Geweke 1986; Pantula 1986).

Moreover, evidence of long memory and persistence (accordingly to the most common definition of long memory: autocovariance function, $\gamma(k)$, decaying at the hypergeometric rate k^{2d-1}, with $0 < d < 0.5$) has been documented in many fields in economics, including volatility of financial series and trading intensity in financial durations data. Baillie et al. (1996) proposed the Fractionally IGARCH(p, d, q), or FIGARCH(p, d, q), in order to accommodate long memory in volatility. The authors started by writing the GARCH(p, q) process as an ARMA(m, p) in X_t^2

$$(1 - a(B) - b(B))X_t^2 = \omega + (1 - b(B))\nu_t,$$

where $m = \max\{p, q\}$ and $v_t = X_t^2 - \sigma_t^2$. When the autoregressive lag polynomial $\phi(B) := 1 - a(B) - b(B)$ contains a unit root, the GARCH(p, q) process is said to be integrated in variance (Engle and Bollerslev 1986). The Integrated GARCH(p, q) or IGARCH(p, q) class of models is given by

$$\phi(B)(1 - B)X_t^2 = \omega + (1 - b(B))v_t.$$

The FIGARCH(p, d, q) class of models is simply obtained by allowing the differencing operator in the above equation to take non-integer values, that is

$$\phi(B)(1 - B)^d X_t^2 = \omega + (1 - b(B))v_t,$$

with $b(B)$ and $\phi(B)$ representing lag polynomials having all their roots lying outside the unit circle. The fractional differencing parameter is denoted as d. The fractional differencing operator $(1 - B)^d$ is most conveniently expressed as

$$(1 - B)^d = \sum_{k=0}^{\infty} \binom{d}{k} (-1)^k B^k.$$

After rearrangement, the FIGARCH(p, d, q) model can be represented as

$$\sigma_t^2 = \frac{\omega}{1 - b(1)} + \lambda(B)X_t^2, \tag{2.37}$$

where

$$\lambda(B) = 1 - \phi(B)(1 - B)^d(1 - b(B))^{-1} = \sum_{i=1}^{\infty} \lambda_i B^i. \tag{2.38}$$

Here, $\lambda(1) = 1$ for every d, with $\lambda_i \geq 0$, for $i = 1, 2, \ldots,$ so that the FIGARCH(p, d, q) model is well-defined and the conditional variance is positive for all t. Conrad and Haag (2006) obtained two sets of sufficient conditions for the conditional variance of the FIGARCH process to be non-negative almost surely. Nonetheless, general conditions are difficult to establish. The simplest version of the FIGARCH(p, d, q) model, which appears to be particularly useful in practice, is the FIGARCH$(1, d, 1)$ for which the volatility σ_t^2 takes the form as in (2.37) with $b(B) = b_1 B$ and $\phi(B) = \phi_1 B$ with $|b_1| < 1$. Necessary and sufficient conditions for the non-negativity of the conditional variance for the FIGARCH$(1, d, 1)$ were obtained by Conrad and Haag (2006). The FIGARCH model has the property that for high lags, say k, the distributed lag coefficients are $\lambda_k \simeq ck^{-d-1}$, with c a positive constant. This implies that the conditional variance can be expressed as a distributed lag of past squared returns with coefficients that decay at a slow, that is hyperbolic, rate which is consistent with the long memory property. Davidson (2004) proposed an alternative definition of the persistence properties of the FIGARCH process in terms of the hyperbolic memory, aiming to make the distinction of the FIGARCH model from the geometric memory cases represented by the GARCH and IGARCH processes more precise.

The statistical properties of the general FIGARCH(p, d, q) process, however, remain unestablished. For example, conditions for the existence of a stationary solution as well as the source of long memory on volatility are not known. For example, Mikosch and Starica (2004) and Granger and Hyung (2004) advocated that spurious long memory can be detected from time series exhibiting structural breaks. As solution to this problem, Baillie and Morana (2009) proposed the Adaptative FIGARCH model, or A-FIGARCH model in short, which simultaneously accounts for long memory and incorporates a deterministic time-varying intercept which allows for breaks, cycles and changes in drift. The A-FIGARCH (p, d, q, k) model can be derived from the FIGARCH (p, d, q) in (2.37) by letting the intercept ω to be time-varying, that is

$$\sigma_t^2 = \omega_t + [1 - \phi(B)(1 - B)^d (1 - b(B))^{-1}]X_t^2,$$

or

$$\sigma_t^2 = \omega_t + \lambda(B)X_t^2,$$

with

$$\omega_t = \omega_0 + \sum_{j=1}^{k} [\gamma_j \sin(2\pi jt/T) + \delta_j \cos(2\pi jt/T)].$$

In practice, k is a small integer often taken as $k = 1$ or 2. An immediate advantage of this model is that it does not require pretesting to determine either the number of structural break points or their locations. Furthermore, this model does not require any smooth transition between volatility regimes. Note that the inclusion of the time-varying intercept component implies that the A-FIGARCH process is neither ergodic nor strictly stationary.

The FIAPARCH(p, d, q) model of Tse (1998) is a special case of (2.31) with

$$\sigma_t^{\delta} = \frac{\omega}{1 - \beta(B)} + \lambda(B)g(X_t), \tag{2.39}$$

where $g(X_t) = (|X_t| - \gamma X_t)^{\delta}$ with $|\gamma| < 1$ and $\delta \geq 0$, and $\lambda(B)$ defined as in (2.38) for every $0 < d < 1$, with $\lambda_i \geq 0$, for $i \in \mathbb{N}$, and $\omega > 0$. Furthermore, in order to allow for long memory, the fractional differencing parameter d is constrained to lie in the interval $0 < d < 1/2$. The FIAPARCH model nests two major classes of ARCH-type models: the APARCH and the FIGARCH models of Ding et al. (1993) and Baillie et al. (1996), respectively. When $d = 0$ the process reduces to the APARCH(p, q) model, whereas for $\gamma = 0$ and $\delta = 2$ the process reduces to the FIGARCH(p, d, q) model. Conrad et al. (2008) pointed out some advantages of the FIAPARCH(p, d, q) class of models, namely (a) it allows for an asymmetric response of volatility to positive and negative shocks, thus being able to traduce the leverage effect. If $\gamma > 0$, negative shocks have stronger impact on volatility than positive shocks as would be expected in the analysis of financial time series. If $\gamma < 0$, the reverse happens; (b) in this particular class of models, it is the data that determines the power of returns for which the predictable structure in the volatility pattern is the strongest, and (c) the models are able to accommodate long memory in volatility, depending on the differencing parameter d.

The simplest version of the FIAPARCH(p, d, q) model, which appears to be particularly useful in practice, is the FIAPARCH$(1, d, 1)$ for which the volatility σ_t takes the form as in (2.39) with $\beta(B) = \beta B$ and $\phi(B) = \phi B$. Necessary and sufficient conditions for the non-negativity of the conditional variance for the FIAPARCH$(1, d, 1)$ resembles the ones obtained by Conrad and Haag (2006) for the FIGARCH$(1, d, 1)$ model.

Volatility, asymmetry and long memory may also be captured using various extensions of the model introduced by Tse (1998) and Davidson (2004) among others. For example, Diongue and Guégan (2007) introduced the so-called seasonal hyperbolic APARCH, in short S-HY-APARCH, model where

$$[1 - b(B)]\sigma_t^\delta = \omega + \{\phi(B)[1 - \tau(1 - (1 - B^s)^d)]\}g(X_t). \tag{2.40}$$

The parameter $\tau \geq 0$ permits to eliminate the non-stationarity of the process. Moreover, by assuming that the roots of $[1 - b(B)] = 0$ lie outside the unit circle, the conditional variance in (2.40) can be expressed as

$$\sigma_t^\delta = \frac{\omega}{1 - b(1)} + \{1 - a(B)(1 - b(B))^{-1}(1 - \tau(1 - (1 - B^s)^d))\}g(X_t).$$

Another popular class of GARCH-type models is the Exponential GARCH, EGARCH in short. Nelson (1991) introduced it in order to overcome some disadvantages exhibited by the GARCH models, namely (a) parameter restrictions that are often violated by estimated coefficients; (b) asymmetric responses of shocks; and (c) interpreting whether shocks to conditional variance persist or not is difficult in GARCH models, since the usual norms measuring persistence often do not agree. The family of EGARCH(p, q) models can be defined as in (2.31) with

$$\ln(\sigma_t^2) = a_0 + \sum_{i=1}^{p} a_i g(Z_{t-i}) + \sum_{j=1}^{q} b_j \ln(\sigma_{t-j}^2). \tag{2.41}$$

For example, setting $g(Z_t) = \theta Z_t + \gamma(|Z_t| - E|Z_t|)$ with non-zero θ and γ in (2.41), we get the EGARCH model of Nelson (1991). Moreover, if in (2.31) and (2.41) we set $g(Z_t) = \theta_i \ln(Z_t^2)$, for $i = 1, \ldots, p$, then we get the logarithmic GARCH (LGARCH) model proposed by Geweke (1986) and Pantula (1986).

As a final class of GARCH-type processes, we mention the model introduced by Liu (2009) which is a generalization of the first-order GARCH processes family introduced in He and Teräsvirta (1999) and further developed by Ling and McAleer (2002). These authors defined the following general class for the GARCH$(1,1)$ model. Assume that in (2.31), σ_t is modeled by

$$\sigma_t^\delta = g(Z_{t-1}) + c(Z_{t-1})\sigma_{t-1}^\delta, \ t \in \mathbb{Z},$$

where $\delta > 0$, (Z_t) is a sequence of i.i.d. non-degenerate r.v's with mean zero. Further, it is assumed that Z_t is independent of $X_{t-1}, X_{t-2}, \ldots$, and $g(\cdot)$ is a positive function whereas $c(\cdot)$ is a non-negative function. This family of GARCH processes includes the GARCH$(1,1)$ model of Bollerslev (1986), the absolute value GARCH$(1,1)$ model of Taylor (1986) and of Schwert (1989), the nonlinear GARCH$(1,1)$ model of Engle (1990), the asymmetric GJR-GARCH$(1,1)$ model of Ding et al. (1993), the TARCH model (Rabemananjara and Zakoïan 1993; Zakoïan 1994), the 4NLGMACH$(1,1)$ model of Yang and Bewley (1995), the generalized quadratic ARCH$(1,1)$ model of Sentana (1995), and the volatility switching GARCH$(1,1)$ model of Fornari and Mele (1997).

Liu (2009) extends He and Teräsvirta (1999) results by allowing for an influence of higher-order past errors and conditional variances on the current conditional variance. Specifically, Liu model for σ_t stands as follows:

$$\sigma_t^\delta = g(Z_{t-1}, \ldots, Z_{t-s}) + \sum_{k=1}^{r} c_k(Z_{t-k})\sigma_{t-k}^\delta, \ t \in \mathbb{Z},$$

where $g(Z,t,s) = g(Z_{t-1}, \ldots, Z_{t-s})$ is a strictly positive function and $c_k(\cdot)$, $k = 1, \ldots, r$, all are nonnegative functions. This new family of GARCH processes includes:

1. The GARCH(p,q) model of Bollerslev (1986) for $\delta = 2$, $g(Z,t,s) \equiv a_0$, $c_k(Z_{t-k}) = b_k + a_k Z_{t-k}^2$ for $k = 1, \ldots, r$ with $r = \max\{p,q\}$, $a_i = 0$ and $b_j = 0$ for $i > p$ and $j > q$, respectively.
2. The absolute value GARCH$(1,1)$ model of Taylor (1986) and of Schwert (1989) for $\delta = 1$, $g(Z,t,s) \equiv a_0$, $c_k(Z_{t-k}) = b_k + a_k|Z_{t-k}|$ for $k = 1, \ldots, r$ with $r = \max\{p,q\}$, $a_i = 0$ and $b_j = 0$ for $i > p$ and $j > q$, respectively.
3. The volatility switching GARCH$(1,1)$ model of Fornari and Mele (1997) for $\delta = 2$, $g(Z,t,s) = a_0 + \sum_{k=1}^{s} \gamma_k \text{sgn}(Z_{t-k})$, $c_k(Z_{t-k}) = b_k + a_k Z_{t-k}^2$ for $k = 1, \ldots, r$ with $r = \max\{p,q\}$, $a_i = 0$ and $b_j = 0$ for $i > p$ and $j > q$, respectively.
4. The nonlinear GARCH(p,q) model of Engle (1990).

   (a) Case $\delta = 1$: $g(Z,t,s) \equiv a_0$, $c_k(Z_{t-k}) = b_k + a_k(1 - 2\eta\text{sgn}(Z_{t-k}) + \eta^2)|Z_{t-k}|$ for $k = 1, \ldots, r$ with $r = \max\{p,q\}$, $a_i = 0$ and $b_j = 0$ for $i > p$ and $j > q$, respectively.

(b) Case $\delta = 2$: $g(Z,t,s) \equiv a_0$, $c_k(Z_{t-k}) = b_k + a_k(1 - 2\eta\text{sgn}(Z_{t-k}) + \eta^2)Z_{t-k}^2$ for $k = 1,\ldots,r$ with $r = \max\{p,q\}$, $a_i = 0$ and $b_j = 0$ for $i > p$ and $j > q$, respectively.

5. The GJR-GARCH(p,q) model of Glosten et al. (1993) for $\delta = 2$ $g(Z,t,s) \equiv a_0$, $c_k(Z_{t-k}) = b_k + (a_k\omega_k I(Z_{t-k}))Z_{t-k}^2$ where $I(Z_{t-k}) = 1$ if $Z_{t-k} < 0$ and $I(Z_{t-k}) = 0$ otherwise, for $k = 1,\ldots,r$ with $r = \max\{p,q\}$, $a_i = 0$ and $b_j = 0$ for $i > p$ and $j > q$, respectively.
6. The APARCH(p,q) model of Ding et al. (1993) for $\delta > 0$, $g(Z,t,s) \equiv a_0$, $c_k(Z_{t-k}) = b_k + a_k(1 - 2\eta\text{sgn}(Z_{t-k}) + \eta^2)|Z_{t-k}|^\delta$ for $k = 1,\ldots,r$ with $r = \max\{p,q\}$, $a_i = 0$ and $b_j = 0$ for $i > p$ and $j > q$, respectively.
7. The threshold GARCH(p,q) model for $\delta > 0$, $g(Z,t,s) \equiv a_0$, $c_k(Z_{t-k}) = b_k + (a_{1k}(1 - I(Z_{t-k})) + a_{2k}I(Z_{t-k}))|Z_{t-k}|^\delta$ for $k = 1,\ldots,r$ with $r = \max\{p,q\}$, $a_i = 0$ and $b_j = 0$ for $i > p$ and $j > q$, respectively. Note that this is generalization of the models introduced by Zakoïan (1994), Hwang and Woo (2001), and Hwang and Basawa (2004).
8. The 4NLGMACH$(1,1)$ model of Yang and Bewley (1995) for $\delta = 2$, $g(Z,t,s) = a_0 + \sum_{k=1}^{s} a_{1k}(Z_{t-k} - d_k)^2 + a_{2k}(Z_{t-1} - d_k)^4$, $c_k(Z_{t-k}) = b_k$ for $k = 1,\ldots,r$. As pointed out by Liu (2009) this is a generalization of the family of moving-average conditional heteroskedasticity models proposed by Yang and Bewley (1995).
9. The first-order GARCH model of He and Teräsvirta (1999) with $r = 1$ and $s = 1$

We refer the reader to Andersen et al. (2009) for the recent developments and applications of this class of models.

Parameter-Driven Models

Parameter driven models for conditional variance are based on representing the variance of the process by a latent stochastic component. A simple example is the log-normal stochastic variance or volatility model

$$X_t|W_t \sim N(0, \exp(W_t)),$$

$$W_{t+1} = \gamma_0 + \gamma_1 W_t + \nu_t,$$

where $\nu_t \sim i.i.d.\ N(0,\sigma^2)$. Here W_t is not observed but can be estimated using the observations. These models lack analytic one-step ahead forecast densities and they need to be approximated through numerical methods. However they extend to higher dimensions and have continuous time analogs; see Sect. 2.2.7 for an extended treatment of parameter-driven models. Recent advances in hierarchical modeling techniques and simulation-based inferential methods make these generalized state space models very attractive.

2.2.6 Mixed Models for the Conditional Mean and Variance

The objective behind these models is to join models for the conditional mean and conditional variance given in the previous sections under a single model. In its simplest form, these composite models can be given as

$$X_t = f(\mathcal{F}_{t-1}, \boldsymbol{\theta}_1) + g(\mathcal{F}_{t-1}, \boldsymbol{\theta}_2) Z_t,$$

Note that

$$E(X_t|\mathcal{F}_{t-1}) = f(\mathcal{F}_{t-1}, \boldsymbol{\theta}_1),$$

$$V(X_t|\mathcal{F}_{t-1}) = g(\mathcal{F}_{t-1}, \boldsymbol{\theta}_2)^2 V(Z_t).$$

Here, the function f and g can be chosen in accordance with the partial models for the conditional mean and variance, discussed in the previous sections. In the simplest case, the conditional mean can be modeled by a linear ARMA model, whereas the conditional variance can be modeled by a GARCH model. Typically, first the model for the conditional mean is fitted, then the conditional variance model is fitted to the residuals from this model. This is the standard procedure in fitting GARCH models.

However, models of the type

$$\begin{aligned} X_t &= \sum_{j=1}^{p} \phi_j X_{t-j} + \sum_{j=1}^{q} \theta_j Z_{t-j} + \sum_{i=1}^{m} \sum_{j=0}^{l} b_{ij} X_{t-i} Z_{t-j} + Z_t \\ &= \sum_{j=1}^{p} \phi_j X_{t-j} + \sum_{j=1}^{q} \theta_j Z_{t-j} + \sum_{i=1}^{m} \sum_{j=1}^{l} b_{ij} X_{t-i} Z_{t-j} \\ &+ \sum_{i=1}^{m} b_{i0} X_{t-i} Z_t + Z_t. \end{aligned}$$

give rise to richer and more complex structures. For example, the model

$$X_t = aX_{t-1} + bX_{t-1}Z_{t-1} + cX_{t-1}Z_t + Z_t,$$

includes nonlinear dynamics both in the mean and the variance, since

$$E(X_t|\mathcal{F}_{t-1}) = ax_{t-1} + bx_{t-1}z_{t-1},$$

and

$$V(X_t|\mathcal{F}_{t-1}) = (1 + cx_{t-1}^2)\sigma^2.$$

Alternatively, we can consider bilinear models given in (2.25), whose innovations are generated by a GARCH model. For example the model

$$X_t = \sum_{i=1}^{r}\sum_{j=1}^{s} b_{ij} X_{t-i} Z_{t-j} + Z_t, \tag{2.42}$$

where $i > j$, and the innovations Z_t are generated by the ARCH(q) process will represent nonlinear dynamics both in the mean and the variance.

The fundamental difference between GARCH and bilinear models is that whereas for GARCH models $E(X_t|\mathcal{F}_{t-1}) = 0$, and $V(X_t|\mathcal{F}_{t-1}) = h_t\sigma_t^2$, for bilinear processes given by (2.25), $E(X_t|\mathcal{F}_{t-1})$ has a nonlinear structure and $V(X_t|\mathcal{F}_{t-1})$ is constant. However, both classes of models can have similar unconditional moments. Often, upon fitting an adequate linear model for the conditional mean, the presence of linear dependence in the squared residuals is tested and this test is used as an indication for the presence of GARCH or bilinear type nonlinear structures in the series. However, these tests cannot provide a guidance in choosing the specific model for the series.

Mixed models of the type described above are quite rich in representing nonlinear dynamics and are seemingly attractive, but conditions of stationarity and invertibility are very difficult if not impossible to verify. Also as described above, there are problems with model identification, thus making these classes of models difficult to manage in practice.

Finite-Order Volterra Series

Infinite-order convergent Volterra series representation is the most general nonlinear representation for stationary time series. This suggests using the finite-order Volterra series

$$\begin{aligned}
Y_t^{(m)} &= \sum_{i_1=0}^{k_1} g_{i_1} Z_{t-i_1} \\
&+ \sum_{i_1=0}^{k_2}\sum_{i_2=0}^{k_3} g_{i_1 i_2} Z_{t-i_1} Z_{t-i_2} \\
&+ \sum_{i_1=0}^{k_4}\sum_{i_2=0}^{k_5}\sum_{i_3=0}^{k_6} g_{i_1 i_2 i_3} Z_{t-i_1} Z_{t-i_2} Z_{t-i_3} \\
&+ \cdots \\
&+ \sum_{i_1=0}^{k_7}\sum_{i_2=0}^{k_8}\cdots\sum_{i_m=0}^{k_{2m}} g_{i_1 i_2 \cdots i_m} Z_{t-i_1} Z_{t-i_2} \cdots Z_{t-i_m},
\end{aligned}$$

as a parametric model. Finite-order Volterra series are used as flexible models for input-output systems where the input process, as well as the output process, are observable. In these models, often $m = 2$, so that second-order approximations are used; see, for example, Mathews and Sicuranza (2000). Within the univariate time series context, it is possible to identify the order m of the series using tail index estimation; see Sect. 4.3 for details. Conditional least square method can be used for parameter estimation. However, the innovation process Z_t is not observed and Granger and Andersen (1978) argue that processes of the form

$$X_t = Z_t + aZ_{t-1}Z_{t-2},$$

where Z_t are i.i.d. r.v's cannot be invertible. Therefore, finite Volterra series as models have limited practical value since they cannot be used for forecasting; see Sect. 4.2 for further discussion on invertibility.

2.2.7 *Generalized State Space Models*

Although all arguments given in this section can be extended to multivariate time series, for the sake of ease in notation, we will consider only univariate time series. A state space model for a linear time series Y_t consists of two equations, denoted by the observation and the state equations, which are given by

$$Y_t = \mathbf{H}_t\mathbf{X}_t + U_t, t = 1, 2, \ldots \tag{2.43}$$

$$\mathbf{X}_{t+1} = \mathbf{G}_t\mathbf{X}_t + \mathbf{V}_t, t = 1, 2, \ldots. \tag{2.44}$$

Here, in the first equation, $\mathbf{H}_t$ is a sequence of matrices whose elements are (constant) parameters and observations Y_t are written as a linear function of the unobserved (latent) v-dimensional state vector $\mathbf{X}_t$, plus a white noise U_t. The second equation determines the evolution of the state process in time in terms of the previous state. Here, $\mathbf{G}_t$ is a sequence of $v \times v$ matrices of parameters and $\mathbf{V}_t$ is a v dimensional white noise process, uncorrelated with U_t. In the simplest case, when Y_t is univariate, we may model the observations as

$$Y_t = m_t + Z_t,$$

where the state process m_t is the mean of the process, and Z_t is white noise. The latent mean m_t of the process can be modeled as a simple random walk

$$m_t = m_{t-1} + \nu_t.$$

It is also possible to add further structure to the model for m_t. The Kalman recursions allow a unified approach to prediction and estimation for state-space models; see Brockwell and Davis (1996) and West and Harrison (1997). Fundamental assumptions behind the state-space representation and the consequent Kalman recursions are linearity and the normality of the error structures. When these assumptions are no longer valid, then observation and state equations are given by

$$Y_t = f(\mathbf{X}_{t-1}, U_t),$$

$$\mathbf{X}_t = g(\mathbf{X}_{t-1}, \mathbf{V}_t),$$

for some nonlinear functions f and g and white noise processes U_t and $\mathbf{V}_t$ independent of each other. Except for some special cases, satisfactory treatment of such a system of difference equations is not possible, and it is more advantageous to work directly with conditional distributions (or densities if they exist) which represent the probability structure of the system. In general terms, these equations are represented by two conditional densities $p(y_t|\mathbf{x}_t, \boldsymbol{\theta})$ and $p(\mathbf{x}_t|\mathbf{x}_{t-1}, \boldsymbol{\theta})$. Here, $\boldsymbol{\theta}$ is the vector of all model parameters of this state space representation. This general state space structure can take several forms depending on different sets of further assumptions on these densities, which we examine below. Typically, there are two sets of fundamental assumptions to facilitate mathematical tractability of these state space models. In parameter-driven models, observations $\mathbf{Y}_t$ are assumed to be independent, conditional on the realization of the state vector $\mathbf{X}_t$, and that the state process $\mathbf{X}_t$ is assumed to be a (latent) Markov process.

In observation-driven models, again observations $\mathbf{Y}_t$ are assumed to be independent conditional on the realizations of the state vector $\mathbf{X}_t$, but rather then assuming a Markovian structure for the state vector, a model is specified directly for $\mathbf{X}_t$ conditional on $\mathbf{Y}_{t-1}$ through the conditional density $p(\mathbf{x}_t|\mathbf{y}_{t-1}, \boldsymbol{\theta})$. These two types of models show fundamental differences, particularly in inferential methods. Parameter driven models, otherwise known as hidden Markov models, are particularly suitable for Bayesian hierarchical modeling and simulation-based inferential techniques. Due to some awkward integrals and updating equations, classical likelihood and least squares methods are not particularly suitable for these models. On the other hand, observation-driven models do not involve such updating equations and difficult integrations and hence permit straight forward likelihood and least square methods. However, it is very difficult to verify stationarity conditions for the observation-driven models; see Brockwell and Davis (1996) for detailed comparison of these models. Here we give a brief summary of these models.

Parameter-Driven Models

For simplicity, assume that we have univariate time series Y_t and the corresponding univariate state X_t. Let $\mathbf{Y}_{t-1} := (Y_{t-1}, Y_{t-2}, \ldots)$ and $\mathbf{X}_{t-1} := (X_{t-1}, X_{t-2}, \ldots)$. Instead of the linear equations (2.43) and (2.44), we define the observation and state equations in terms of the conditional densities, assuming they exist, in the following manner:

Assume that Y_t conditional on X_t, is independent of $(\mathbf{X}_{t-1}, \mathbf{Y}_{t-1})$, so that the density of Y_t conditional on $(\mathbf{X}_t, \mathbf{Y}_{t-1})$ can be written as

$$p(y_t|x_t, \mathbf{x}_{t-1}, \mathbf{y}_{t-1}, \boldsymbol{\theta}) = p(y_t|x_t, \boldsymbol{\theta}). \tag{2.45}$$

We also assume that X_{t+1} conditional on X_t is independent of $(\mathbf{X}_{t-1}, \mathbf{Y}_t)$ so that we can write

$$p(x_{t+1}|x_t, \mathbf{x}_{t-1}, \mathbf{y}_t, \boldsymbol{\theta}) = p(x_{t+1}|x_t, \boldsymbol{\theta}). \tag{2.46}$$

For linear Gaussian state space equations, (2.43), (2.44), and the conditional densities (2.45), (2.46) represent the same probability model, with $p(y_t|x_t, \boldsymbol{\theta})$ and $p(x_{t+1}|x_t, \boldsymbol{\theta})$ being normal densities. The joint density of the n observations $\mathbf{Y}_n$ and the state $\mathbf{X}_n$ at each time point $t = 1, \ldots, n$ can be written as

$$\begin{aligned} p(\mathbf{y}_t, \mathbf{x}_t|\boldsymbol{\theta}) &= p(y_n|x_n, \mathbf{x}_{n-1}, \mathbf{y}_{n-1}, \boldsymbol{\theta})p(x_n, \mathbf{x}_{n-1}, \mathbf{y}_{n-1}|\boldsymbol{\theta}) \\ &= p(y_n|x_n, \boldsymbol{\theta})p(x_n|\mathbf{x}_{n-1}, \mathbf{y}_{n-1}, \boldsymbol{\theta})p(\mathbf{x}_{n-1}, \mathbf{y}_{n-1}|\boldsymbol{\theta}) \\ &\vdots \\ &= \left(\prod_{i=1}^{n} p(y_i|x_i, \boldsymbol{\theta})\right)\left(\prod_{i=2}^{n} p(x_i|x_{i-1}, \boldsymbol{\theta})\right) p(x_1). \end{aligned}$$

Note that

$$p(\mathbf{y}_t|\mathbf{x}_t, \boldsymbol{\theta}) = \prod_i p(y_i|x_i, \boldsymbol{\theta}),$$

hence observations are independent, conditional on the state of the process, and the time series Y_t inherits the dependence structure of the state process X_t, which is often called the latent process. Note also that from (2.46), the state X_t is a Markov process. These are indeed strong assumptions but are necessary to bring in some mathematical tractability to nonlinear, non Gaussian structures.

Conditional densities $p(x_t|\mathbf{y}_t, \boldsymbol{\theta})$ and $p(y_{t+1}|\mathbf{y}_t, \boldsymbol{\theta})$ are particularly relevant in the study of the system from which one can calculate the conditional expectations $E(X_t|\mathbf{y}_t)$ and $E(Y_{t+1}|\mathbf{y}_t)$. The former and the latter conditional expectations are the best predictors for X_t and Y_{t+1} in terms of the observation $\mathbf{y}_t$ and are respectively called the filtering and prediction problem. With the above assumptions and using the Bayes' Theorem

$$\begin{aligned} p(x_t|\mathbf{y}_t, \boldsymbol{\theta}) &= \frac{p(x_t, y_t, \mathbf{y}_{t-1}|\boldsymbol{\theta})}{p(\mathbf{y}_t|\boldsymbol{\theta})} \\ &= \frac{p(y_t|x_t, \mathbf{y}_{t-1}, \boldsymbol{\theta})p(x_t|\mathbf{y}_{t-1}, \boldsymbol{\theta})p(\mathbf{y}_{t-1}|\boldsymbol{\theta})}{p(y_t|\mathbf{y}_{t-1}, \boldsymbol{\theta})p(\mathbf{y}_{t-1}|\boldsymbol{\theta})} \\ &= \frac{p(y_t|x_t, \boldsymbol{\theta})p(x_t|\mathbf{y}_{t-1}, \boldsymbol{\theta})}{p(y_t|\mathbf{y}_{t-1}, \boldsymbol{\theta})}. \end{aligned} \tag{2.47}$$

Here, the conditional density $p(x_t|\mathbf{y}_{t-1}, \boldsymbol{\theta})$ has to be calculated from the integral

$$\begin{aligned} p(x_t|\mathbf{y}_{t-1}, \boldsymbol{\theta}) &= \int p(x_t, x_{t-1}|\mathbf{y}_{t-1}, \boldsymbol{\theta})dx_{t-1} \\ &= \int p(x_t|x_{t-1}, \boldsymbol{\theta})p(x_{t-1}|\mathbf{y}_{t-1}, \boldsymbol{\theta})dx_{t-1}. \end{aligned} \tag{2.48}$$

For non-Gaussian and nonlinear processes, this updating equation for X_t in terms of the state equations $p(x_t|x_{t-1}, \boldsymbol{\theta})$ is not immediately available and can be computationally complicated; hence $p(x_t|\mathbf{y}_t, \boldsymbol{\theta})$ in (2.47) does not admit closed form expression.

In order to solve recursive relation in t, one assumes that $p(x_1|\mathbf{y}_0, \boldsymbol{\theta}) = p(x_1|\boldsymbol{\theta})$. The density $p(x_{t+1}|\mathbf{y}_t, \boldsymbol{\theta})$ and the corresponding conditional expectation give the prediction for the future value of the state equation, whereas the predictions for the future observation $\hat{y}_{t+1}$ can be obtained as the expected value of the conditional density $p(y_{t+1}|\mathbf{y}_t, \boldsymbol{\theta})$. In the classical approach, where $\boldsymbol{\theta}$ are unknown but fixed model parameters to be estimated from data, the unknown parameters are substituted by their estimates $\hat{\boldsymbol{\theta}}$ and the *plug-in* predictions are obtained from $p(y_{t+1}|\mathbf{y}_t, \hat{\boldsymbol{\theta}})$. In the Bayesian context the parameters are r.v's and the predictions are obtained from the predictive density $p(y_{t+1}|\mathbf{y}_t)$ through the relationship

$$p(y_{t+1}|\mathbf{y}_t) = \int p(y_{t+1}|\mathbf{y}_t, \boldsymbol{\theta})p(\boldsymbol{\theta}|\mathbf{y}_t)d\boldsymbol{\theta}$$

and

$$p(y_{t+1}|\mathbf{y}_t, \boldsymbol{\theta}) = \int p(y_{t+1}|x_{t+1}, \boldsymbol{\theta})p(x_{t+1}|\mathbf{y}_t, \boldsymbol{\theta})dx_{t+1}.$$

The key expression for the classical and the Bayesian inferential methods is the likelihood function $L(\boldsymbol{\theta}|\mathbf{y}_t)$ which can be computed from the relation

$$\begin{aligned} L(\boldsymbol{\theta}|\mathbf{y}_t) &= \int \cdots \int p(\mathbf{x}_t|\boldsymbol{\theta})p(\mathbf{y}_t|\mathbf{x}_t, \boldsymbol{\theta})dx_1 \cdots dx_n \\ &= \int \cdots \int p(x_1|\boldsymbol{\theta}) \prod_{i=2}^{n} p(x_i|x_{i-1}, \boldsymbol{\theta})p(y_i|x_i, \boldsymbol{\theta})dx_1 \cdots dx_n. \end{aligned} \tag{2.49}$$

The computation of the likelihood given in (2.49) requires the computation of n-dimensional integrals. Except for few special cases, calculation of such integrals are very difficult. Thus one relies on approximate solutions based on numerical methods on Monte Carlo methods. Recent advances in Bayesian simulation-based inferential methods and composite likelihood methods permit efficient simulation based estimation techniques and approximations. In a Bayesian hierarchical setup, upon defining a prior density $p(\boldsymbol{\theta})$ for the (random) model parameters $\boldsymbol{\theta}$,

Bayesian inference relies on the joint density $p(\boldsymbol{\theta}, \mathbf{x}_t | \mathbf{y}_t)$ which is proportional to $p(\mathbf{x}_t, \mathbf{y}_t | \boldsymbol{\theta}) p(\boldsymbol{\theta})$. This joint density does not have closed form expressions, and Monte Carlo methods, in particular recent sequential Monte Carlo methods and particle filters (see, e.g. Andrieu et al. 2010) provide a flexible computational framework to carry out inference for these data sets with complex time dependence structures. In Sect. 4.5 we give a very brief introduction to these simulation-based methods. In Sect. 4.4.3 we also give a brief introduction to composite likelihood methods which are used as alternative pseudo-likelihood method for the observation-driven generalized state space models.

However, it may be possible to escape from such computational difficulties by specifying in (2.47) a model for $p(x_t | \mathbf{y}_t, \boldsymbol{\theta})$, thus eliminating the need for the updating Eq. (2.48). This strategy simplifies inference for generalized state-space models and the resulting models are called the observation-driven models.

Observation-Driven Models

In observation-driven models, the observation equation is the same as in parameter-driven models; namely it is assumed that

$$p(y_t | x_t, \mathbf{x}_{t-1}, \mathbf{y}_{t-1}, \boldsymbol{\theta}) = p(y_t | x_t, \boldsymbol{\theta}). \tag{2.50}$$

However, the representation of the state is done through the densities

$$p(x_t | \mathbf{y}_{t-1}, \boldsymbol{\theta}), t = 1, 2, \ldots. \tag{2.51}$$

Here, the updating equation for the state

$$p(x_t | x_{t-1}, \boldsymbol{\theta}),$$

is not specified, since the conditional density of the state vector given the data $p(x_t | \mathbf{y}_t, \boldsymbol{\theta})$ and the predictive density can be directly calculated from (2.47) and (2.48) respectively, with the estimated value of the parameter $\hat{\boldsymbol{\theta}}$. Within the Bayesian framework, when $\boldsymbol{\theta}$ is random with prior specification $p(\boldsymbol{\theta})$, this predictive density is calculated from

$$p(y_{t+1} | \mathbf{y}_t) = \int p(y_{t+1} | x_{t+1}, \boldsymbol{\theta}) p(x_{t+1} | \boldsymbol{\theta}, \mathbf{y}_t) p(\boldsymbol{\theta} | \mathbf{y}_t) dx_{t+1} d\boldsymbol{\theta},$$

where $p(\boldsymbol{\theta} | \mathbf{y}_t)$ is the posterior density.

The state equation (2.51) without specifying precisely how x_t translates from x_{t-1}, simplifies the calculation of the posterior and the predictive distributions, but observations are no longer Markovian and y_t depend on the whole $\mathbf{y}_t$ rather than y_{t-1}, so that

$$p(y_1, \ldots, y_n|\boldsymbol{\theta}) = \prod_{t=1}^{n} p(y_t|\mathbf{y}_{t-1}, \boldsymbol{\theta}). \tag{2.52}$$

The lack of Markovian property particularly makes it more difficult to verify stationarity conditions. Also, due to the lack of Markovian structure, observation-driven models are not suitable for Bayesian hierarchical modeling. The specification given by (2.50) and (2.51) is not unique, in the sense that it can hold for two different state equations having different transitions, resulting in the same likelihood (2.52) for the data. This model miss-specification can be overcome by assuming that

$$p(x_{t+1}|\mathbf{x}_t, \mathbf{y}_t) = p(x_{t+1}|\mathbf{y}_t),$$

that is assuming that x_t, conditional on $\mathbf{y}_{t-1}$, is independent of $\mathbf{x}_{t-1}$. In this case

$$\begin{aligned} p(\mathbf{x}_n, \mathbf{y}_n) &= p(y_n|x_n)p(x_n|\mathbf{y}_{n-1})p(\mathbf{x}_{n-1}, \mathbf{y}_{n-1}) \\ &\vdots \\ &= \prod_{t=1}^{n} p(y_n|x_n)p(x_t|\mathbf{y}_{t-1}). \end{aligned}$$

We give an example to highlight the difference between the two modeling strategies.

Example: State Space Models for Count Data

In this example, we follow Brockwell and Davis (1996) and Davis et al. (2003a). Assume that Y_t is a time series of counts. Let $\mathcal{F}_{t-1}$ be the σ-field generated by the observation $(Y_s, s \leq t_1)$, and let $\boldsymbol{W}_t$ be a vector of explanatory variables with dimension p, observed at time t.

1. **Parameter-driven model**:
   We assume that observations, conditional on the intensity function λ_t, are independent, having the observation equation

   $$Y_t|\lambda_t \sim \text{Po}(\lambda_t),$$

   so that the likelihood for the data $y_1, \ldots, y_n$ conditional on the realization of the state process or the intensity process λ_t is given by

   $$p(y_1, \ldots, y_n|\lambda_1, \ldots, \lambda_n) = \prod_{t=1}^{n} \frac{e^{-\lambda_t}(\lambda_t)^{y_t}}{y_t!}. \tag{2.53}$$

   The dependence structure is then introduced into the model through the state equation (or the link function)

$$\log \lambda_t = \boldsymbol{\beta}' \boldsymbol{W}_t + U_t, \tag{2.54}$$

where $\boldsymbol{\beta}$ is a p dimensional vector of regression coefficients and U_t a latent time-dependent process. In the simplest case, U_t is assumed to follow an AR(1) process of the form

$$U_t = \phi U_{t-1} + Z_t,$$

where (Z_t) is a sequence of i.i.d. $N(0, \sigma^2)$, independent of the Y_t process. In this case, the state equation can equivalently be written in terms of the conditional density $p(\lambda_t|\lambda_{t-1})$ by

$$p(\lambda_t|\lambda_{t-1}) \sim N(\mu_t, \sigma^2),$$

where

$$\mu_t = \boldsymbol{\beta}' \boldsymbol{W}_t + \phi(\lambda_{t-1} - \boldsymbol{\beta}' \boldsymbol{W}_{t-1}).$$

The above model expressed in terms of Eqs. (2.53) and (2.54) can be implemented in Bayesian context as a hierarchical model upon defining appropriately the prior specifications of parameters and hyper-parameters. The posterior density $p(\lambda_t|\mathcal{F}_{t-1})$ and the predictive density $p(Y_{t+1}|\mathcal{F}_t)$, as well as the posterior densities of all other model parameters can be obtained by applying proper simulation-based inferential methods; see Brockwell and Davis (1996) for an alternative Monte Carlo-based estimation method. Implementation of this model, using standard maximum likelihood estimation is not straightforward and can be difficult, since the closed form for the unconditional likelihood $p(y_1, \ldots, y_n)$ is obtained by integrating the conditional likelihood (2.53) with respect to the joint density of $p(\lambda_1 \ldots, \lambda_t)$. Brockwell and Davis (1996) suggest a simulation-based estimation based on the EM algorithm.

2. **Observation-driven model for the counts:**
Assume for the time-being that there are no explanatory variables available in modeling the counts and that only information available are the counts themselves $(y_1, \ldots, y_n)$. In this case, the observation-driven model can be written as

$$Y_t|\lambda_t \sim \text{Po}(\lambda_t),$$

where λ_t is written as a positive function of the observations $y_{t-1}, \ldots, y_1$. The class of INGARCH(p, q) processes is constructed by assuming a specific linear function for λ_t, where

$$\lambda_t = \mu + \sum_{i=1}^{p} a_i \lambda_{t-i} + \sum_{j=1}^{q} b_j Y_{t-j}, \tag{2.55}$$

where $\mu > 0$, $a_i \geq 0$, $b_j \geq 0$ for every $i = 1, \ldots, p$ and $j = 1, \ldots, q$, so that λ_t is strictly positive for every t. If we further assume that all the roots of the polynomial $A(B) = 1 - \sum_{i=1}^{p} a_i B^i$ lie outside the unit circle (for non-negative a_i this is equivalent to the condition $\sum_{i=1}^{p} a_i < 1$), then λ_t can be written in terms of the $(Y_s, s < t)$ as

$$\lambda_t = A^{-1}(B)\mu + \sum_{j=1}^{\infty} \pi_j Y_{t-j}.$$

Note that (Y_t) are no longer conditionally independent and the joint density of $(Y_1, \ldots, Y_n)$ is written as

$$p(y_1, \ldots, y_n) = \prod_{t=1}^{n} p(y_t|\mathcal{F}_{t-1})p(y_1),$$

where

$$p(y_t|\mathcal{F}_{t-1}) = p(y_t|\lambda_t).$$

INGARCH(p, q) processes are restricted, so that the state equation λ_t can be written as a strictly positive, linear function of the observations. Such restriction simplifies the conditions of existence of stationary solutions, as well as estimation procedures. For example, the INGARCH(p, q) process defined above with $\sum_{i=1}^{p} a_i + \sum_{j=1}^{q} b_j < 1$, is strictly stationary with finite second-order moments (see Ferland et al. 2006, for the case $p = q = 1$ and Weiß 2009 for the general one). The classical (conditional) likelihood-based inference is also relatively easy. Set $\boldsymbol{\beta} := (a_1, \ldots, a_p, b_1, \ldots, b_q)$, then the conditional log-likelihood is written in the form

$$L(\boldsymbol{\beta}|\mathbf{y}_n) = \sum_{t=\max(p,q)}^{n} [-\lambda_t(\boldsymbol{\beta}) + y_t \log \lambda_t(\boldsymbol{\beta}) - \log y_t!],$$

from which we get the score function

$$\frac{\partial L(\boldsymbol{\beta}|\mathbf{y}_n)}{\partial \boldsymbol{\beta}} = (\frac{\partial L(\boldsymbol{\beta}|\mathbf{y}_n)}{\partial \beta_i}, i = 1, \ldots, p + q),$$

where

$$\frac{\partial L(\boldsymbol{\beta}|\mathbf{y}_n)}{\partial \beta_i} = \sum_{t=\max(p,q)}^{n} \frac{\partial \lambda_t(\boldsymbol{\beta})}{\partial \beta_i} \left(\frac{y_t}{\lambda_t(\boldsymbol{\beta})} - 1 \right).$$

The elements of the Hessian matrix $H_n(\boldsymbol{\beta})$ are calculated as

$$\frac{\partial^2 L(\boldsymbol{\beta})}{\partial \boldsymbol{\beta}_j \partial \boldsymbol{\beta}_i} = \sum_{t=\max(p,q)}^{n} \left[\frac{\partial^2 \lambda_t(\boldsymbol{\beta})}{\partial \boldsymbol{\beta}_j \partial \boldsymbol{\beta}_i} \left(\frac{y_t}{\lambda_t(\boldsymbol{\beta})} - 1 \right) - \frac{y_t}{\lambda_t^2(\boldsymbol{\beta})} \frac{\partial \lambda_t(\boldsymbol{\beta})}{\partial \boldsymbol{\beta}_j} \right] \tag{2.56}$$

from which a proper numerical optimization method can be constructed.

Extensions of (2.55) for the simplest case $p = q = 1$ have been recently proposed by Fokianos et al. (2009) by considering a more general representation for λ_t, namely

$$\lambda_t = f(\lambda_{t-1}) + g(Y_{t-1}), \ t \geq 1, \tag{2.57}$$

where $f, g : \mathbb{R}^+ \to \mathbb{R}^+$ are known functions up to an unknown finite dimensional parameter vector. The initial values Y_0 and λ_0 are assumed to be fixed. Special models for λ_t in (2.57) include the model in (2.55) upon defining $f(x) = c + dx$ and $g(x) = bx$ with $c, d, g > 0$ and $x \geq 0$, and the so-called exponential autoregressive model with

$$\lambda_t = (a_1 + c_1 \exp\{-\gamma_1 \lambda_{t-1}\})\lambda_{t-1} + bY_{t-1}.$$

Fokianos et al. (2009) proved that under geometric ergodicity the maximum likelihood estimators of the parameters are asymptotically Gaussian in the linear model (2.55); see also Tjøstheim (2012) and Fokianos (2011) for further details.

If we have explanatory variables to account for the variations in the latent intensity λ_t of the counts, then the statistical and probabilistic properties of the model get more complicated. In this case, in order to satisfy the positivity of the intensity process λ_t, we model $\log \lambda_t$ by a linear function, giving rise to

$$Y_t | \mathcal{F}_{t-1} \sim \text{Po}(\lambda_t),$$

$$\log \lambda_t = \boldsymbol{\beta}' \boldsymbol{W}_t + \sum_{i=1}^{p} a_i \lambda_{t-i} + \sum_{j=1}^{q} b_j Y_{t-j}.$$

In this case, it is not clear under what conditions this process may be stationary. For example, the simpler process with

$$\log \lambda_t = \boldsymbol{\beta}' \boldsymbol{W}_t + \sum_{j=1}^{q} b_j Y_{t-j},$$

cannot be stationary unless some normalization is applied to the observations. Davis et al. (2003a) suggest using the model

$$\log \lambda_t = \boldsymbol{\beta}' \boldsymbol{W}_t + \sum_{j=1}^{q} \theta_i Z_{t-j},$$

where

$$Z_t = \frac{Y_t - \lambda_t}{\lambda_t^{\eta}}, \quad \eta \geq 0.$$

Note that Y_t is not Markov process, but the intensity process λ_t is pth-order Markov. Existence of a stationary solution depends on the value of η. For example, for the simpler first-order model and assuming that $\boldsymbol{\beta}' \boldsymbol{W}_t = \boldsymbol{\beta}'$,

$$\log \lambda_t = \boldsymbol{\beta}' + \frac{Y_{t-1} - \lambda_{t-1}}{\lambda_t^{\eta}}.$$

Davis et al. (2003a) proved the existence of a stationary solution for $\eta \in [1/2, 1]$, showing that this solution is unique when $\eta = 1$.

Estimation of the parameters using likelihood is relatively easy and the likelihood is maximized by using the Newton-Raphson method; see Davis et al. (2003b) for details.

In Chap. 5, we will study alternative models for integer-valued time series, which have linear representations similar to ARMA models but are constructed with thinning operations.

2.2.8 *Max-Stable Moving Average Processes*

Max-stable moving average processes are introduced as models for heavy-tailed data by Davis and Resnick (1993). This class is defined as follows: X_t is said to be max-stable moving average process if

$$X_t = \bigvee_{j=0}^{\infty} \psi_j Z_{t-j},$$

where $\bigvee_j \psi_j \equiv \max_j \psi_j$ and (Z_t) are i.i.d. r.v's with distribution $\exp[-\sigma z^{-1}]$. Analogous finite parameter version of these models are also defined. The reason why the authors suggest such classes for modeling heavy-tailed data is that their sample paths very much resemble the sample paths of corresponding linear models formed from the same residuals, and the predictions and estimation of parameters for these models can be done by an optimality criterion which minimizes the probability of large errors, that is likely to give better fit to sudden burst. The optimal predictor can be explicitly written for several models. However, since second-order moments

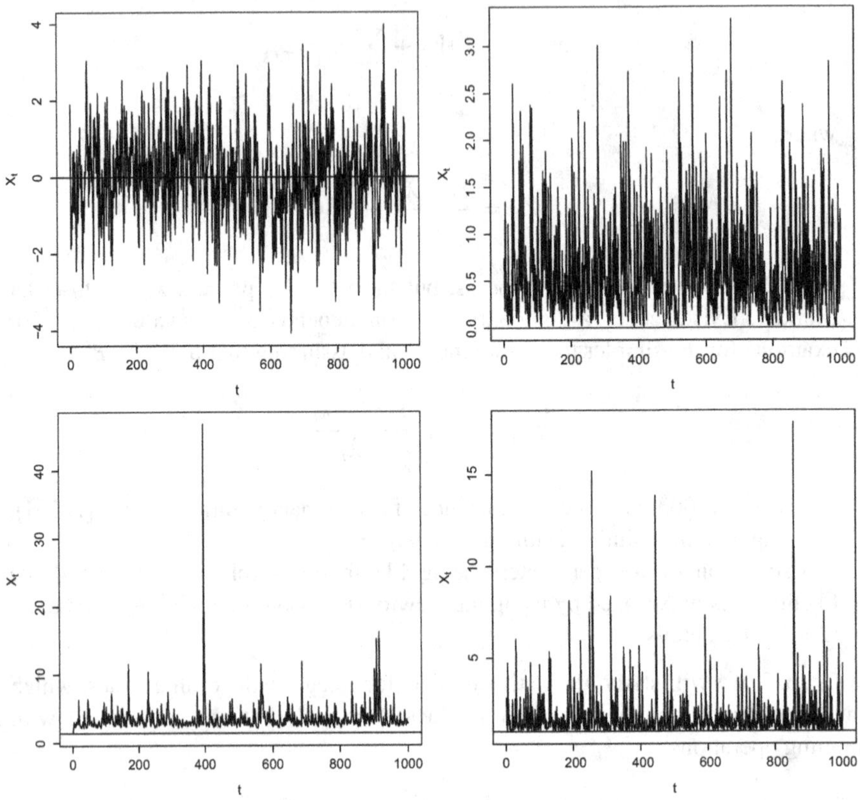

Fig. 2.8 Sample paths – AR(1) and max-stable models with N(0,1) errors (*top row*) and Pareto($\alpha = 2.5$) errors (*bottom row*)

cannot be used for identification and estimation, such classes are not very frequently used in practice. Figure 2.8 shows the sample path of $n = 1{,}000$ observations generated from an AR(1), $X_t = 0.5X_{t-1} + Z_t$, and the corresponding max-stable process $X_t = \max(0.5X_{t-1}, Z_t)$, where Z_t is a N(0, 1) sequence. The same models are represented in the bottom row for Pareto($\alpha = 2.5$) residuals.

2.2.9 *Nonparametric Methods*

In the class of parametric models, the main emphasis was on building parametric models for the conditional mean and variance of the process, either separately or jointly. If the emphasis is on prediction rather than on explaining how these conditional means and variances change in time, then a plausible alternative is to estimate them using nonparametric methods. This would be quite flexible, since one is not restricted by a specific parametric model. The most common way is to use kernel estimators. For example, for a given time series (X_t) the conditional mean

$$M(x_1, \ldots, x_p) := E(X_t | X_{t-1} = x_1, \ldots, X_{t-p} = x_p),$$

can be estimated by

$$\hat{M}(x_1, \ldots, x_p) = \frac{(n-p)^{-1} \sum_{t=p+1}^{n} X_t \prod_{i=1}^{p} K_h(X_{t-i} - x_i)}{(n-p+1)^{-1} \sum_{t=p+1}^{n+1} \prod_{i=1}^{p} K_h(X_{t-i} - x_i)},$$

where

$$K_h(x) = \frac{1}{h} K(\frac{x}{h})$$

and K representing a kernel function. A similar expression can be obtained for the conditional variance. The drawback with these models is that one needs large sample sizes for a reasonable fit and the sample needs to increase drastically with the increase in p, a typical case of the curse of dimension. The curse of dimension can be reduced by simplifying the model. In the simplest case, one can model the conditional mean function by

$$M(x_1, \ldots, x_p) = \sum_j f_j(x_p),$$

where the $f_j(\cdot)$ are unknown functions to be estimated. Each of these functions are one-dimensional, thus simplifying the problem. A similar model can be written for the conditional variance. Such additive models can further be extended to include linear combinations of past values. Such models are known as projection pursuit models. In general, these models are taken from regression context and adopted to the time series context. Further models and references can be found in Tjøstheim (1994) and Gao (2007). Another possibility is to use splines in estimating these conditional means and variances. We refer the reader to Hardle et al. (1997) for a general review of nonparametric methods in time series analysis.

Bayesian Nonparametric Methods

Bayesian nonparametric methods have been one of the fastest growing topics in statistics. Bayesian methods inherently are likelihood based; therefore they need specification of a parametric model. Indeed, what is usually called nonparametric Bayesian method, in fact corresponds to models with priors defined on an infinite dimensional parameter space. Suppose that X_t is a stochastic process with probability measure F defined through its finite dimensional distributions. In ordinary Bayesian inferential methods, one assumes a parametric model for the probability distribution and expresses prior belief on the parameters and then the inference concentrates on deriving the posterior distribution of the model parameters and the predictive distribution for the future values of the process. In nonparametric Bayesian methods, no parametric form is assumed for the probability distribution; instead prior beliefs are assigned to the probability distributions (i.e. models) which are now random elements themselves belonging to some measure space. Hence consider the probability space $(\Omega, \mathcal{B}, P)$ where the random variable (or the stochastic process) resides. Typically $\Omega = \mathbb{R}^d$, $\mathcal{B}$ the Borel σ-algebra over Ω and P is the probability measure of the random variable. Assume that P is a random measure residing in a space of probability measures $(\mathcal{P}, \mathcal{C}, \mathcal{Q})$ so that the probability measures P of the random variable (or the stochastic process) is a simple element of $\mathcal{P}$. Often $(\Omega, \mathcal{B}, P)$ is called the base space and $(\mathcal{P}, \mathcal{C}, \mathcal{Q})$ the distributional space. The Dirichlet process is a probability measure on $(\mathcal{P}, \mathcal{C})$ and is often used as the prior distribution for the random measure P. A Dirichlet process (DP) is defined by a concentration parameter α and base distribution P_0. Random measure P is said to follow a DP prior if for any measurable partition $(A_1, A_2, \ldots, A_k)$ of the sample space of the random variable, the vector $(P(A_1), \ldots, P(A_k))$ has a Dirichlet distribution with parameters $(\alpha P_0(A_1), \ldots, \alpha P_0(A_k))$. The DP is centered at P_0, so that $E(P(A)) = P_0(A)$ for any measurable set $A \in \mathcal{B}$. The inferential problem is then given in terms of a hierarchical representation. For example, in the simplest form, when the observed data are i.i.d. with common marginal distribution F, the hierarchical model is given as

$$x_1, x_2, \ldots, x_n | F \sim \text{ i.i.d. } F,$$

$$F | \alpha, F_0 \sim DP(\alpha, F_0),$$

whereas the classical Bayesian parametric modeling paradigm would result in the following hierarchical representation;

$$x_1, x_2, \ldots, x_n | \boldsymbol{\theta} \sim \text{ i.i.d. } F(x|\boldsymbol{\theta}),$$

$$\boldsymbol{\theta} \sim \pi(\boldsymbol{\theta}),$$

where, $\pi(\boldsymbol{\theta})$ is the prior distribution of the model parameters $\boldsymbol{\theta}$. The difference in these alternative approaches is evident.

For time-dependent data, the specification of the hierarchical model needs the notion of dependent Dirichlet processes and is beyond the scope of this book. We refer the reader to Rodriguez (2007) and Hjort et al. (2010) for excellent accounts of Bayesian nonparametric modeling.

In Sect. 4.5 we will give a detailed summary of Bayesian inferential methods for nonlinear time series based on parametric likelihood methods.

References

Andersen TG, Davis RA, Kreib J-P, Mikosch T (eds) (2009) Handbook of financial time series. Springer, Berlin/Heidelberg

Andrews B, Davis RA, Breidt FJ (2006) Maximum likelihood estimation for all-pass time series models. J Multivar Anal 97:1638–1659

Andrieu C, Doucet A, Holenstein R (2010) Particle Markov chain Monte Carlo methods. J R Stat Soc B 72:269–342

Baillie RT, Morana C (2009) Modelling long memory and structural breaks in conditional variances: an adaptive FIGARCH approach. J Econ Dyn Control 33:1577–1592

Baillie RT, Bollerslev T, Mikkelsen HO (1996) Fractionally integrated generalized autoregressive conditional heteroskedasticity. J Econom 74:3–30

Bauwens L, Laurent S, Rombouts JVK (2006) Multivariate GARCH models: a survey. J Appl Econom 21:79–109

Berkes I, Horváth L, Kokoszka P (2003) GARCH processes: structure and estimation. Bernoulli 9:201–227

Bollerslev T (1986) Generalized autoregressive conditional heteroskedasticity. J Econom 31:307–327

Bollerslev T, Chou RY, Kroner KF (1992) ARCH modelling in finance: a review of the theory and empirical evidence. J Econom 52:5–59

Bollerslev T, Engle RF, Nelson DB (1994) ARCH models. In: Engle RF, McFadden DL (eds) Handbook of econometrics. North Holland, Amsterdam, pp 2959–3038

Brandt A (1986) The stochastic equation $Y_{n+1} = A_n Y_n + B_n$ with stationary coefficients. Adv Appl Probab 18:211–220

Brockett RW (1976) Non-linear systems and differential geometry. Automatica 12:167–176

Brockwell PJ, Davis RA (1991) Time series: theory and methods. Springer, New York

Brockwell PJ, Davis RA (1996) Introduction to time series and forecasting. Springer, New York

Chan KS, Tong H (1986) On estimating thresholds in autoregressive models. J Time Ser Anal 7:179–190

Conrad C, Haag BR (2006) Inequality constraints in the fractionally integrated GARCH model. J Financ Econom 4:413–449

Conrad C, Karanasos M, Zeng N (2008) Multivariate fractionally integrated APARCH modeling of stock market volatility: a multi-country study. Discussion paper no. 472, University of Heidelberg

Cont R (2001) Empirical properties of asset returns: stylized facts and statistical issues. Quant Finance 1:223–236

Davidson JEH (2004) Conditional heteroskedasticity models and a new model. J Bus Econom Stat 22:16–29

Davis RA, Resnick SI (1993) Prediction of stationary max-stable processes. Ann Appl Probab 3:497–525

Davis RA, Dunsmuir WTM, Streett SB (2003a) Observation-driven models for Poisson counts. Biometrika 90:777–790

Davis RA, Dunsmuir WTM, Streett SB (2003b) Maximum likelihood estimation for an observation driven model for Poisson counts. Methodol Comput Appl Probab 7:149–159

Davis RA, Lee TCM, Rodriguez-Yam GA (2008) Break detection for a class of nonlinear time series models. J Time Ser Anal 29:834–867

Ding Z, Granger CWJ, Engle RF (1993) A long memory property of stock market returns and a new model. J Empir Finance 1:83–106

Diongue AK, Guégan D (2007) The stationary seasonal hyperbolic asymmetric power ARCH model. Stat Probab Lett 77:1158–1164

Engle RF (1982) Autoregressive conditional heteroskedascity with estimates of the United Kingdom inflation. Econometrica 50:987–1008

Engle RF (1990) Discussion: stock market volatility and the crash of 87. Rev Financ Stud 3: 103–106
Engle RF (2004) Nobel lecture. Risk and volatility: econometric models and financial practice. Am Econ Rev 94:405–420
Engle RF, Bollerslev T (1986) Modelling the persistence of conditional variances. Econom Rev 5:1–50
Fan J, Yao Q (2003) Nonlinear time series. Springer, New York
Ferland R, Latour A, Oraichi D (2006) Integer-valued GARCH processes. J Time Ser Anal 27: 923–942
Fokianos K (2011) Some recent progress in count time series. Stat Pap 45:49–58
Fokianos K, Rahbek A, Tjøstheim D (2009) Poisson autoregression. J Am Stat Assoc 104:1430–1439
Fornari F, Mele A (1997) Sign- and volatility-switching ARCH models: theory and applications to international stock markets. J Appl Econom 12:49–65
Franses PH, van Dijk D (2000) Non-Linear Time Series Models in Empirical Finance. Cambridge University Press, New York
Friedman M (1977) Nobel lecture: inflation and unemployment. J Polit Econ 85:451–472
Gao J (2007) Nonlinear time series: semiparametric and nonparametric methods. Chapman and Hall, Boca Raton
Geweke J (1986) Modeling the persistence of conditional variances: a comment. Econom Rev 5:57–61
Glosten L, Jagannathan R, Runkle D (1993) On the relation between the expected value and the volatility of the nominal excess return on stocks. J Finance 48:1779–1801
Goldfeld SM, Quandt R (1973) The estimation of structural shifts by switching regressions. Ann Econ Soc Meas 2:475–85
Granger CWJ, Andersen A (1978) On the invertibility of time series models. Stoch Process Appl 8:87–92
Granger CWJ, Hyung N (2004) Occasional structural breaks and long memory with an application to the S&P 500 absolute stock returns. J Empir Finance 11:399–421
Granger CWJ, Teräsvirta T (1993) Modelling nonlinear economic relationships. Oxford University Press, New York
Hamilton JD (1989) A new approach to the economic analysis of nonstationary time series and the business cycle. Econometrica 57:357–384
Hamilton JD (2008) Regime switching models. In: Durlauf SN, Blume LE (eds) The new Palgrave dictionary of economics, 2nd edn. Palgrave Macmillan, Basingstoke/New York
Härdle W, Lütkepohl H, Chen R (1997) A review of nonparametric time series analysis. Int. Stat. Rev. 65:49–72
He C, Teräsvirta T (1999) Properties of moments of a family of GARCH processes. J Econom 92:173–192
Higgins ML, Bera AK (1992) A class of nonlinear ARCH models. Int Econ Rev 33:137–158
Hjort N, Holmes C, Mueller P, Walker S (2010) Bayesian Nonparametrics: Principles and Practice. Cambridge University Press, Cambridge
Hwang SY, Basawa IV (2004) Stationarity and moment structure for Box-Cox transformed threshold GARCH(1, 1) processes. Stat Probab Lett 68:209–220
Hwang SY, Woo MJ (2001) Threshold ARCH(1) processes: asymptotic inference. Stat Probab Lett 53:11–20
Kashikar AS, Rohan N, Ramanathan TV (2013) Integer autoregressive models with structural breaks. J. Appl. Statist. 40:2653–2669
Ling S, McAleer M (2002) Stationary and the existence of moments of a family of GARCH processes. J Econom 106:109–117
Liu J-C (2009) Stationarity of a family of GARCH processes. Econom J 12:436–446
Mathews VJ, Sicuranza GL (2000) Polynomial signal processing. Wiley, New York
Meyn S, Tweedie RL (2009) Markov chains and stochastic stability. Cambridge University Press, Cambridge

Mikosch T, Starica C (2004) Changes of structure in financial time series and the GARCH model. REVSTAT 2:41–73

Nelson DB (1991) Conditional heteroskedasticity in asset returns: a new approach. Econometrica 2:347–370

Nisio M (1960) On polynomial approximation for strictly stationary processes. J. Math. Soc. Japan 12:207–226

Pagan A (1996) The econometrics of financial markets. J Empir Finance 3:15–102

Palm F (1996) GARCH models of volatility. In: Rao CR, Maddala GS (eds) Handbook of statistics, vol 14. North Holland, Amsterdam, pp 209–240

Pantula SG (1986) Modeling the persistence of conditional variances: a comment. Econom Rev 5:71–74

Pham DT, Tran TL (1981) On the first-order bilinear time series model. J Appl Probab 18:617–627

Priestley MB (1981) Spectral analysis and time series. Academic, London

Rabemananjara R, Zakoïan JM (1993) Threshold ARCH models and asymmetries in volatility. J Appl Econom 8:31–49

Resnick SI, Van den Berg E (2000) Sample correlation behaviour for the heavy tailed general bilinear process. Stoch Models 16:233–258

Robinson PM (1991) Testing for strong serial correlation and dynamic conditional heteroskedasity in multiple regression. J Econom 47:67–78

Rodriguez A (2007) Some advances in Bayesian nonparametric modeling. Unpublished doctoral thesis, Institute of Statistics and Decision Science, Duke University

Schwert GW (1989) Why does stock market volatility change over time? J Finance 45:1129–1155

Schwert GW (1990) Stock volatility and the crash of '87. Rev Financ Stud 3:77–102

Sentana E (1995) Quadratic ARCH models. Rev Econ Stud 62:639–661

Shephard N (1996) Statistical aspects of ARCH and stochastic volatility. In: Cox DR, Barndorff-Nielsen OE (eds) Likelihood, time series with econometric and other applications. Chapman and Hall, London

Silvennoinen A, Teräsvirta T (2009) Multivariate GARCH models. In: Andersen TG, Davis RA, Kreiss J-P, Mikosch T (eds) Handbook of financial time series. Springer, New York, pp 201–229

Taylor S (1986) Modeling financial time series. Wiley, New York

Teräsvirta T (2009) An introduction to univariate GARCH models. In: Andersen TG, Davis RA, Kreiss J-P, Mikosch T (eds) Handbook of financial time series. Springer, New York, pp 17–42

Terdik G (1999) Bilinear stochastic models and related problems of nonlinear time series analysis. Springer, New York

Tjøstheim D (1994) Non-linear time series: a selective review. Scand J Stat 21:97–130

Tjøstheim D (2012) Some recent theory for autoregressive count time series. Test 21:413–438. (With discussion)

Tong H (1990) Non-linear time series. Oxford Science Publications, Oxford

Tse Y (1998) The conditional heteroskedascity of the Yen-Dollar exchange rate. J Appl Econom 13:49–55

Weiß CH (2009) Modelling time series of counts with overdispersion. Stat Methods Appl 18: 507–519

West M, Harrison J (1997) Bayesian forecasting and dynamic models. Springer, New York

Yang M, Bewley R (1995) Moving average conditional heteroskedastic processes. Econ Lett 49:367–372

Zakoïan JM (1994) Threshold heteroskedastic models. J Econ Dyn Control 18:931–955

Zivot E, Wang J (2006) Modeling financial time series with S-PLUS. Springer, New York

Chapter 3
Extremes of Nonlinear Time Series

3.1 Tail Behavior

We have seen in Sect. 2.1.4 that nonlinear processes, due to their dependence on initial conditions, often magnify error causing unstable behavior. Even when stationary solutions exist, this noise magnification and dependence on initial conditions reflects on the tails of the stationary distribution, as well as on how large values cluster.

In this chapter we show why and how nonlinear processes tend to create clusters of large values. Specifically, we look at the stationary solutions of fairly general classes of stationary nonlinear processes and study their extremal properties. In order to exhibit the heavy-tailed phenomena of nonlinear processes, consider the following linear process

$$X_t = -0.1X_{t-1} + Z_t \tag{3.1}$$

and the bilinear process

$$X_t = (-0.1 + 0.9Z_{t-1})X_{t-1} + Z_t, \tag{3.2}$$

where Z_t are i.i.d. $\sim N(0, 1)$. Two realizations of size 500 of these processes are given in Fig. 3.1. This figure indicates that such a deviation from linearity can cause heavy-tailed data, in the sense that the output series has more variability than the input series. In fact, the stationary distribution of the linear model (3.1) has Gaussian tails, whereas the stationary distribution of the bilinear model (3.2) has regularly varying tails with polynomial decay. In this chapter, we make these statements precise by studying the extremal properties of such processes.

Extreme value theory for stochastic processes is well-known and can be found in many good references, for example Leadbetter et al. (1983) and Embrechts et al. (1997). In this chapter, we give a very brief summary of the basic notions and then some specific results for nonlinear time series.

K.F. Turkman et al., *Non-Linear Time Series*, DOI 10.1007/978-3-319-07028-5_3,

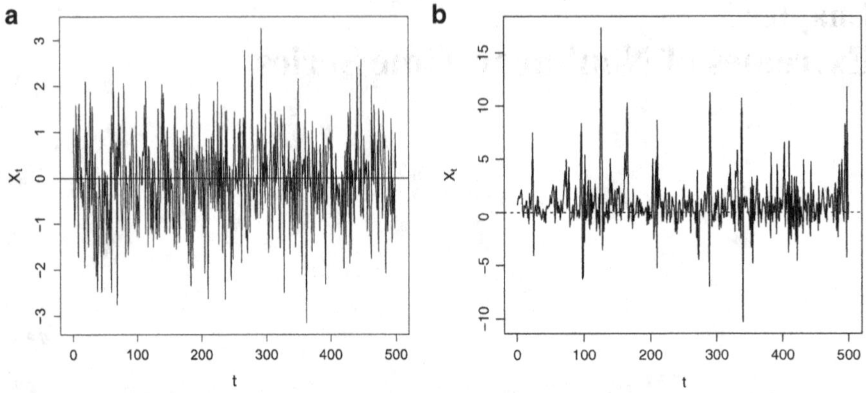

Fig. 3.1 AR model (**a**) and bilinear model (**b**) with $n = 500$ and Normal residuals

3.1.1 Extreme Value Theory

Extreme value theory deals with the probabilistic mechanisms of a process (X_t) that generate large values, such as the partial maximum $M_n(X) := \max(X_1, \ldots, X_n)$ and in particular its behavior as $n \to \infty$. If (X_t) is an i.i.d. sequence with distribution function F and right endpoint $x_{sup} := \sup\{x : F(x) < 1\}$, then

$$P(M_n(X) \le x) = F^n(x),$$

so that for all $x < x_{sup}$, $P(M_n(X) \le x) \to 0$, as $n \to \infty$. Therefore, in order to obtain meaningful limiting results, we look at normalized limits

$$P(M_n(X) \le a_n x + b_n) = F^n(a_n x + b_n) =: G(x),$$

for some suitable linear normalization (a_n, b_n) such that $a_n x + b_n \to x_{sup}$ in a controlled fashion as $n \to \infty$. Without loss of generality, for ease of notation, we assume that $x_{sup} = \infty$, unless we need to specifically refer to cases with finite endpoint.

The major result of the extreme value theory says that if this limit exists, then $G(x)$ has to be one of three types, called Fréchet, Gumbel and Weibull extreme value distributions given respectively by

$$G(x) := G_1(x) = \begin{cases} 0, & x \le 0; \\ e^{-x^{-\alpha}}, & x > 0, \alpha > 0\,. \end{cases}$$

$$G(x) := G_0(x) = \exp\{-e^{-x}\} \quad , -\infty < x < \infty$$

$$G(x) := G_2(x) = \begin{cases} e^{-(-x)^{\alpha}}, & x < 0, \alpha > 0; \\ 1, & x \ge 0. \end{cases}$$

These three distributions can be written in a single parametric family called the *Generalized Extreme Value* (GEV) *distribution*

$$G(x) \equiv G_\alpha(x) := \exp -(1+\alpha x)^{-1/\alpha}, \tag{3.3}$$

where $1 + \alpha x > 0$ and the parameter α, taking real values, is called the extreme value index (or simply tail index). When $\alpha = 0$, the right-hand side of (3.3) is interpreted as $\exp\{-e^{-x}\}$. When (X_t) is a stationary but dependent sequence with distribution function F, if for suitable normalization

$$F^n(a_n x + b_n) \to G(x),$$

then under fairly general conditions

$$P(M_n(X) \leq a_n x + b_n) \to G^\theta(x),$$

where $\theta \in (0, 1]$ is called the *extremal index*, and is a measure of how local dependence affects the extremal behavior of the process. When X_t is an independent sequence, then $\theta = 1$, whereas when the local dependence in X_t is strong, θ approaches 0.

The class of distributions F such that $F^n(a_n x + b_n) \to G(x)$ for some suitable normalization (a_n, b_n) is called the domain of attraction of $G(x)$. It is the behavior of the tail $P(X_t > x) = 1 - F(x)$, as $x \to \infty$ that decides which domain of attraction F belongs to. Extreme value distributions have the property that $G^\theta(x) = G(ax + b)$, for some a and b, therefore the local dependence structure of X_t does not influence the domain of attraction of the stationary distribution of X_t.

As **a general rule**, the Fréchet domain of attraction embraces heavy-tailed distributions, with polynomially decaying tails, that is

$$P(X_t > x) \sim x^{-\alpha} L(x) \;,\; x > 0$$

for $\alpha > 0$ and some slowly varying function $L(x)$, that is

$$L(cx)/L(x) \to 1,$$

as $x \to \infty$, for every $c > 0$. All cdf's belonging to the Weibull domain of attraction are light-tailed with finite right endpoint. The intermediate case $\alpha = 0$ is of particular interest in many applied sciences where extremes are relevant, not only because of the simplicity of inference within the Gumbel domain of attraction, but also for the great variety of distributions that belong to this domain. Such distributions may have finite or infinite endpoints. Tails of distributions that belong to the domain of attraction of the Gumbel distribution are typically products of polynomial and exponential functions of the form (Rootzén 1986)

$$P(X_t > x) \sim Cx^\alpha e^{-x^p}, \tag{3.4}$$

as $x \to +\infty$, where $p > 0$, $C > 0$ and α is a real number. Such tails exhibit different rates of decay depending on the values of the parameters in (3.4). For example, the normal and the lognormal, as well as the exponential distributions belong to this class. We refer the reader to Embrechts et al. (1997) or Beirlant et al. (2004) for an overview of extreme value theory and its applications.

3.1.2 Tail Behavior of Linear Processes

Gaussian r.v's have exponentially bounded tails and belong to the domain of attraction of the Gumbel distribution. For example, the standard Gaussian random variable $X \sim N(0,1)$ with density $\phi(x)$ satisfies

$$P(X > x) \sim \frac{\phi(x)}{x} \sim \frac{1}{x\sqrt{2\pi}} e^{-x^2/2}, \quad x \to \infty.$$

Linear Gaussian time series have exponentially bounded tails, producing moderately large values. Many types of data, particularly coming from finance, insurance, telecommunications, as well as from environmental sciences, show behavior often inconsistent with this observation. These data sets often exhibit an erratic behavior producing clusters of large values, consistent with distributions having much heavier tails and dependence structures that go beyond covariances. How we deal with such data sets depends very much on our understanding of the stochastic mechanisms that generate such large values and the dependence structures that cluster these large values.

A random variable X has heavy tails if $E|X|^p = \infty$ for some $p > 0$. Naturally, the behavior of $P(X > x)$ and $P(X < -x)$ as $x \to \infty$ will define the behavior of integrals $E|X|^p$. The class of regularly varying distributions is particularly attractive for modeling heavy-tailed distributions.

Definition 3.1.1. The probability distribution of a random variable is regularly varying with tail index $\alpha \geq 0$, if there exist constants $p, q \geq 0$ such that $p = 1 - q$ with $p \in (0, 1]$ and as $x \to \infty$,

$$P(X > x) = px^{-\alpha} L(x), \tag{3.5}$$

and

$$P(X < -x) = qx^{-\alpha} L(x), \tag{3.6}$$

for some slowly varying function $L(x)$.

An equivalent way of defining a regularly varying distribution is the following: the probability distribution of a random variable is regularly varying with (tail) index $\alpha \geq 0$, if there exist constants $p, q \geq 0$ such that $p = 1 - q$

$$\lim_{t\to\infty} \frac{P(X > tx)}{P(|X| > t)} = px^{-\alpha} \tag{3.7}$$

and

$$\lim_{t\to\infty} \frac{P(X < -tx)}{P(|X| > t)} = qx^{-\alpha}. \tag{3.8}$$

Here, α is often called the tail index or the index of regular variation, whereas p and q represent respectively the proportion of left and right tails on the overall tail behavior of the distribution.

Examples of distributions which have regularly varying tails are the t, F, Cauchy, Pareto and the stable distributions. There is strong empirical evidence for using such distributions in many real life data sets, particularly in internet communications, insurance, finance and environment; see for example, Resnick (2007) and Embrechts et al. (1997). Within this class, distributions with $\alpha \in (0, 2)$ stand out as truly heavy-tailed, since, if $\alpha < 2$, then the second-order moment of X is infinite, whereas, when $\alpha < 1$, X will not have finite mean. In general,

$$E|X|^p = \begin{cases} \infty, & p \geq \alpha; \\ < \infty, & p < \alpha. \end{cases}$$

Another class of distributions which is often used in modeling the heavy-tailed data is called the class of *sub-exponential distributions*. The class of sub-exponential distributions can be defined through the asymptotic relation of the partial sums and maxima of i.i.d. random sequences.

Definition 3.1.2. The probability distribution of a positive random variable X is called sub-exponential if for all $n \geq 2$ and i.i.d. replicates $X_1, \ldots, X_n$,

$$\lim_{x\to\infty} \frac{P(X_1 + \cdots + X_n > x)}{P(M_n(X) > x)} = \lim_{x\to\infty} \frac{P(X_1 + \cdots + X_n > x)}{nP(X_1 > x)} = 1. \tag{3.9}$$

Basically, (3.9) says that sums of i.i.d. r.v's are dominated by one extreme observation. Such behavior clearly cannot be displayed by a Gaussian sequence, since in this case the limit (3.9) would be infinite. Positive, regularly varying functions are sub-exponential, and therefore the class of regularly varying r.v's (as models for the stationary solutions of time series, when they exist), seems to be a more flexible way of modeling heavy-tailed time series.

Can a linear model with moderately light-tailed innovations generate large values? In other words, can the distribution of

$$X_t = \sum_{j=-\infty}^{\infty} \psi_j Z_{t-j}, \tag{3.10}$$

be heavy-tailed when the innovations Z_t have lighter tails? The answer is **no**. We know that when the innovations are Gaussian, the stationary distribution of a linear model is also Normal and therefore will have exponentially bounded tails. This result is fairly general for other types of innovations having regularly varying tails. The following theorem which is due to Cline (1983) (see also Resnick 1987) tells us that the linear time series models will have the same tail behavior as the input series but may have clusters of large values depending on the model parameters.

Theorem 3.1.1. *If Z_t satisfies* (3.5) *and* (3.6) *and* (ψ_j) *satisfy*

$$\sum_{j=-\infty}^{\infty} |\psi_j|^{\delta} < \infty,$$

for some $0 < \delta < \min\{1, \alpha\}$, *then X_t in* (3.10) *is almost surely convergent and*

$$\lim_{x\to\infty} \frac{P(\sum_j |\psi_j||Z_j| > x)}{P(|Z_1| > x)} = \sum_j |\psi_j|^{\alpha}.$$

Moreover,

$$P(a_n^{-1}(M_n(X) - b_n) \le x) \to e^{-x^{-\alpha}},$$

where $a_n = (p\phi_+ + q\phi_-)^{-1/\alpha} n^{-1/\alpha}$, $b_n = 0$ *and* $\phi_+ := \max_j(0, \phi_j)$, $\phi_- := \max_j(0, -\phi_j)$.

Note that the maximum of the i.i.d. noise process, $\hat{M}_n(Z)$, has the same asymptotic behavior, in the sense that

$$P(a_n^{-1}(\hat{M}_n(Z) - b_n) \le x) \to e^{-x^{-\alpha}},$$

with normalization $a_n = n^{1/\alpha}$ and $b_n = 0$.

Theorem 3.1.1 above clearly indicates that

- If the innovations have regularly varying tails with index $\alpha > 0$, then the stationary distribution of X_t and the distribution function of Z_t belong to the same domain of attraction, having regularly varying tails with the same tail index;
- An extreme value of X_t is caused by just one large noise term Z_t.

In general, such results may not be true for nonlinear processes. We will see that Volterra series expansions and stationary solutions of certain types of nonlinear difference equations have heavier tails than the input noise and may not belong to the same domain of attraction as the i.i.d. input sequence. Large values of these processes also tend to cluster in a different manner. This partially explains why nonlinear relationships generate sudden bursts (clusters) of large values.

Similar, albeit much more complicated results, exist for linear processes with lighter tailed input noise process. One such general class is studied by Rootzén (1986), where the input noise Z_t satisfies

$$P(Z_t > x) \sim Cx^{\alpha}e^{-x^p}, x \to \infty \tag{3.11}$$

where $p > 0$, $C > 0$ and $\alpha \in \mathbb{R}$. Note that polynomial tails given in Theorem 3.1.1 are a special case with $p = 0$ and $\alpha < 0$, whereas Gaussian noise input is also another special case with $\alpha = -1$ and $p = 2$. For such input noise processes, the asymptotic behavior of extremes of the linear process is much more complicated, depending on (ϕ_j) and p in a very intricate manner. In general, the stationary solution of the linear process is in the same domain of attraction of the i.i.d. noise sequence, but the normalization and the clustering of large values will be different depending on three separate cases: when $0 < p < 1$, $p = 1$ and $p > 1$. In all cases, $M_n(X)$ is in the domain of attraction of the Gumbel distribution, that is

$$P(a_n^{-1}(M_n(X) - b_n) \leq x) \to e^{-e^{-x}}.$$

Therefore, the extremes are not as large as the extremes of linear processes with regularly varying input noise process. In general, the case $0 < p < 1$ is similar to the regularly varying case where an extreme value of the linear process is caused by one large noise term Z_t. For $p > 1$, an extreme value of the linear process depends on many large and moderately large values of the input process in a very complex way. The linear process and the input process no longer need to be tail equivalent in the sense that, as $x \to \infty$,

$$\frac{P(X_t > x)}{P(Z_t > x)} \sim \textit{constant},$$

and therefore they need different orders of normalization, although they both belong to the domain of attraction of the Gumbel distribution. Again, in general this may not be the case for nonlinear processes. We will see that some nonlinear processes can have regularly varying tails even when the input noise process has lighter tails of the form (3.11) with $p > 1$. Rootzén's results can be found in Klüppelberg and Lindner (2005).

3.1.3 Bivariate Regular Variation and the Coefficient of Tail Dependence

Clusters of large values of time series can also be characterised by the coefficient of tail dependence which in essence characterizes the nature of the limiting dependence between the margins of a bivariate extreme value distributions. The basic idea is to look at the bivariate extremes of the pair (X_t, X_{t+h}), for some positive lag h.

The usual approach is to assume that the bivariate distribution $F_{X_t,X_{t+h}}(x, y)$ lies in the domain of attraction of the bivariate extreme value distribution with Fréchet margins. However, in order to allow for more general models, the data are transformed marginally to obtain Fréchet margins, or alternatively to standard Pareto margins using the transformation

$$Y_t = \frac{1}{1 - F(X_t)}. \tag{3.12}$$

Consider a stationary time series X_t with marginal distribution function F, assumed to be continuous. For any fixed time lag $h > 0$, the pair (X_t, X_{t+h}) is said to be regularly varying on $(0, \infty)^2$ if as $u \to \infty$,

$$\frac{P(X_t > ux, X_{t+h} > uy)}{P(X > u, X_{t+h} > u)} \to d(x, y), \tag{3.13}$$

for some non-degenerate function d.

Assume that upon the standardization (3.12), the transformed time series satisfies the bivariate regular variation condition given in (3.13). Then it follows that (Ledford and Tawn 1996) as $u \to \infty$,

$$P(X_t > u, X_{t+h} > u) = P(\min(X_t, X_{t+h}) > u) \sim u^{-1/\eta} L(u), \tag{3.14}$$

for some slowly varying function $L(u)$ and for some constant $\eta = \eta_h \in (0, 1]$. Here, the constant η describes the type of limiting dependence between the extreme values of X_t and X_{t+h} and is called the coefficient of tail dependence. The perfect negative and positive dependence respectively correspond to $\eta \to 0$ and $\eta = 1$, whereas, asymptotic independence in the tails corresponds to $\eta = 1/2$. The degree of dependence in general is determined by the combination of η and the slowly varying function $L(u)$. However, increased values of η can be interpreted as stronger association between the large values of X_t and X_{t+h}. These results can be used to compare competing time series models. For example, it is known that for $GARCH(p, q)$ models, $\eta = 1$, whereas for stochastic volatility models, $\eta = 1/2$. If we can estimate the value of η of an observed time series, then we can check if the observed time series is compatible with a GARCH model, at least in the tails. However, the estimation of η is not straightforward.

If $T_i := \min(Y_i, Y_{i+h})$, then from (3.12), $1 - F_T(u) = u^{-1/\eta} L(u)$. If T_i are observable, η can be estimated by standard methods such as the Hill estimator: for suitably chosen k_n and the corresponding upper order statistics $Tn{-}k_n{:}n, \ldots, T_{n:n}$,

$$\hat{\eta} = \frac{1}{k_n} \sum_{j=1}^{k_n} \log \frac{T_{n-j+1:n}}{T_{n-k_n:n}}. \tag{3.15}$$

However, the stationary marginal distribution of the time series in general is not known, thus the transformation in (3.12) can not be performed. Draisma et al. (2004) suggest substituting the marginal distribution by the empirical distribution function to obtain the corresponding estimator. Drees and Rootzén (2010) study the asymptotic behavior of such estimator by proving limit theorems for suitable empirical processes.

The coefficient of tail dependence can be a powerful tool in model choice by calculating the theoretical η for the model under study and comparing it with the estimated η from the observed time series.

3.2 Connection Between Nonlinearity and Heavy Tails

Consider the general difference equation

$$X_t = g(X_s, Z_s, s < t),$$

where Z_t are i.i.d. r.v's with distribution function F, and g is some measurable function, such that the stationary solution for X_t with distribution function G exists. We have seen that for a very general class of noise input processes, under fairly general conditions, G and F belong to the same domain of attraction when g is a linear function, although how the large values cluster and what normalization to choose will depend on the constants in the linear representation, as well as on the tail of the noise distribution F. When g is nonlinear, G and F in general do not belong to the same domain of attraction and the nonlinear relationship g magnifies the large values of the input noise. To highlight this point, we will report on the extremes of Volterra expansions, as well as on the extremes of certain types of general nonlinear difference equations which give raise to many known classes of nonlinear processes.

Extremes of Volterra expansions

Assume that the time series X_t has Volterra series expansion

$$X_t = \sum_{i=1}^{\infty} \left(\sum_{|j_1|\le\infty,\ldots,|j_i|\le\infty} h_{j_1,j_2,\ldots,j_i} \prod_{r=1}^{i} Z_{t-j_r} \right), \tag{3.16}$$

where (Z_t) is a sequence of i.i.d. innovations with distribution function satisfying (3.5) and (3.6) and $(h_{j_1}), (h_{j_1,j_2}), \ldots,$ are such that (3.16) converges to a well-defined random variable. In order to understand the effect of nonlinearity on the tail behavior of X_t, we look at mth-order Volterra approximation to X_t, namely

$$X_t = \sum_{|j_1|\le\infty} h_{j_1} Z_{t-j_1} + \cdots + \sum_{|j_1|\le\infty,|j_2|\le\infty,\ldots,|j_m|\le\infty} h_{j_1,j_2,\ldots,j_m} \prod_{r=1}^{m} Z_{t-j_r}. \tag{3.17}$$

In order to ensure the absolute almost sure convergence of the mth-order Volterra series given in (3.17), the sequence of kernels $(h_{j_1}), (h_{j_1,j_2}), \ldots$, has to satisfy the following summability condition

$$\sum_{|j_1|\leq\infty,|j_2|\leq\infty,\ldots,|j_m|\leq\infty} |h_{j_1,j_2,\ldots,j_m}|^{\delta} < \infty, \tag{3.18}$$

for some $\delta < \min(1, \alpha/m)$, where α is the tail index given in (3.5) and (3.6). The following result (Scotto and Turkman 2005) essentially shows how the order of nonlinearity affects the tail behavior of the process X_t.

Theorem 3.2.1. *If Z_1 satisfies* (3.5) *and* (3.6), *and the sequence of kernels satisfies* (3.18), *then*

$$\lim_{x\to\infty} \frac{P(|X_t| > x)}{P(|Z_1|^m > x)} = \sum_{|j|<\infty} |h_{j,j,\ldots,j}|^{\alpha/m} < \infty$$

and

$$\lim_{x\to\infty} \frac{P(X_t > x)}{P(|Z_1|^m > x)} = \begin{cases} \sum_{|j|\leq\infty} (h^+)^{\alpha/m}, & \text{if } m \text{ is even;} \\ p\sum_{|j|\leq\infty} (h^+)^{\alpha/m} + q\sum_{|j|\leq\infty} (h^-)^{\alpha/m}, & \text{if } m \text{ is odd,} \end{cases}$$

where $h^+ := \max(h_{j,j,\ldots,j}, 0)$, $h^- := \max(-h_{j,j,\ldots,j}, 0)$ and p, q are given respectively in (3.7) *and* (3.8).
Moreover, as $n \to \infty$

$$P(M_n(X) \leq a_n^m x) \to \begin{cases} \exp\left\{-(h^{(+)})^{\alpha/m} x^{-\alpha/m}\right\}, & \text{if } m \text{ is even;} \\ \exp\left\{-[p(h^{(+)})^{\alpha/m} + q(h^{(-)})^{\alpha/m}]x^{-\alpha/m}\right\}, & \text{if } m \text{ is odd,} \end{cases} \tag{3.19}$$

and the sequence has extremal index

$$\theta = \begin{cases} \frac{(h^{(+)})^{\alpha/m}}{\sum_{|j|\leq\infty}(h^+)^{\alpha/m}}, & \text{if } m \text{ is even;} \\ \frac{p(h^{(+)})^{\alpha/m} + q(h^{(-)})^{\alpha/m}}{p\sum_{|j|\leq\infty}(h^+)^{\alpha/m} + q\sum_{|j|\leq\infty}(h^-)^{\alpha/m}}, & \text{if } m \text{ is odd.} \end{cases}$$

Note that if Z_t has regularly varying tails with tail index α, then X_t has also regularly varying tails with tail index α/m. One consequence of this result is that, while the i.i.d. noise sequence has finite moments up to α, the output process X_t has finite moments up to α/m. This result also shows how large values of the process cluster through the extremal index θ. Note that the tail of the process is tail equivalent to $|Z|^m$, and the kernels $h_{j,\ldots,j}$ corresponding to $|Z_j|^m$ influence how the clusters are formed in an intricate manner. At present there are no comparable results when the noise sequence has lighter tails satisfying (3.11). This is due to the fact that when Z_t has regularly varying tails, an extreme value of X_t is caused by just one large noise variable $|Z_1|^m$, whereas when Z_t has lighter tails, clusters of $\prod_{r=1}^m Z_{t-j_r}$ affect the tail behavior X_t, and these clusters are very difficult to quantify. Since Volterra expansions are the most general representation for nonlinear processes, Theorem 3.2.1 gives us a very general characterization of the extremal behavior of nonlinear processes. However, it is not particularly useful and in order to get a better description, it is important to look at some specific nonlinear structures.

3.2.1 Extremal Properties of Certain Types of Nonlinear Difference Equations

Consider the stochastic difference equation (alternatively, we will call this a stochastic recurrence equation) first introduced in (2.16)

$$X_t = A_t X_{t-1} + B_t, \tag{3.20}$$

where (A_t, B_t) is a sequence of r.v's. We will call (3.20) a random difference equation (or stochastic recurrence equation). It is possible to define (3.20) in a more general form

$$\mathbf{X}_t = \mathbf{A}_t \mathbf{X}_{t-1} + \mathbf{B}_t, \tag{3.21}$$

where $\mathbf{X}_t$ and $\mathbf{B}_t$ are random vectors in $\mathbb{R}^d$, $\mathbf{A}_t$ are $d \times d$ random matrices and $(\mathbf{A}_t, \mathbf{B}_t)$ is a strictly stationary ergodic process. For the sake of simplicity and completeness of presentation, we give most of the results for the scalar difference equation given in (3.20) but we will also mention the existing ones for the vector representation (3.21).

The difference equations given in (3.20) and (3.21) are quite general and many known families of nonlinear processes can be given in these representations. For example, consider the first-order bilinear process

$$X_t = aX_{t-1} + bX_{t-1}Z_{t-1} + Z_t, \tag{3.22}$$

where Z_t are i.i.d. r.v's and a and b are constants satisfying certain conditions to be specified later. The representation given in (3.22) can be written as

$$X_t = (a + bZ_{t-1})X_{t-1} + Z_t.$$

This representation is not Markovian and the random pair $(a + bZ_{t-1}, Z_t)$ forms a 1-dependent, identically distributed pair. However, by setting

$$V_t = (a + bZ_t)X_t,$$

we see that

$$\begin{aligned} V_t &= (a + bZ_t)X_t \\ &= (a + bZ_t)(a + bZ_{t-1})X_{t-1} + (a + bZ_t)Z_t \\ &= (a + bZ_t)V_{t-1} + (a + bZ_t)Z_t, \end{aligned} \tag{3.23}$$

and

$$X_t = V_{t-1} + Z_t. \tag{3.24}$$

Note that V_t given in (3.23) is a Markov process, and hence X_t in (3.24) has the standard state space representation; see Sect. 2.2.7 for details. See also Subba Rao and Gabr (1984) and Pham (1985) for the existence of such Markovian representations for a general bilinear process. Note that

$$V_t = A_t V_{t-1} + B_t,$$

where $(A_t, B_t) = (a + bZ_t, (a + bZ_t)Z_t)$ forms an i.i.d. random sequence.

Another well-known process which can be put in the form (3.20) is the first order ARCH process

$$X_t = (\beta + \lambda X_{t-1}^2)^{1/2} Z_t, \tag{3.25}$$

where $\beta > 0$, $0 < \lambda < 1$. Here, X_t^2 satisfies (3.20) with $A_t = \lambda Z_t^2$, $B_t = \beta Z_t^2$. Note that (A_t, B_t) forms an i.i.d. pair. Higher-order bilinear processes and ARCH processes can similarly be put in the form (3.21) in the higher dimension. Apart from this example, it is quite clear that many other nonlinear processes, such as random coefficient autoregressive models (RCA) and threshold models, can similarly be put either in the form (3.20) or (3.21), depending on the dimension.

In this section we show that the stationary solutions of (3.20) when they exist, under fairly general conditions, have regularly varying tails. For example, the simple bilinear process given in (3.22) and the ARCH model given in (3.25) have stationary solutions with regularly varying tails even when the input innovation process Z_t has Gaussian or similar lighter tailed distributions. This difference in the tails of input-output processes is due to the nonlinear nature of the relationship. We also show that the extremal behavior of these processes differs significantly from those of linear Gaussian models. The following result (due to Vervaat 1979) establishes the conditions for the existence of a stationary solution for (3.20).

Theorem 3.2.2. *Assume that* (A_t, B_t), $t \in \mathbb{N}$, *in* (3.20) *is an i.i.d. sequence of random pairs. Let*

$$R = \sum_{k=1}^{\infty} \prod_{j=1}^{k-1} A_j B_k. \tag{3.26}$$

1. *If* $E(\log|A|) \in (-\infty, 0)$, *then the sum in* (3.26) *converges almost surely and* (3.20) *has a strictly stationary solution with distribution equivalent to the distribution of* R *in* (3.26) *iff* $E(\log^+|B|) < \infty$.
2. *If* $E(\log|A|) = -\infty$ *then the sum in* (3.26) *converges almost surely and* (3.20) *has a strictly stationary solution with distribution equivalent to the distribution of* R *in* (3.26) *either if* $E(\log^+|B|) < \infty$ *or if* $P(A = 0) > 0$.

Here, for simplicity, we drop the index of A_i *and* B_i, *since they are identically distributed.*

The assumption that (A_t, B_t) is an i.i.d. sequence can be relaxed and the existence of stationary solutions can be given in a very general set-up in higher dimensions. Let $|\cdot|$ be the Euclidian norm in $\mathbb{R}^d$ and $||\cdot||$ be the corresponding operator norm in $\mathbb{R}^d$ such that for any $d \times d$ matrix $\mathbf{M}$

$$||\mathbf{M}|| = \sup\left\{\frac{||\mathbf{Mx}||}{||\mathbf{x}||}, \quad \mathbf{x} \in \mathbb{R}^d, \mathbf{x} \neq 0\right\}.$$

Theorem 3.2.3 (Brandt 1986). *If* (A_t, B_t), $t \in \mathbb{N}$, *is a strictly stationary ergodic process such that both* $E(\log^+||A||)$, $E(\log^+||B||)$ *are finite and the top Lyapunov exponent* γ *defined by*

$$\gamma = \inf\left\{E\left(\frac{1}{n+1}\log||A_0\cdots A_n||\right), \quad n \in \mathbb{N}\right\},$$

is strictly negative, then for every t *the series*

$$X_t = \sum_{k=1}^{\infty} A_t A_{t-1}\cdots A_{t-k+1}\mathbf{B}_{t-k},$$

converges almost surely and the process X_t *is the unique strictly stationary solution of* (3.21).

If we assume that (A_t, B_t) is an i.i.d. sequence, then the converse of Theorem 3.2.3 is also true.

Theorem 3.2.4 (Bougerol and Picard 1992). *If* (A_t, B_t) *in* (3.21) *is an i.i.d. sequence and both* $E(\log^+||A||)$, $E(\log^+||B||)$ *are finite then* (3.21) *has a strictly stationary solution, which is independent of* (A_t, B_t) *for every* t *if and only if the top Lyapunov exponent* γ *is strictly negative.*

The top Lyapunov exponent referred to in theorem 3.2.4 is always less than $E(\log||A||)$, with equality in dimension $d = 1$. That is, $\gamma = E(\log|A|)$. Although it is difficult to get an explicit form for the distribution of R in (3.26), it is possible to say something about the tail behavior of R. Here we will consider the case when (A_t, B_t) are i.i.d. scalar r.v's. Results on the tail behavior of the stationary solution of (3.21) need the notion of multivariate regular variation and will be given later in the section. In the scalar case, there are basically two cases, depending on A and B (we use the notation A and B representing the common probability law for each of the A_t and B_t).

1. Light-tailed stationary solutions.
   Provided that B is sufficiently light-tailed, the case $P(|A| \leq 1) = 1$ results in a light-tailed R, in the sense that all moments are finite (Goldie and Grübel 1996). In this case neither B nor $\left(\prod_{k=1}^{n} A_k\right)$ will have dominant contribution towards the tails of R.[1]
2. Heavy-tailed stationary solutions.
   This is the most common situation and most of the literature deals with this case. Heavy tails can arise either when B is heavy-tailed (Grincevičius 1975; Grey 1994) or when neither A nor B is heavy-tailed, but A can take values outside the interval $[-1, 1]$ (Kesten 1973; Goldie 1991; Embrechts and Goldie 1994). In the former case, individual large B values dominate the tail of R, whereas in the latter case, clusters of large values of the sequence $\left(\prod_{k=1}^{n} A_k\right)$ dominate the large values of R.

We now give in detail the conditions (Goldie 1991) under which R has heavy- or light-tailed tails.

Theorem 3.2.5. *Suppose that the stationary solution R in* (3.26) *is non-degenerate and satisfies*

$$R =^d B + AR, \tag{3.27}$$

with R independent of (A, B). If $P(|A| > 1) > 0$, then $|R|$ has regularly varying tails. If $|R|$ is non-degenerate, has all moments finite and satisfies (3.27) *then $P(|A| \leq 1) = 1$.*

Therefore, if $|A|$ takes values larger then 1 with positive probability, then the stationary solution of (3.20) cannot have light tails.

Theorem 3.2.6. *Let R satisfy* (3.27)*, where $P(|A| \leq 1) = 1$ and $P(|A| < 1) > 0$. If $Ee^{\epsilon|B|} < \infty$ for some $\epsilon > 0$, then $P(|R| \geq x)$ decreases at least exponentially fast, in the sense that there is a positive constant ρ such that*

$$\limsup_{x\to\infty} \frac{1}{x} \log P(|R| \geq x) \leq -\rho.$$

In fact, it is possible to say more about the value of ρ; see Goldie and Grübel (1996) for details. Trivially, if $P(|B| \leq b) = 1$ for some $b < \infty$, and $P(|A| \leq c) = 1$ for some $c < 1$ then $P(|R| \geq x) = 0$, for $x > a/(1-c)$. Examples

[1] Extensions of Goldie and Grübel's results, namely the connection between the tails of R and the behavior of A near 1, can be found in Hitczenko and Wesolowski (2009).

of processes for which the tail decay of $|R|$ is faster than exponential are given in Goldie and Grübel (1996), with exact rate of decay.

We first start with the case when neither A nor B have heavy tails (Kesten 1973; Vervaat 1979; Goldie 1991).

Theorem 3.2.7. *Assume that Z_t satisfies the difference equation* (3.20) *with i.i.d.* (A_t, B_t) *such that*

- $E \log |A| < 0$;
- *For some* $\alpha > 0$,

$$E|A|^\alpha = 1,$$
$$E|A|^\alpha \log^+ |A| < \infty,$$
$$0 < E|B|^\alpha < \infty;$$

- $\log |A|$ *does not have a lattice distribution and for every* $r \in R$, $P(B = (1 - A)r) < 1$.

Then

$$R = \sum_{k=1}^{\infty} \prod_{j=1}^{k-1} A_j B_k,$$

converges almost surely and moreover, as $x \to \infty$,

$$P(R > x) \sim c_+ x^{-\alpha},$$
$$P(R < -x) \sim c_- x^{-\alpha}, \tag{3.28}$$

for some constants c_+, c_- *such that at least one of them is strictly positive. Further, if* $P(A < 0) > 0$, *then* $c_+ = c_- > 0$.

Note that in the above theorem, it is the distribution of A that is important in determining the tail behavior of R through the determination of the tail index α. However, α does not always exist. Consider the case when A is nonnegative such that $EA^\kappa < 1$ and $EA^\beta < \infty$, for some $\beta > \kappa > 0$ and $P(B > x) \sim x^{-\kappa} L(x)$, where $L(x)$ is a slowly varying function. In this case α satisfying $EA^\alpha = 1$ cannot exist, since if it does, it needs to be greater that κ, and for such a case $E|B|^\alpha < \infty$ cannot hold. In this case, it is the distribution of B that is important in defining the tail of R:

Theorem 3.2.8. *Let* (A_t, B_t) *be an i.i.d. sequence of r.v's with* $E \log^+ |B| < \infty$, $P(A \geq 0) = 1$, $EA^\alpha < 1$ *and* $EA^\beta < \infty$ *for some* $\beta > \alpha > 0$. *Then as* $x \to \infty$,

$$P(B > x) \sim x^{-\alpha} L(x),$$

if and only if

$$P(R > x) \sim \frac{1}{1 - EA^{\alpha}} x^{-\alpha} L(x),$$

for some slowly varying function $L(x)$.

Theorem 3.2.8 above is stated for non-negative A. Similar results, however, exist when A also takes negative values (Grey 1994). In that case, simultaneous effects of the right and the left tails of A on those of R have to be taken into account complicating slightly the result. For this case, the result is as follows:

Theorem 3.2.9. *Let* (A_t, B_t) *be jointly distributed r.v's such that for some* $\beta > \alpha > 0$, $E(|A|^{\alpha}) < 1$ *and* $E(|A|^{\beta}) < \infty$. *In addition, assume that the tails of the distribution of* B *satisfy*

$$P(B > x) \sim x^{-\alpha} L(x) \text{ and } P(-B > x) \sim s x^{-\alpha} L(x), x \to \infty,$$

for some slowly varying function $L(x)$ *at infinity, and* $s \geq 0$. *Then*

$$P(R > x) \sim d_+ x^{-\alpha} L(x) \text{ and } P(-R > x) \sim d_- x^{-\alpha} L(x), x \to \infty,$$

where

$$d_+ = \frac{1}{2}\left(\frac{1+s}{1 - E[(A^+)^{\alpha}] - E[(A^-)^{\alpha}]} + \frac{1-s}{1 - E[(A^+)^{\alpha}] + E[(A^-)^{\alpha}]}\right)$$

and

$$d_- = \frac{1}{2}\left(\frac{1+s}{1 - E[(A^+)^{\alpha}] - E[(A^-)^{\alpha}]} - \frac{1-s}{1 - E[(A^+)^{\alpha}] + E[(A^-)^{\alpha}]}\right).$$

Under fairly general conditions, the stationary solution of (3.21) also has regularly varying tails. In order to write precisely the results, we need the concept of multivariate regular variation; see Basrak et al. (2002) for details.

Definition 3.2.1. The d-dimensional random vector $\mathbf{X}$ is said to be regularly varying with index $\alpha > 0$ if there exists a sequence of constants a_n and a random vector $\boldsymbol{\theta}$ with values in the unit sphere $\mathcal{S}^{d-1} \subset \mathbb{R}^d$ such that for all $t > 0$ and for every Borel measurable $A \subset \mathcal{S}^{d-1}$, as $n \to \infty$

$$nP(|\mathbf{X}| > t a_n, \mathbf{X}/|\mathbf{X}| \in A) \to t^{-\alpha} P(\boldsymbol{\theta} \in A). \tag{3.29}$$

Here, the convergence is vague convergence on the unit sphere $\mathcal{S}^{d-1}$. Condition (3.29) is equivalent to the condition that for all $t > 0$, as $x \to \infty$,

$$\frac{P(|\mathbf{X}| > tx, \mathbf{X}/|\mathbf{X}| \in A)}{P(|\mathbf{X}| > x)} \to t^{-\alpha} P(\boldsymbol{\theta} \in A). \tag{3.30}$$

One particular consequence of the convergence in (3.30) is that linear combinations

$$\sum_{i=1}^{d} \phi_i X_i,$$

for any $\boldsymbol{\phi} \in \mathbb{R}^d \backslash \{\mathbf{0}\}$ are also regularly varying with the same index $\alpha > 0$. Specifically,

$$\lim_{u \to \infty} P\left(\sum_{i=1}^{d} \phi_i X_i > u\right) = w(\boldsymbol{\phi}) L(u) u^{-\alpha}, \tag{3.31}$$

for some function $w(\boldsymbol{\phi})$ such that

$$w(t\boldsymbol{\phi}) = t^{\alpha} w(\boldsymbol{\phi}).$$

Conditions (3.30) and (3.31) are essentially equivalent.

Theorem 3.2.10. *Assume that* $E \ln^+ ||\mathbf{A}|| < \infty$, $E \ln^+ |\mathbf{B}| < \infty$ *and the top Lyopunov exponent* $\gamma < 0$. *Then*

$$\mathbf{X}_n = \mathbf{B}_n + \sum_{j=1}^{\infty} \mathbf{A}_n \times \cdots \times \mathbf{A}_{n-j+1} \mathbf{B}_{n-j},$$

converges almost surely and is the unique stationary solution of (3.21). *Moreover, under further technical conditions on the i.i.d. random sequence* $(\mathbf{A}_t, \mathbf{B}_t)$, *(see Basrak et al. 2002), the stationary solution* $\mathbf{X}$ *is regularly varying with index* κ, *where* κ *is the unique solution to the equation*

$$\lim_{n \to \infty} \frac{1}{n} \ln E ||\mathbf{A}_n \times \cdots \times \mathbf{A}_1||^{\kappa} = 0.$$

As we have stressed, the stochastic difference equation (3.21) is quite general and many known nonlinear processes can be put in this form. Bilinear and GARCH are two classes of nonlinear processes which can be written in this form. Thus, Theorem 3.2.10 clearly says that under fairly general (but often complex) conditions, starting with moderately light-tailed innovations, we get heavy-tailed outputs from general nonlinear systems. We now write the conditions and the results specifically for these two classes of nonlinear processes.

3.2.2 Tails of Bilinear Processes

The bilinear process $BL(p, q, m, l)$

$$X_t = \sum_{j=1}^{p} \phi_j X_{t-j} + \sum_{j=1}^{q} \theta_j Z_{t-j} + \sum_{i=1}^{m} \sum_{j=1}^{l} b_{ij} X_{t-i} Z_{t-j} + Z_t, \tag{3.32}$$

can be written in the form (Resnick and Van den Berg 2000)

$$\mathbf{X}_t = \mathbf{A}_{t-1}\mathbf{X}_{t-1} + \mathbf{B}_t,$$

where $\mathbf{X}_t$ and $\mathbf{Z}_t$ are respectively $(p-1) \times 1$ and $q \times 1$ column vectors

$$\mathbf{X}_t := (X_t,\ X_{t-1}, \cdots, X_{t-p+1})',$$
$$\mathbf{Z}_t := (Z_t,\ Z_{t-1}, \cdots, Z_{t-q})'.$$

Here,

$$\mathbf{B}_t = \mathbf{\Theta}\mathbf{Z}_t,$$

where $\mathbf{\Theta}$ is a $p \times (1+q)$ matrix given by

$$\mathbf{\Theta} = \begin{pmatrix} 1 & \theta_1 & \ldots & \theta_{q-1} & \theta_q \\ 0 & 0 & \ldots & 0 & 0 \\ 0 & 0 & \ldots & 0 & 0 \\ \vdots & \ldots & \ddots & \ldots & \vdots \\ 0 & 0 & \ldots & 0 & 0 \end{pmatrix}$$

and $\mathbf{A}_{t-1}$ is a $p \times p$ matrix given by

$$\mathbf{A}_{t-1} = \begin{pmatrix} \phi_1 + \sum_{j=1}^{l} b_{1j} Z_{t-j} & \phi_2 + \sum_{j=1}^{l} b_{2j} Z_{t-j} & \ldots & \ldots & \phi_p + \sum_{j=1}^{l} b_{pj} Z_{t-j} \\ 1 & 0 & \ldots & 0 & 0 \\ \vdots & \ldots & \ddots & \ldots & \vdots \\ 0 & 0 & \ldots & 1 & 0 \end{pmatrix}$$

Theorem 3.2.11. *Assume that the i.i.d. noise sequence* (Z_t) *has regularly varying tails with tail index* $\alpha > 0$*, satisfying the conditions* (3.5) *and* (3.6).

1. *Assume further that*

   (a) *For* $l \geq 1$ *and* $0 < \alpha \leq l + 1$,

$$\sum_{i=1}^{p}\left(|\phi_i|^{\alpha/(l+1)} + \sum_{j=1}^{l}|b_{ij}|^{\alpha/(l+1)}E(|Z_1|^{l\alpha/(l+1)})^{1/l}\right) < 1, \tag{3.33}$$

(b) For $l \geq 1$ and $\alpha > l+1$,

$$\sum_{i=1}^{p}\left(|\phi_i| + \sum_{j=1}^{l}|b_{ij}|E(|Z_1|^{l\alpha/(l+1)})^{(l+1)/l\alpha}\right) < 1. \tag{3.34}$$

If $\mathbf{e}_1 := (1,0,\ldots,0)$ is the p-dimensional unit vector,

$$X_t = \mathbf{e}_1\boldsymbol{\Theta}\mathbf{Z}_t + \sum_{n=1}^{\infty}\mathbf{e}_1\prod_{j=1}^{n}\mathbf{A}_{t-j}\boldsymbol{\Theta}\mathbf{Z}_{t-n},$$

is a almost surely convergent stationary solution of (3.32).

2. *Further if $\prod_{i=1}^{n-l} b_{1i} \neq 0$ then $|X_t|$ has regularly varying tails with tail index $\alpha/(l+1)$, satisfying*

$$\frac{P(|X_t| > x)}{P(|Z_1|^{l+1} > x)} = c, \tag{3.35}$$

for some constant c, depending on the expectation of products of the noise process, as well as the constants b_{ij}.

The proof of this technical theorem, as well as the exact expression for the constant c, can be found in Resnick and Van den Berg (2000). The effect of nonlinearity on the tail of X_t is evident. In the bilinear difference equation, the cross product terms $X_{t-i}Z_{t-j}$, and in particular, the order l of the product term $X_{t-i}Z_{t-l}$ determines the degree of nonlinearity. The order of these product terms l directly influences how heavy the tail of X_t gets relatively to the input process Z_t, whereas the model parameters, in particular the cross product term kernels b_{ij}, determine the constant c in (3.35).

So far we have looked at the tail behavior of the stationary solution of (3.20). The results indicate that under quite general conditions the tails of this stationary solution will be regularly varying. This would partially explain why some nonlinear models have sudden bursts of large values. The formation of clusters of large values is not only a consequence of the heavy tails of the stationary solution, but it is also a result of the local dependence structure of the process defined in (3.20). Quantifying the effect of local dependence on the formation of clusters of large values, particularly when Z_t have lighter tails given in (3.11), is difficult and there are no known general results on the tail behavior of the BL(p,q,m,l) process with light-tailed innovations, except for some special simple cases such as the first order bilinear model

$$X_t = aX_{t-1} + bX_{t-1}Z_{t-1} + Z_t.$$

Theorem 3.2.12 (Turkman and Amaral Turkman 1997). *Assume that the conditions of Theorem* 3.2.7 *hold true. Assume further that* $P(A < 0) > 0$ *so that* c_+ *and* c_- *in* (3.28) *are strictly positive. Then,* X_t *has an extremal index* θ *given by*

$$\theta = \int_1^{\infty} P(\max_{1\le j\le\infty} \prod_{i=1}^{j} A_i \le y^{-1})\alpha y^{-\alpha-1}\, dy$$

and

$$\lim_{n\to\infty} P(M_n(X) < n^{1/\alpha}x) = \exp\{-c_+\theta x^{-\alpha}\},$$

where α *is the solution of*

$$E|a + bZ_t|^{\alpha} = 1.$$

3.2.3 *The Relation Between the Extremes of Finite-Order Volterra Series and Bilinear Processes*

We now compare the extremes of bilinear processes and finite-order Volterra series expansions with heavy-tailed input innovations. Consider the first-order bilinear process

$$X_t = cX_{t-1}Z_{t-1} + Z_t, \tag{3.36}$$

where (Z_t) satisfies conditions (3.5) and (3.6). If $c(c > 0)$ is such that

$$c^{\alpha/2}EZ_1^{\alpha/2} < 1,$$

then X_t has an almost surely convergent infinite order Volterra series expansion

$$X_t = Z_t + \sum_{j=1}^{\infty} c^j \left(\prod_{i=1}^{j-1} Z_{t-i}\right) Z_{t-j}^2. \tag{3.37}$$

Extremal properties of the model (3.36) were first studied by Davis and Resnick (1996) who proved that

1.

$$\lim_{x\to\infty} \frac{P(X_t > x)}{P(Z_1^2 > x)} = \frac{c^{\alpha/2}}{1 - c^{\alpha/2} E Z_1^{\alpha/2}}.$$

2. If a_n is the $1 - n^{-1}$ quantile of $|Z_1|$,

$$a_n = \inf\{x : P(|Z_1| > x) < n^{-1}\},$$

then

$$P(a_n^{-2} M_n(X) \le x) \to \exp\{-E(V_1^{\alpha/2}) x^{-\alpha/2}\},\ x > 0 \tag{3.38}$$

with $V_1 := \max_{1 \le k \le \infty}(c^k W_k)$, and

$$W_k = \begin{cases} \prod_{i=1}^{k-1} U_i, & k > 1; \\ 1, & k = 1; \\ 0, & k < 1. \end{cases}$$

Here, (U_i) are i.i.d. r.v's with the same distribution F of Z_1.

Although the series given in (3.37) has infinite order Volterra series expansion, the highest order of Z_t appearing in the expansion is two, and the random coefficients in each term are bounded; hence we would expect the extremal behavior of this process to be well approximated by a second-order expansion. Indeed, if we approximate this bilinear process by its second-order Volterra series (take $\prod_{i=1}^{0} Z_{k-i} = 1$)

$$X_t = Z_t + cZ_{t-1}^2,$$

we see from Theorem 3.2.11 that

$$\lim_{x\to\infty} \frac{P(X_t > x)}{P(Z_1^2 > x)} = c^{\alpha/2}$$

and, as $n \to \infty$

$$P(a_n^{-2} M_n(X) \le x) \to \exp\{-c^{\alpha/2} x^{-\alpha/2}\},\ x > 0. \tag{3.39}$$

Thus, the asymptotic distribution of the maximum of this bilinear process differs from the asymptotic distribution of the maximum of the second-order approximation by a constant which represents the (average) aggregate weight of the random coefficients of the Z^2 terms in the higher-order expansions.

We consider now the general bilinear process. Let

$$X_t = \sum_{i=1}^{p} \phi_i X_{t-i} + \sum_{j=1}^{q} \theta_j Z_{t-j} + \sum_{i=1}^{m} \sum_{j=1}^{l} b_{ij} X_{t-i} Z_{t-j} + Z_t, \tag{3.40}$$

where $\prod_{j=1}^{l} b_{1j} \neq 0$, and (Z_t) satisfies the conditions (3.5) and (3.6). Assume that the parameters of the model (3.40) satisfy the conditions (3.33) and (3.34), so that the process in (3.40) is almost surely convergent. We can write the process (3.40) as

$$\Phi(B)X_t = \Theta(B)Z_t + C(B, B)[X_t, Z_t],$$

where

$$\Phi(B) = (1 - \sum_{i=1}^{p} \phi_i B^i),$$
$$\Theta(B) = (1 + \sum_{j=1}^{q} \theta_j B^j),$$

and

$$C(B, B)[X_t, Z_t] = \sum_{i=1}^{p} \sum_{j=1}^{q} b_{ij} (B^i X_t)(B^j Z_t)$$
$$= \sum_{i=1}^{p} \sum_{j=1}^{q} b_{ij} X_{t-i} Z_{t-j}.$$

Let

$$\Phi(B)X_t^{(1)} = \Theta(B)Z_t,$$

and for $j = 2, 3, \ldots,$

$$\Phi(B)X_t^{(j)} = C(B, B)[X_t^{(j-1)}, Z_t].$$

Then

$$X_t = \sum_{j=1}^{\infty} X_t^{(j)}, \tag{3.41}$$

is the unique solution of (3.40) (cf. Mathews and Sicuranza 2000). Note also that the representation given in (3.41) is the Volterra series expansion of the process (3.40). For example, consider the two processes

$$X_t = b_{12} X_{t-1} Z_{t-2} + Z_t$$

and

$$X_t = b_{11} X_{t-1} Z_{t-1} + b_{12} X_{t-1} Z_{t-2} + Z_t.$$

Using the representation (3.41), we see that for the first process, we have

$$\begin{aligned} X_t^{(j)} &= b_{12}^{j-1} Z_{t-j-1} \prod_{i=2}^{j} Z_{t-i} \\ &= b_{12}^{j-1} Z_{t-j+1}^2 \prod_{\substack{i=2 \\ i \neq j-1}}^{j} Z_{t-i}, \end{aligned}$$

whereas for the second process, we have

$$X_t^{(1)} = Z_t,$$

$$X_t^{(2)} = b_{11} Z_{t-1}^2 + b_{12} Z_{t-1} Z_{t-2},$$

$$X_t^{(3)} = b_{11} b_{12} Z_{t-3}^3 + b_{12}^2 Z_{t-3} Z_{t-1}^2 + b_{11}^2 Z_{t-2}^2 + b_{11} b_{12} Z_{t-1} Z_{t-2} Z_{t-3}.$$

Higher-order terms are complicated, however no power of Z_t higher than 3 appears in these expansions. Note that while for the first process $b_{11} b_{12} = 0$, for the second process $b_{11} b_{12} \neq 0$.

In general, the infinite Volterra series expansions of the process (3.40) will involve products of innovations Z_t, with the highest power of Z_t being equal to $l + 1$, provided that $\prod_{i=1}^{l} b_{1i} \neq 0$. In this case the diagonal kernels $h_{j,j,\ldots,j}$ in the $(l + 1)$th Volterra term

$$\sum_{j_1} \cdots \sum_{j_{l+1}} h_{j_1, j_2, \ldots, j_{l+1}} \prod_{i=1}^{l+1} Z_{t-j_i},$$

are given by

$$h_{j,j,\ldots,j} = \begin{cases} \prod_{i=1}^{l} b_{1i}, & i = l + 1; \\ 0, & \text{otherwise.} \end{cases}$$

If, however, $b_{1i} = 0$ for some $i = 1, \ldots, l$ then the infinite Volterra series expansions will involve powers of Z_t less than $l + 1$. If we approximate the general bilinear process by its $(l + 1)$th order Volterra series expansion, then from (3.19) we see that, as $n \to \infty$

$$P(M_n(X) \leq a_n^{-(l+1)}x) \to$$
$$\begin{cases} \exp\left\{-(h^{(+)})^{\alpha/(l+1)}x^{-\alpha/(l+1)}\right\}, & \text{if } l+1 \text{ is even;} \\ \exp\left\{-[p(h^{(+)})^{\alpha/(l+1)} + q(h^{(-)})^{\alpha/(l+1)}]x^{-\alpha/(l+1)}\right\}, & \text{if } l+1 \text{ is odd,} \end{cases}$$

where $h^{(+)} := \max_{-\infty \leq j \leq \infty}(\prod_{j=1}^{l} b_{1j}, 0)$, and $h^{(-)} := \max_{-\infty \leq j \leq \infty}(-\prod_{j=1}^{l} b_{1j}, 0)$. If we compare this result with Theorem 4.1 of Resnick and Van den Berg (2000) we see that such approximation characterizes the asymptotic distribution of the maximum of the general bilinear process given in (3.40) up to a constant in the same way as (3.38) is related to (3.39), although in a much more complicated manner.

The above results indicate a possible way to determine a finite-order Volterra (polynomial) model for a given observed input and nonlinear output time series: Let $z_1, z_2, \ldots, z_n$ and $x_1, x_2, \ldots, x_n$ respectively be observations from an i.i.d. regularly varying input series (we assume for the time being that the input process is observed) and stationary nonlinear output time series. In this case, when it exists, the stationary distribution $P(X_t \leq x)$ has regularly varying tails (see for example, Goldie 1991 and Davis and Resnick 1985), and we assume that, as $x \to \infty$

$$P(|X_t| > x) = x^{-\alpha^*} L(x),$$

for some $\alpha^* > 0$. The tail indexes α of Z_t and α^* of X_t can be estimated by one of the many methods suggested, for example by the Hill estimator (Hill 1975; Embrechts et al. 1997). Consequently, these results suggest using a Volterra series approximation of order m, where m is taken to be the integer part of α/α^*.

In the classical time series setup the input innovation process Z_t is not observed, therefore α cannot be estimated directly. In this case, one possible way is to over-fit a high order bilinear or a GARCH model to the data and then estimate the residuals from this fitted model. The tail index of Z_t, and hence the order $\hat{m} = \hat{\alpha}/\hat{\alpha}^*$, can then be estimated from these residuals. However, when the input series is not observed or reliable estimates of residuals do not exist, we suggest leaving α as well as m to be unknown model design parameters and choose a combination (α, m) in such a way that m is the integer part of α/α^*. Then the corresponding finite-order Volterra series, although arbitrary and not unique, would be a consistent model with the observed nonlinearities of the process X_t. Note that the choice of a high value for α would permit us, for example, to work with innovations with desired finite moments, but at the cost of higher-order polynomial approximations. Similar arguments can be used to identify the order l in the bilinear specification given in (3.32).

In some cases, it may not be possible to find a finite-order Volterra series approximations. Consider, for example, the bilinear process (3.36) with normally distributed innovations. In this case, when it exists, the stationary distribution of the process X_t has regularly varying tail and is in the max-domain of attraction of the Fréchet distribution (see Turkman and Amaral Turkman 1997). However, under quite general conditions, the low order Volterra terms

$$c^m \prod_{i=1}^{m-1} Z_{t-i} Z_{t-j}^2,$$

will have comparatively lighter tails; hence the maximum of finite-order Volterra expansions would be in the domain of attraction of Gumbel distribution. Hence the extremal behavior of the process and its finite-order Volterra approximation are quite different. For example, in the second-order approximation

$$X_t = Z_t + cZ_{t-1}^2$$

with Gaussian innovations, Z_t^2 are i.i.d. r.v's having a χ^2-distribution, and the process $X_t = Z_t + cZ_{t-1}^2$ is 1-dependent. Hence the asymptotic distribution of the maximum of this process will be a Gumbel distribution and the second-order expansions in this case will not capture all the nonlinearities of the bilinear process given in (3.36). This is due to the fact that, when the residuals have regularly varying tails, the tail behavior of the Volterra expansion is dominated by individual large innovations, whereas in the case of Gaussian distributed innovations, the clusters of large values of $\prod_{i=1}^{n} Z_{t-i}$ dominate the large values of the process.

3.3 Linear Models at the Presence of Heavy Tails

Suppose that we observe a time series and estimate the tail index $\hat{\alpha} < 2$, suggesting that this time series has infinite variance. In the previous section, we made a case for using an appropriate nonlinear time series model for such data set depending on the pair (m, α). One possible choice is $m = 1$ with $\alpha = \alpha^*$, that is, to assume a linear model

$$X_t = \sum_{j=1}^{\infty} \psi_j Z_{t-j} + Z_t, \tag{3.42}$$

where the innovations are i.i.d., having regularly varying tails with index $\alpha < 2$, satisfying conditions (3.5) and (3.6). We may then be tempted to model the data using ARMA(p, q) models with heavy-tailed innovations. The tools of identification and estimation are readily available for using such models; see for example, Brockwell and Davis (1991), Andrews et al. (2009) and Calder and Davis (1998).

Theorem 3.3.1. *Let Z_t be an i.i.d. sequence of r.v's satisfying* (3.5) *and* (3.6) *and* (ψ_j) *is a sequence of constants such that*

$$\sum_{j=0}^{\infty} |\psi_j|^{\delta} < \infty,$$

for some δ such that $\delta \in (0, \alpha) \cap [0, 1]$. Then X_t converges almost surely.

Moreover, if $\theta(z)$ *and* $\phi(z)$ *are moving average and autoregressive polynomials of orders q and p such that*

$$\sum_{j=0}^{\infty} \psi_j z^j = \theta(z)/\phi(z)$$

and

$$\phi(z) \neq 0,$$

for $|z| \leq 1$*, then the difference equation*

$$\phi(B)X_t = \theta(z)Z_t,$$

has the unique stationary solution (3.42)*. Further, if* $\phi(z)$ *and* $\theta(z)$ *have no common zeroes, then the process is invertible iff* $\theta(z) \neq 0$ *for* $|z| \leq 1$.

The sample autocorrelation function, which is fundamental in linear models for identification as well as estimation, is given by

$$\hat{\rho}(k) = \frac{\sum_{t=1}^{n-k}(X_t - \overline{X})(X_{t+k} - \overline{X})}{\sum_{t=1}^{n}(X_t - \overline{X})^2}.$$

This function may not make much sense in the presence of heavy tails, especially when the data are positive, due to the centering at the sample average. A heavy-tailed version without centering is given by

$$\hat{\rho}_{hev}(k) = \frac{\sum_{t=1}^{n-k} X_t X_{t+k}}{\sum_{t=1}^{n} X_t^2}. \tag{3.43}$$

When $\alpha < 2$ there is no finite variance for X_t, hence the mathematical autocovariance function does not exist, but the sample covariance function given in (3.43) still has many desirable properties. For example,

$$\hat{\rho}_{hev}(k) \to \rho(k), \tag{3.44}$$

in probability, where

$$\rho(k) = \frac{\sum_{j=0}^{\infty} \psi_j \psi_{j+k}}{\sum_{j=0}^{\infty} \psi_j^2},$$

and this consistency result leads to consistency of Yule-Walker estimates of autoregressive coefficients in an AR(p) model with non-Gaussian innovations. Therefore, it is possible to make identification and estimation based on the adjusted sample autocovariance function $\hat{\rho}_{hev}(k)$. If a general ARMA(p,q) model is identified then estimation methods other than Yule-Walker estimation are needed. The likelihood methods implicitly require a specific innovation distribution and may take a very complicated form. Least squares methods may not be suitable, as they require the finiteness of the second-order moments. The method of M-estimation suggested by Calder and Davis (1998) may be much more suitable for this case. Although inference for nonlinear models will be given in Chap. 4 in detail, here we give a brief description of the M-estimation method since it is specific to linear models with heavy-tailed innovations.

Let $x_1, \ldots, x_n$ be observations generated from the heavy-tailed linear model given above. Let $\boldsymbol{\theta}$ be the vector of parameters of this model. An M-estimate $\hat{\boldsymbol{\theta}}$ of $\boldsymbol{\theta}$ is obtained by minimizing the objective function

$$\sum_{t=1}^{n} L(Z_t(\boldsymbol{\theta})), \tag{3.45}$$

where $L(\cdot)$ is a suitably chosen loss function and $Z_t(\boldsymbol{\theta})$ is an estimate of the innovation sequence Z_t. Innovations of the linear model can be obtained via

$$\begin{aligned} Z_1(\boldsymbol{\theta}) &= x_1, \\ Z_2(\boldsymbol{\theta}) &= x_2 - \phi_1 x_1 - \theta_1 Z_1(\boldsymbol{\theta}), \\ &\vdots \end{aligned}$$

Many known estimation procedures are special cases of M-estimation. For example, in least squares, $L(x)$ is chosen as x^2. Other options are least absolute deviations with $L(x) = |x|$, and maximum likelihood estimation with $L(x) = -\log f(x)$, where f is the density function of the innovations. For heavy-tailed linear ARMA(p,q) models, Davis (1996) suggests first defining a centered and scaled parameter vector

$$\boldsymbol{\beta} = a_n(\boldsymbol{\theta} - \boldsymbol{\theta}_0),$$

where the scaling constants are chosen as

$$a_n = \inf\{x : P(|Z_1| > x) \leq 1/n\}.$$

Davis (1996) then shows that minimizing the objective function

$$\sum_{t=1}^{n} [L(Z_t(\boldsymbol{\theta}_0 + a_n^{-1}\boldsymbol{\beta})) - L(Z_t(\boldsymbol{\theta}_0))],$$

with respect to $\boldsymbol{\beta}$ is equivalent to minimizing (3.45) with respect to $\boldsymbol{\theta}$. The reason for rescaling and centering is that convergence results are more tractable under such transformations. However, the asymptotic distribution of $\hat{\beta}$ which depends on the particular choice of the loss function and innovation distribution, is typically intractable and resampling techniques are used to approximate these distributions. For detailed study of these estimators, as well as examples, see Calder and Davis (1998). Maximum likelihood estimation for such heavy-tailed distributions are in general difficult, as the limiting distributions of these estimators are generally intractable. Andrews et al. (2009) suggest using bootstrapping to study asymptotic properties.

Asymptotic consistency of the adjusted sample covariance function (3.44) holds under the specific linear model given above. However, this property may not exist when data is generated by a nonlinear process. For example, if the data X_t is generated by a bilinear BL(p, q, m, l) process satisfying the conditions of Theorem 3.2.11, then

$$(\hat{\rho}_{hev}(k), k = 1, \ldots, p) \to (Y_1, \ldots Y_p), \tag{3.46}$$

where $(Y_1, \ldots, Y_p)$ are r.v's depending on the innovations in a rather complicated and nonlinear manner; see Resnick and Van den Berg (2000) for details.

Figure 1.24 shows the autocovariance function of three independent and identical samples of 100 observations from a bilinear process with tail index $\alpha < 1$. Hence, these observations are from a process with infinite mean. The serious implication of the result given in (3.46) is that if we have data truly coming from a nonlinear process, then the sample covariance function, estimated from different samples or from different portions of the same sample, will not be consistent with the autocorrelation function estimated from the whole sample; thus identification methods based on autocovariance function will give misleading results. Therefore the modeler runs not only the risk of using a inferior linear model when the data come from a nonlinear process but also of using the wrong linear model; see Resnick (1998) for further details. It is our belief that if the observed data is heavy-tailed with the estimated tail index $\hat{\alpha} < 2$, then nonlinear models should be considered as better alternatives to heavy-tailed linear models.

References

Andrews B, Calder M, Davis RA (2009) Maximum likelihood estimation for α-stable autoregressive processes. Ann Stat 37:1946–1982

Basrak B, Davis RA, Mikosch T (2002) Regular variation of GARCH processes. Stoch Process Appl 99:95–115

Beirlant J, Goegebeur Y, Segers J, Teugels J (2004) Statistics of extremes: theory and applications. Wiley, Chichester

Bougerol P, Picard N (1992) Strict stationarity of generalized autoregressive processes. Ann Probab 20:1714–1730

Brandt A (1986) The stochastic equation $Y_{n+1} = A_n Y_n + B_n$ with stationary coefficients. Adv Appl Probab 18:211–220

Brockwell PJ, Davis RA (1991) Time series: theory and methods. Springer, New York

Calder M, Davis RA (1998) Inference for linear processes with stable noise. In: Adler RJ, Feldman RE, Taqqu MS (eds) A practical guide to heavy tails. Birkhäuser, Boston, pp 159–176

Cline D (1983) Estimation and linear prediction for regression, autoregression and ARMA with infinite varaiance data. Ph.D. Theses, Dept. of Statistics, Colorado State University, Fort Collins

Davis RA (1996) Gauss-Newton and M-estimation for ARMA processes with infinite variance. Stoch Process Appl 63:75–95

Davis RA, Resnick SI (1985) Limit theory for moving averages of random variables with regularly varying tail probability. Ann Probab 13:179–195

Davis RA, Resnick SI (1996) Limit theory for bilinear processes with heavy tailed noise. Ann Appl Probab 6:1191–1210

Draisma G, Drees H, Ferreira A, de Haan L (2004) Bivariate tail estimation; dependence in asymptotic independence. Bernoulli 10:251–280; 25:721–736

Drees H, Rootzen H (2010) Limit theorems for empirical processes of cluster functionals. Ann Stat 38:2145–2186

Embrechts P, Goldie C (1994) Perpetuities and random equations. In: Mandl P, Huskova M (eds) Asymptotic statistics: proceedings of fifth Prague symposium. Physica, Heidelbeg, pp 75–86

Embrechts P, Klüppelberg C, Mikosch T (1997) Modelling extremal events for insurance and finance. Springer, Heildelberg

Goldie CM (1991) Implicit renewal theory and tails of solutions of random equations. Ann Appl Probab 1:126–166

Goldie CM, Grübel R (1996) Perpetuities with thin tails. Adv Appl Probab 28:463–480

Grey DR (1994) Regular variation in the tail behaviour of solutions of random difference equations. Ann Appl Probab 4:169–183

Grincevičius AK (1975) One limit distribution for a random walk on the line. Lithuanian Math J 15:580–589. (English translation)

Hill BM (1975) A simple general approach to inference about the tail of a distribution. Ann Stat 3:1163–1174

Hitczenko P, Wesolowski J (2009) Perpetuities with thin tails revisited. Ann. Appl. Probab. 19:2080–2101

Kesten H (1973) Random difference equations and renewal theory for products of random matrices. Acta Math 131:207–248

Klüppelberg C, Lindner A (2005) Extreme value theory for moving average processes with light-tailed innovations. Bernoulli 11:381–410

Leadbetter MR, Lindgren G, Rootzén H (1983) Extremes and related properties of random sequences and processes. Springer, New York

Ledford AW, Tawn JA (1996) Statistics for near dependence in multivariate extreme values. Biometrika 83:169–187

Mathews VJ, Sicuranza GL (2000) Polynomial signal processing. Wiley, New York

Pham DT (1985) Bilinear Markovian representation and bilinear model. Stoch Proc Appl 20:295–306

Resnick SI (1987) Extreme values, regular variation and point processes. Springer, New York

Resnick SI (1998) Why non-linearities can ruin the heavy-tailed modeler's day. In: Adler RJ, Feldman RE, Taqqu MS (eds) A practical guide to heavy tails. Birkhäuser, Boston, pp 219–239

Resnick SI (2007) Heavy-tail phenomena: probabilistic and statistical modeling. Springer, New York

Resnick SI, Van den Berg E (2000) Sample correlation behaviour for the heavy tailed general bilinear process. Stoch Models 16:233–258

Rootzén H (1986) Extreme value theory for moving average processes. Ann Probab 14:612–652

Scotto MG, Turkman KF (2005) Extremes of Volterra series expansions with heavy-tailed innovations. Nonlinear Anal Theory Methods Appl 63:106–122

Subba Rao T, Gabr MM (1984) An introduction to bispectral analysis and bilinear time series models. Springer, New York

Turkman KF, Amaral Turkman MA (1997) Extremes of bilinear time series models. J Time Ser Anal 18:305–320

Vervaat W (1979) On a stochastic difference equation and a representation of non-negative infinitely divisible random variables. Adv Appl Probab 11:750–783

Chapter 4
Inference for Nonlinear Time Series Models

4.1 Identification of Nonlinearity

Suppose we have an observed time series $x_1, x_2, \ldots, x_n$ and want to know if a linear time series model is adequate for the data, or an alternative nonlinear model should be considered. Linear models are often taken as the null hypotheses against a nonlinear alternative due to the simplicity of inference. Often we know much about the underlying process which generate the data set. Therefore it is possible to decide if a linear model will be adequate and if not, what aspects of nonlinearity should be modeled as alternative. For example, if the data is on population dynamics, as explained before, limit cycle behavior can be expected, and an adequate model such as a self-exciting threshold model should seriously be considered as an alternative. On the other hand, bilinear models may be more adequate for telecommunication data which often exhibit heavy-tailed behavior. However, if we know little about the underlying process, we will have to rely only on the information contained in the data, and empirical methods for testing nonlinearity are needed. In this case, formulation of formal tests is quite difficult due to the fact that there are many different ways a process can be nonlinear, and often it is difficult to specify alternative hypotheses to a linear model. Naturally, the power of tests constructed will depend on how well the alternative hypotheses are constructed, and complicated tests of hypotheses may not worth the effort put into devising such tests. Therefore, quick graphical methods and portmanteau tests are often used for checking nonlinearity.

Subba Rao and Gabr (1980) were possibly the first to propose a linearity test, which is based on the characteristics of the third-order cumulant spectrum of a linear process, and later this test was improved by Hinich (1982). McLeod and Li (1983) proposed a portmanteau test and Keenan (1985) devised an easy to use test which is based on arranged auto-regressions. Petruccelli and Davies (1986) constructed a CUSUM test which is also based on the arranged auto-regression approach. Brock et al. (1996) (see also Luukkonen et al. 1988) proposed a test for linearity against a STAR model alternative. Tsay (1989) devised an F-type test for assessing a

K.F. Turkman et al., *Non-Linear Time Series*, DOI 10.1007/978-3-319-07028-5_4,

TAR/SETAR model alternative hypothesis that used the arranged auto-regression methodology. Tsay (1991) modified his previously proposed test in order to be both simple to use and general enough in order to be able to identify several types of nonlinearity, and consequently a wide range of possible nonlinear models such as Bilinear, STAR, EXPAR and SETAR were included as alternative hypotheses. These tests were also used as basis for model selection. Peat and Stevenson (1996) slightly modified the test proposed by Tsay (1991). Bera and Higgins (1997) proposed two stage tests for GARCH and bilinear type models. In the first stage, a joint test is devised to see if existing nonlinearity can be attributed to ARCH or bilinearity. Then in the next stage, GARCH and bilinear models are taken as null and alternative hypotheses. In the next section we look at some simple tests for checking the presence of nonlinearity in the data. Detailed discussion of tests for linearity can be found in Tong (1990).

4.1.1 Graphical Methods and Other Informal Tests for Types of Nonlinearity

Checking for the Gaussian structure of the data is the first step to see possible deviations from linearity. Apart from the QQ-normal plots and histograms, one can check other deviations from Gaussian assumption by verifying;

- **Reversibility**: Reversibility is more a Gaussian property then linearity, but it is a strong indication of nonlinearity. The reverse data plots are the simplest way of checking reversibility. By comparing the plots of $x_1, x_2, \ldots, x_n$ and $x_n, x_{n-1}, \ldots, x_1$, time irreversibility can be revealed.
- **Sample autocorrelation function of squared residuals**: If X_t is a Gaussian stationary time series, then the autocorrelation function of the squared process X_t^2 and the square of the autocorrelation function of X_t should be identical, namely

$$\rho_{X^2}(k) = (\rho_X(k))^2,$$

for every $k = 1, 2, \ldots$. A departure from this observation indicates the possibility of nonlinearity. Therefore a comparison of the plots of the respective autocorrelation functions, can reveal deviations of nonlinearity.

Often the autocorrelation functions of the absolute values and the squares of residuals from a fitted linear model also indicate nonlinearity. Suppose we fit an adequate linear model to data, and let $(\hat{Z}_1, \ldots, \hat{Z}_n)$ be the fitted residuals. If Z_t are i.i.d. r.v's, then

$$\rho_{Z^2}(k) = \rho_Z(k)$$

and

$$\rho_{|Z|}(k) = \rho_Z(k),$$

but these results need not hold if Z_t are identical but uncorrelated r.v's.

We can compare the plots of sample autocorrelation functions $\rho_{\hat{Z}^2}(k)$ and $\rho_{\hat{Z}}(k)$ (similarly $\rho_{|\hat{Z}|}(k)$ with $\rho_{\hat{Z}}(k)$). Any deviation may be used as an indication that there is structure in the series other than the linear model. This graphical check is particularly valuable for assessing linearity against GARCH-type alternatives. The above idea can be formalized as a portmanteau test. If

$$r(k) := \frac{\sum_{t=1}^{n-k}(\hat{Z}_t^2 - \hat{\sigma}^2)(\hat{Z}_{t+k}^2 - \hat{\sigma}^2)}{\sum_{t=1}^{n}(\hat{Z}_t^2 - \hat{\sigma}^2)},$$

where

$$\hat{\sigma}^2 = \sum_{t=1}^{n} \hat{Z}_t^2/n,$$

then under the assumption that the innovations are i.i.d.,

$$Q := n(n+2)\sum_{k=1}^{m} r^2(k)/(n-k), \tag{4.1}$$

has asymptotic distribution χ_m^2. Another alternative is to look at the cross correlations between $\hat{Z}_t$ and $\hat{Z}_t^2$. For linear models, these cross correlations should be zero, and the plot of this cross correlation function may reveal deviations from this assumption. The expression given in (4.1) is the widely known McLeod-Li (1983) test statistics. Here, m has to be chosen as a function of the sample size and $m = n/4$ is often seen as a good compromise.

- **Index of Linearity**

If X_t is a linear process, then for every $j = 1, 2, \ldots$, $E(X_t|X_{t-j})$ is a linear function of X_{t-j}. Therefore, if we have good, nonparametric estimates of the regression functions of X_t on X_{t-j} for various values of j, and plot them as function of t, we may get an understanding of the degree of nonlinearity that exists in the series. The degree of nonlinearity between X_t and X_{t-j} can be quantified in the following way:

Let X_t be a stationary time series with $\mu := E(X_t)$, $\sigma^2 := E(X_t^2) < \infty$ and ACF $\rho(k)$. Consider the best mean-square predictor of X_t in terms of X_{t-j}, for $j = 1, 2, \ldots$ which is given by $E(X_t|X_{t-j})$. If we restrict ourselves to the best linear predictor then

$$E_L(X_t|X_{t-j}) = \mu + \rho(j)(X_{t-j} - \mu).$$

The variance about the best predictor $E(X_t|X_{t-j})$ is given by

$$E(X_t - E(X_t|X_{t-j}))^2 = \sigma^2 - V(E(X_t|X_{t-j})),$$

and in particular, if we have the best linear predictor, the variance about this best linear predictor is given by

$$E(X_t - E_L(X_t|X_{t-j}))^2 = \sigma^2 - \sigma^2\rho^2(j).$$

For a nonlinear stationary time series, clearly

$$0 \leq E(X_t - E(X_t|X_{t-j}))^2 \leq E(X_t - E_L(X_t|X_{t-j}))^2,$$

hence

$$\sigma^2 - V(E(X_t|X_{t-j})) \leq \sigma^2 - \sigma^2\rho(j),$$

so that

$$0 \leq \frac{\sigma^2\rho(j)}{V(E(X_t|X_{t-j}))} \begin{cases} = 1, \text{ if } X_t \text{ and } X_{t-j} \text{ are related linearly;} \\ \leq 1, \text{ if not.} \end{cases}$$

Therefore, we can define the index of linearity as

$$L := \max_j \frac{\sigma^2\rho(j)}{V(E(X_t|X_{t_j}))}.$$

This index may give some indication of the degree of deviation of the data from linearity.

4.1.2 Statistical Tests for Linearity and Gaussianity

Tests Based on Polyspectra

The most important property of polyspectra that has relevance to nonlinear models is that all polyspectra higher than the second-order vanish if X_t is a Gaussian process. Existence of higher-order polyspectra thus can be used as a measure of deviation from normality and to some extend, deviation from linearity. Hence, tests for Gaussianity and linearity can be devised based on polyspectra. Two tests are due to Subba Rao and Gabr (1980).

1. If the process is Gaussian then all the polyspectra, in particular bispectrum, of the process should be 0 for all frequencies. Hence a test statistic can be constructed from bispectral estimates over a grid of frequencies based on this observation. The distribution of the test statistic will depend on the region over which the

grid is defined, and as a result the test is performed in two stages depending on the regions. The details of the test can be found in Subba Rao and Gabr (1984) or Priestley (1981). The problem with this test is that it is possible to have a non-Gaussian process with zero bispectrum. However, it is reasonable to expect that the bispectrum should be a good indicator of deviation from the Gaussian assumption.

2. If the data is generated by a linear model of the form

$$X_t = \sum_{j=0}^{\infty} \psi_j Z_{t-j},$$

then the cumulant $C(s_1, s_2)$ is given by

$$C(s_1, s_2) = E(Z_t^3) \sum_{j=0}^{\infty} \psi_j \psi_{j+s_1} \psi_{j+s_2}.$$

Taking the Fourier transforms, the bispectrum for such a process is given by

$$\begin{aligned} h_3(\omega_1, \omega_2) &= \frac{E(Z_t^3)}{4\pi^2} [\sum_j \psi_j e^{i(\omega_1+\omega_2)}][\sum_j \psi_j e^{-i\omega_1 j}][\sum_j \psi_j e^{-i\omega_2 j}] \\ &= \frac{E(Z_t^3)}{4\pi^2} \Gamma[-(\omega_1 + \omega_2)]\Gamma(\omega_1]\Gamma(\omega_2). \end{aligned}$$

Since the spectral density function of the process is given by

$$h(\omega) = \frac{\sigma_Z^2}{2\pi} |\Gamma(\omega)|^2,$$

the function

$$X(\omega_1, \omega_2) = \frac{|h_3(\omega_1, \omega_2)|^2}{h(\omega_1)h(\omega_2)h(\omega_1 + \omega_2)}, \tag{4.2}$$

is constant for all ω_1 and ω_2. A test of linearity can then be constructed to test the null hypothesis that the function $X(\omega_1, \omega_2)$ is zero for all ω_1, ω_2 by substituting the functions $h_3(\omega_1, \omega_2)$ and $h(\omega)$ by their sample estimates and testing the constancy of the $X(\omega_1, \omega_2)$ over a fine grid of frequencies ω_1, ω_2. This naturally suggests the use of the statistics

$$\sum_{\omega_1} \sum_{\omega_2} \frac{|\hat{h}_3(\omega_1, \omega_2)|^2}{\hat{h}(\omega_1)\hat{h}(\omega_2)\hat{h}(\omega_1 + \omega_2)}.$$

Again the details of the test can be found in Subba Rao and Gabr (1984) and Priestley (1981). We note that this is not a full-proof test of linearity. It is possible to find nonlinear processes for which (4.2) is zero. However, it is reasonable to expect that comparing the bispectrum with the spectrum should indicate the presence of nonlinearity. Criticism to these tests are given in Tong (1990). His general criticism is basically due to lack of power and possible problems due to smoothing the spectral estimates.

Tests Based on Residuals

We have seen how to use the sample ACF of squared residuals obtained from a linear model to check graphically the presence of possible nonlinearity. These arguments can be formalized by devising a formal test based on Tukey's one degree of freedom test for nonadditivity (Keenan 1985). Suppose that $x_1, \ldots, x_n$ is an observed time series and we start by assuming that

$$H_0 : X_t = \mu + \sum_{j=1}^{M} \phi_j X_{t-j} + Z_t,$$

where Z_t are i.i.d. zero-mean, with finite variance and finite fourth moment. The test is constructed by fitting linear models for X_t and X_t^2 on

$$(X_{t-1}, \ldots, X_{t-M}),$$

for some large, fixed value M. Let Z_t and η_t, $t = M+1, \ldots, n$ be the respective residuals obtained from these regressions. Now, regress Z_t on η_t with

$$Z_t = a_0 + a_1 \eta_t + \xi_t.$$

Note that, under H_0, there should be no linear relationship between Z_t and η_t. If

$$\hat{a} = \hat{a}_1 \left(\sum_{t=M+1}^{N} \xi_t^2 \right)^{1/2},$$

then under the null hypothesis H_0, the test statistic

$$F := \frac{\hat{a}^2 (N - 2M - 2)}{\sum_{t=M+1}^{N} Z_t^2 - \hat{a}^2},$$

has the asymptotic F-distribution with $(1, N - 2M - 2)$ degrees of freedom; see Tong (1990) for the proof. Large value of the test statistics F should indicate deviation from linearity. See also Tong (1990) for slightly modified tests to improve the power of the above F test.

Lagrange Multiplier Tests or Score Tests

In this approach, a specific nonlinear model is indicated as alternative hypothesis to the null hypothesis of a linear model. Such tests have more power than the nonparametric approach based on bispectra due to having a specific alternative hypothesis. These tests depend on a variant of the likelihood approach, and the basic idea is to assume a model of the form

$$X_t = \mathbf{a}'\mathbf{X}_{t-1} + f(\boldsymbol{\theta}, \mathbf{X}_{t-1}) + Z_t,$$

where $\mathbf{X}'_{t-1} := (X_{t-1}, \ldots, X_{t-p})$ and $\mathbf{a}' := (a_1, \ldots, a_p)$ are unknown model parameters. The function f is assumed to be known and $\boldsymbol{\theta}$ is a parameter vector such that $f(\boldsymbol{\theta}, \mathbf{X}_{t-1}) = 0$, when $\boldsymbol{\theta} = 0$. Under such model, the test of nonlinearity can be formulated as

$$H_0 : \boldsymbol{\theta} = 0 \text{ vs } H_1 : \boldsymbol{\theta} \neq 0.$$

The problem with these tests, as expected, is that they are very sensitive to the choice of f, that is, to the choice of alternative hypotheses.

Consider for example, the hypothesis H_0 versus H_1 defined as

$$H_0 : X_t \sim \mathrm{AR}(p) \text{ vs } H_1 : X_t \sim \mathrm{BL}(p, 0, m, l),$$

specifically defined as

$$H_0 : X_t + \phi_1 X_{t-1} + \cdots + \phi_p X_{t-p} = \mu + Z_t$$

versus

$$H_1 : X_t + \phi_1 X_{t-1} + \cdots + \phi_p X_{t-p} = \mu + Z_t + \sum_{i=1}^{m} \sum_{j=1}^{l} \phi_{ij} Z_{t-i} X_{t-j},$$

where (Z_t) are i.i.d. $N(0, \sigma^2)$. Therefore the test of nonlinearity can be reduced to the test

$$H_0 : \phi_{ij} = 0, \text{ vs } H_1 : \text{not all are zero.}$$

Here $\boldsymbol{\Phi} := (\phi_{ij}, i = 1, \ldots, m, j = 1, \ldots, l)$ are the parameters of interest and $\boldsymbol{a} := (\phi_1, \ldots, \phi_p, \mu)$ are the nuisance parameters. The Lagrange multiplier test statistic L in its general form is given by (see Tong 1990)

$$L = S(\boldsymbol{\Phi}_0, \hat{\boldsymbol{a}}_0)^T (I_{11} - I_{12} I_{22}^{-1} I_{21})^T S(\boldsymbol{\Phi}_0, \hat{\boldsymbol{a}}_0),$$

where

$$I(\mathbf{\Phi}_0, \hat{\boldsymbol{a}}_0) = \begin{bmatrix} I_{11} & I_{12} \\ I_{21} & I_{22} \end{bmatrix},$$

is the Fisher Information matrix, partitioned according to the parameters $(\mathbf{\Phi}_0, \hat{\boldsymbol{a}}_0)$ and $S(\mathbf{\Phi}_0, \hat{\boldsymbol{a}}_0)$ is the score statistic for $(\mathbf{\Phi}, \boldsymbol{a})$, calculated under the null hypothesis. Here, $\hat{\boldsymbol{a}}_0$ are the maximum likelihood estimates of $\boldsymbol{a}$ under the null hypothesis, conditional on the fixed values of $(x_1, \ldots, x_p)$. Tong (1990) also gives the specific expression of the test statistic L for the above bilinear alternative hypothesis. Under H_0, the test statistic L has asymptotic χ^2_{ml} distribution and can be used to test deviation of linearity from bilinear BL$(p, 0, m, l)$ processes; see Tong (1990) for details.

Petruccelli and Davies Test

Other tests with different alternative hypotheses are also available. For example, Petruccelli and Davies (1986) present a linearity test which is commonly classified in the literature as a CUSUM test. This test is devised with the purpose of testing a null hypothesis of AR(p) model vs a SETAR model as alternative. The test is based on the arranged autoregression methodology.

Bera and Higgins Test

It is possible to device test of hypotheses to see if nonlinear dynamics in the mean and variance are present in an observed time series. Based on the model (2.42), Bera and Higgins (1997) suggest a test

$$H_0 : \mathbf{b} = 0, \boldsymbol{\alpha} = 0 \text{ versus } H_1 : \text{not all are equal to } 0,$$

where $\mathbf{b} := (b_{ij}, c_{ij})$ and $\boldsymbol{\alpha} := (\alpha_0, \alpha_1, \ldots, \alpha_q)$. This test has power against both GARCH and bilinear type nonlinearity, but if the null hypothesis is rejected, it cannot give any guidance as to the origin of nonlinearity.

4.2 Checking for Stationarity and Invertibility

For linear processes, conditions of stationarity and invertibility are given elegantly in terms of the roots of AR and MA characteristic polynomials and are relatively easy to check. These conditions depend heavily on the linear representation and cannot be extended to nonlinear processes. However, some nonlinear processes can be given in a Markovian representation, and one can use the general theory of Markov processes to find conditions for the existence of stationary solutions. We have seen in Sect. 3.2.1 that some of the well-known nonlinear processes can be put in the Markov representation

$$\mathbf{X}_t = \mathbf{A}_t\mathbf{X}_{t-1} + \mathbf{B}_t,$$

and conditions for the existence of stationary solutions for such representations are well studied; see Meyn and Tweedie (2009), Tjøstheim (1990) and Nummelin (1984) for details.

Let $(\mathbf{X}_t; t \geq 0)$ be a homogeneous discrete time Markov process taking values in $(\mathbb{R}^k, \mathcal{B})$, where $\mathcal{B}$ is the Borel σ-algebra of subsets of $\mathbb{R}^k$. Denote by $p^n(x, B)$, the n-step transition probabilities $P(X_n \in B|X_0 = x)$, so that $p(x, B)$ is the one-step transition probability.

Theorem 4.2.1. *Existence of a stationary solution.*

- *If for any $\epsilon > 0$, there exists a compact set $B \subset \mathbb{R}^k$ such that*

$$p(x, B^c) < \epsilon,$$

for all $x \in \mathbb{R}^k$, then $\mathbf{X}_t$ is bounded in probability and hence there exists at least one stationary solution for the chain.

- *If further the chain is aperiodic and satisfies Doeblin's condition given below, then it is uniformly ergodic and hence it has an unique stationary distribution.*

Definition 4.2.1 (Doeblin's condition). The sequence $\mathbf{X}_t$ is said to satisfy the Doeblin's condition, if there exists a probability measure ν on σ-field $\mathcal{B}$ and $\epsilon < 1$, $\delta > 0$ such that whenever $\nu(B) > \epsilon > 0$, then

$$\inf_x p^m(x, B) \geq \delta,$$

for some integer $m \geq 1$.

In general, it is not easy to verify the tightness of the m-step transition probabilities and the Doeblin's condition. Another problem is to write these conditions in terms of the model parameters, which may create further difficulties. Yet, even with these difficulties, the verification of stationarity is relatively easier than verifying the condition of invertibility. Conceptually stationarity and invertibility are quite similar. If we have a nonlinear model of the form

$$X_t = f(X_{t-j}, Z_{t-k}), \; j = 1, \ldots, p, \; k = 0, 1 \ldots, q,$$

then stationarity imposes conditions on the function f so that X_t is measurable with respect to the σ-field $\mathcal{B}_Z$ generated by $(Z_{t-j}; j = 0, 1, \ldots)$. On the other hand, invertibility imposes conditions on f so that Z_t is measurable with respect to the σ-field $\mathcal{B}_X$ generated by $(X_{t-j}; j = 0, 1, \ldots)$. Since in a nonlinear model the innovations are not observed and are recovered from the data, for the purpose of prediction and model diagnosis, invertibility is crucial in obtaining the residuals from the observed data. Unfortunately, apart from few special cases it is very difficult to find conditions of invertibility.

Consider first the general univariate class of processes of the form

$$X_t = f(X_{t-i}, Z_{t-i}) + Z_t, \; i = 1, 2, \ldots, q, \tag{4.3}$$

for some i.i.d. unobserved sequence Z_t and model parameters $\boldsymbol{\theta}$. If such a model is chosen to represent the data with the objective of making forecasts, it is necessary to be able to estimate the noise process Z_t from the observed data. Typically this can be done by assuming initial values for $\hat{Z}_1, \hat{Z}_2, \ldots, \hat{Z}_q$ and the innovations can be estimated through

$$\hat{Z}_t = x_t - f(x_{t-j}, \hat{Z}_{t-j}), \; j = 1, 2, \ldots, q,$$

iteratively for $t = q + 1, \ldots$. The error attached to this procedure can be quantified by

$$e_t := Z_t - \hat{Z}_t.$$

See Mauricio (2008) for alternative ways of defining residuals in linear models. Granger and Andersen (1978) define invertibility in terms of these errors. The model in (4.3) is said to be invertible if $E(e_t^2) \to 0$ as $t \to \infty$. This definition implicitly assumes that the functional form $f(\cdot)$ as well as the parameters $\boldsymbol{\theta}$ are completely known. If the model parameters are not known but estimated by an earlier set of data, then Granger and Andersen (1978) suggest relaxing this condition by assuming that $E(e_t^2) \to c$, as $t \to \infty$, for some constant $c < \infty$. Equipped with this simpler definition of invertibility, Tong (1990) suggests testing invertibility based on simulated data. The above definition of invertibility is slightly different from the definition of invertibility given for the linear MA(q) processes. Consider, for example, the MA(1) process

$$X_t = bZ_{t-1} + Z_t.$$

Assuming that $Z_0 = 0$, the innovations can be conditionally generated from

$$\hat{Z}_t = Z_t - b\hat{Z}_{t-1},$$

so that

$$\begin{aligned} e_t &= Z_t - \hat{Z}_t \\ &= -be_{t-1} \\ &= (-b)^t e_0. \end{aligned}$$

Clearly, if $|b| < 1$ then $E(e_t^2) \to 0$ and the process is invertible. However, if the parameter b is not known but estimated from the data, then

$$\hat{e}_t = (b - \hat{b})Z_{t-1} - \hat{b}e_{t-1}.$$

The solution for this difference equation is given in two parts of which the first component is

$$e_t = (-\hat{b})^t e_0,$$

whereas the second component has the following expression

$$e_t = (b - \hat{b}) \sum_{i=1}^{t} \hat{b}^{(i-1)} Z_{t-i}.$$

For large t,

$$V(e_t) = \frac{(b - \hat{b})^2}{1 - \hat{b}^2} V(Z_t),$$

hence condition $|\hat{b}| < 1$ is required for the first component to converge and for the second component to have a finite variance. In order to achieve these results, often b is assumed to be estimated once, but not updated as more data becomes available. When X_t follows a simple bilinear process

$$X_t = Z_t + bX_{t-1}Z_{t-1},$$

assuming that $Z_0 = 0$, the innovations can be calculated iteratively from

$$\hat{Z}_t = X_t - bX_{t-1}\hat{Z}_{t-1},$$

and

$$\begin{aligned} e_t &= Z_t - \hat{Z}_t \\ &= -bX_{t-1}e_{t-1} \\ &= (-b)^t \left(\prod_{i=1}^{t} X_{t-i} \right) e_0. \end{aligned}$$

A necessary and sufficient condition for $e_t \to 0$ in probability as $t \to \infty$ is given by

$$E(\ln |X_t|) = \ln |b| \leq 0,$$

(see e.g., Quinn 1982) but this condition depends on the unknown stationary distribution of X_t. However, if Z_t are zero-mean, i.i.d. r.v's with variance σ_Z^2 and if $b^2\sigma_Z^2 < 1$ then $E(X_t^2) = \sigma_Z^2/(1 - b^2\sigma_Z^2)$ and

$$\begin{aligned}\ln|b| + E(\ln|X_t|) &= \frac{1}{2}(\ln b^2 + E\ln(X_t^2)) \\ &\leq \frac{1}{2}(\ln b^2 + \ln(EX_t^2)) \\ &= \frac{1}{2}b^2\sigma_Z^2/(1 - b^2\sigma_Z^2).\end{aligned}$$

Hence, in this case, a sufficient condition for the invertibility of this model is given by

$$\sigma_Z|b| < 1/\sqrt{2}.$$

However, such conditions are far from being necessary. For example, when Z_t are further assumed to be Gaussian, then $E(\ln|X_t|)$ can be calculated to give (see, Quinn 1982)

$$E(\ln|X_t|) = \ln(\sigma_Z) - \frac{1}{2}(\gamma + \ln 2) + \frac{1}{2}E(\ln(1 + b^2X_t^2)),$$

where γ denotes the Euler's constant and a sharper sufficient condition is given by

$$\sigma_Z|b| < 2\exp(\gamma)/(1 + 2\exp(\gamma))^{1/2} \approx 0.8836.$$

Although, it is relatively easy to obtain sufficient conditions for the invertibility of this simple bilinear model, conditions for higher-order models are much more difficult if not impossible; see Subba Rao and Gabr (1980) for a set of sufficient conditions for the invertibility of the BL$(p, 0, p, 1)$ model. Assuming invertibility, it is possible to obtain conditional least squares estimators (alternatively conditional likelihood if the (Z_t) are assumed to be i.i.d. Normal r.v's). However, asymptotic distributions of these estimators are not known. Therefore, except for very simple cases such as the model BL$(1, 0, 1, 1)$, their practical use is very limited.

Tong (1990) suggests the following practical procedure for checking invertibility: assume that the model

$$X_t = f(X_{t-j}, Z_{t-j}) + Z_t,$$

was fitted to the data $x_j, j = 1, \ldots, n$ conditional on $x_0 = x_1 = \cdots = x_p = Z_0 = \cdots = Z_p = 0$. Here, the choice of p depends on the model chosen. Using this fitted model, simulate further m residuals $\hat{Z}_{n+1}, \ldots, \hat{Z}_{n+m}$ from the model. We can also simulate m innovations $Z_1, \ldots, Z_m$ from the fitted distribution of the innovations. Note that if (Z_t) are i.i.d. $N(0, \sigma^2)$ r.v's then the fitted distribution is

$N(0, \hat{\sigma}^2)$. Tong (1990) suggests calculating the sample mean of $(Z_t - \hat{Z}_t)^2$, for $t = 1, \ldots, m$. Explosive tendency suggests noninvertibility. However, if the data indicates noninvertibility, it is not clear what diagnostic tools can be used to adjust the model towards invertibility.

4.3 Tail Index Estimation

We have seen in Sect. 3.2 that heavy-tailed data are a good indicator of the presence of nonlinearity. Naturally, heavy-tailed data are also consistent with linear models with heavy-tailed innovations. However, in order to employ standard inferential methods, one often needs to assume finite fourth-order moments. Therefore, if there is evidence in the data of heavy-tailed behavior, and in particular if there is an indication that the variance is infinite, then a proper nonlinear model, rather than a linear model with heavy-tailed innovations, should seriously be considered. There are several ways we can check for the presence of heavy tails;

1. *The kurtosis* of a random variable X with distribution F is defined to be

$$k := \frac{E(X-\mu)^4}{(E(X-\mu)^2)^2}.$$

   Kurtosis can also be defined in terms of the quantiles x_p

$$k_p := \frac{\frac{1}{2}(x_{0.75} - x_{0.25})}{x_{0.9} - x_{0.1}},$$

   where $0 < p < 1$ and $x_p = F^{-1}(p)$ is the pth quantile of F. For the normal distribution, $k = 3$ and $k_p = 0.263$. A distribution function is called leptokurtic, if either $k > 3$ or $k_p > 0.263$. Hence, estimated kurtosis in terms of empirical quantiles may give an indication of tail heaviness.
2. *Exploratory data analysis for extremes*. Probability and QQ-plots are often used to compare several distributions which may be a good model for the data, and they are also appropriate for indicating tail heaviness. For a given observed stationary time series $x_1, \ldots, x_n$, with marginal distribution F, let us define the ordered sample as

$$x_{1,n} \leq x_{2,n} \leq \cdots \leq x_{n,n}.$$

   The QQ-plot is obtained by plotting $x_{k,n}$ against $F^{-1}(p_{k,n})$, where $p_{k,n}$ are the plotting positions. Typical choices of $p_{k,n}$ are given by

$$p_{k,n} = \frac{n-k+0.5}{n},$$

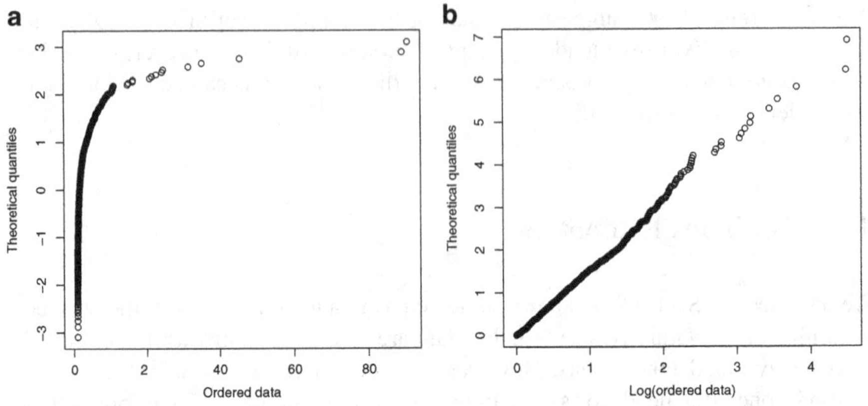

Fig. 4.1 QQ-plots of Pareto data vs theoretical Gaussian quantiles (**a**); and (**b**) theoretical Pareto quantiles

or by

$$p_{k,n} = \frac{n-k+1}{n+1} = 1 - \frac{k}{n+1}.$$

If the data are generated from the reference distribution F, the plot should look linear. On the other hand, if the data were generated from a distribution with heavier tails than the reference distribution, then the plot should curve upwards. These features can be seen in the plots presented in Fig. 4.1. The plots show the QQ-plot of a Pareto (1.5) sample of size $n = 1{,}000$ against the reference standard Gaussian distribution (a) and a Pareto distribution (b) (Fig. 4.1).

There are other exploratory plots that may be useful in understanding the tail heaviness, such as the mean excess function that was referred to in Chapter 1. The theoretical MEF of a positive random variable X with cdf F is defined as

$$e(u) := E(X - u \mid X > u) = \frac{\int_u^{x_F} (1 - F(y))dy}{1 - F(u)}$$

for a certain threshold u and $x_F \leq \infty$ is the end-point of the distribution. For an observed sample $x_1, x_2, \ldots, x_n$ of size n, the MEF is estimated by

$$\hat{e}_n(u) = \frac{\sum_{i=1}^{n} (x_i - u) I(x_i > u)}{\sum_{i=1}^{n} I(x_i > u)},$$

where $I(\cdot)$ is the indicator function.

Although it might be interesting to analyze the behavior of the MEF for several statistical models, it is particularly useful in the case of the Generalized Pareto distribution (GPD). The GPD plays a very important role in the extreme value theory (Pickands 1975). GPD has been widely used to model the asymptotic distributional behavior of the excesses above a sufficiently high threshold. This approach is known in the literature as the Peaks Over Threshold (POT), and was first used in a hydrological framework. Nowadays, the POT has been applied to many other areas such as insurance, finance, environment and meteorology (see Embrechts et al. 1997; Coles 2001; Beirlant et al. 2004).

Let F_u be the conditional cdf of the excesses above u defined as

$$F_u(y) = P(X - u \leq y \mid X > u) = \frac{F(u+y) - F(u)}{1 - F(u)},$$

then

$$\lim_{u \to x_F} \sup_{0 < x < x_F - u} \mid F_u(x) - F(x \mid k, \sigma) \mid = 0,$$

where F is the cdf of the GPD(k,σ) given by

$$F(x \mid k, \sigma) = \begin{cases} 1 - \left(1 + \frac{kx}{\sigma}\right)^{-1/k}, & k \neq 0 \\ 1 - \exp(-\frac{x}{\sigma}), & k = 0 \end{cases},$$

where k and σ are respectively the shape and scale parameters satisfying $k \geq 0$ when $x > 0$, and $0 < x < -\sigma/k$ when $k < 0$. Heavy-tailed distributions are obtained for $k > 0$, whereas light-tailed behavior is observed when k is negative. The theoretical MEF of the GPD given as

$$e(u) = \frac{\sigma}{1-k} + u\frac{k}{1-k}, k < 1 \text{ and } \sigma + ku > 0,$$

is clearly a straight line with intercept and slope equal to $\frac{\sigma}{1-k}$ and $\frac{k}{1-k}$, respectively. If the tail of the underlying distribution is exponential then the line is constantly equal to σ whereas it increases (decreases) for $k > 0$ ($k < 0$) reflecting a heavier tailed (lighter tailed) than the exponential distribution. Figure 4.2 shows the graphical representation of the MEF of an Exponential, a Pareto and an Uniform distributions. These three distributions correspond to special cases of the GPD(k,σ) for $k = 0$, $k > 0$ and $k = -1$, respectively. In practice choosing an adequate threshold above which the GPD model assumptions hold is frequently a difficult task. We refer the reader to Embrechts et al. (1997) for a detailed study of exploratory data analysis for extremes.

3. *Estimation of tail index.* We have seen in Chap. 3 that under fairly general conditions, nonlinear relations $X_t = f(\mathbf{Z}_t)$ cause heavy tails in the sense that even for moderately heavy-tailed input Z_t, the output process X_t, will have a stationary solution (when exists) with a regularly varying distribution

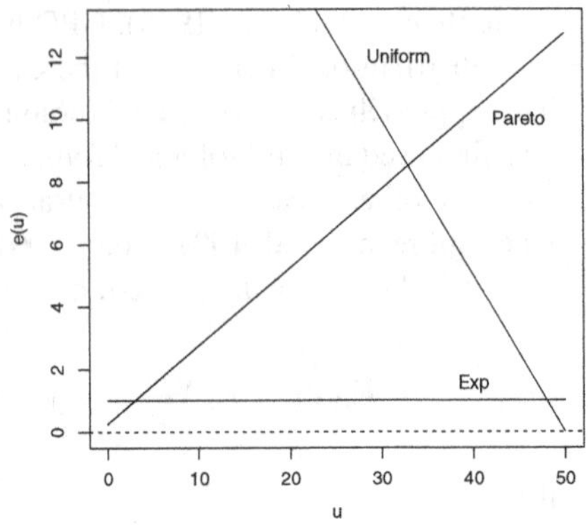

Fig. 4.2 Theoretical MEF of some common distributions

$$P(X_t > x) \sim x^{-\alpha}.$$

The tail index parameter α, as explained in the previous chapter, quantifies the degree of tail heaviness and is also a good indicator of nonlinearity in the data. The tail index α can be estimated from data either by using likelihood or semi-parametric methods. One such estimator is the Hill estimator.

Let $X_{1,n} \leq \cdots \leq X_{n,n}$ be the order statistics corresponding to an i.i.d. sample of size n coming from a distribution with regularly varying tail with tail index α. The Hill estimator of the tail index α is given by

$$\hat{\alpha}^{-1} = \frac{1}{k}\sum_{j=1}^{k} \log X_{n-j+1,n} - \log X_{n-k,n},$$

where $k = k(n)$ is the number of upper order statistics used in the estimator. Here k is a design parameter chosen as $k(n) \to \infty$ as $n \to \infty$, in a manner that $k(n)/n \to 0$. The proper choice of k affects the bias and variance of the resulting estimator and there are many suggestions in the literature as to how it should be chosen. There are also several variants of Hill estimator resulting in more efficient estimators. Hill estimator is a natural and versatile estimator for the tail index and different estimators based on maximum likelihood and other semi and nonparametric methods give asymptotically equivalent versions. Hill estimator can still be used for dependent data. The general argument is that serial dependence among high upper order statistics in general is expected to be low. Under fairly general dependence conditions, Hill estimators are still consistent, but their asymptotic bias may increase. See for example Embrechts et al. (1997) or Beirlant et al. (2004) for a complete treatment of tail index estimation for i.i.d. and dependent observations.

4.4 Methods of Inference

There are two principal methods of parameter estimation commonly used for time series analysis, the maximum likelihood and the least squares methods. These two methods coincide when the time series have a Gaussian likelihood. It is possible to give an unified approach to these alternative methods in a very general setting which is often called quasi-likelihood method (e.g., Heyde 1997). This theory is developed in terms of estimating functions, as compared to estimators. Estimators are functions of data only, ideally having desired properties such as unbiasness and consistency, whereas estimating functions are functions of the data, as well as the model parameters, satisfying similar characteristics. When we deal with linear time series models with finite variances, likelihood and least squares methods give good results and there is no need to look further for other methods of parameter estimation. Likelihood functions of most nonlinear time series are quite complicated, often analytical expressions for these likelihoods are not available and least squares methods cannot be used with simple conditioning. Similarly, the sampling properties of the parameter estimators obtained by these methods are quite diverse and are often very hard to get. Quasi-likelihood methods, based on estimating functions, although by no means numerically easier to obtain, give a better unified approach to parameter estimation for nonlinear time series. In Sect. 4.4.1, we give a brief introduction to parameter estimation for linear time series using least squares and likelihood approaches under the Gaussian assumption. Under this assumption, the likelihood can be written in terms of the one-step predictors and their variances. This representation is a natural gateway to estimating functions and their use in nonlinear time series estimation. In Sect. 4.4.2, we give a brief introduction to estimating functions.

4.4.1 Least Squares and Likelihood Methods

Complete and precise treatment of parameter estimation of linear time series can be found in Brockwell and Davis (1991); see also Brockwell and Davis (1996, 2006) for a lighter treatment. Here we will only give a summary to tie the likelihood-based methods to estimating functions. Details can be found in Brockwell and Davis (1996).

Assume that we have n observations $x_1, \ldots, x_n$ coming from the stationary and invertible linear model

$$X_t = \sum_{i=1}^{p} a_i X_{t-i} + \sum_{j=0}^{q} b_j Z_{t-j},$$

where (Z_t) are i.i.d. Normal r.v's, with zero-mean and variance σ_Z^2. In this case, the joint density of $Z_1, \ldots, Z_n$ is given by

$$f(z_1, \ldots, z_n) = (2\pi)^{-n/2}(\sigma_Z)^{-n} \exp(-\sum_{t=1}^{n} z_t^2/2\sigma_Z^2).$$

Since

$$Z_t = X_t - a_1 X_{t-1} - \cdots - a_p X_{t-p} - b_1 Z_{t-1} - \cdots - b_q Z_{t-q},$$

we can write the likelihood for the model parameters $(\boldsymbol{\theta}, \sigma_Z^2)$, where

$$\boldsymbol{\theta} := (a_1, \ldots, a_p, b_1, \ldots, b_q),$$

as

$$L(\boldsymbol{\theta}, \sigma_Z^2 | \mathbf{x}_t, \mathbf{Z}_t) = (2\pi)^{-n/2}(\sigma_Z)^{-n} \exp(-\frac{1}{2\sigma_Z^2} \sum_{t=\max(p,q)+1}^{n} z_t^2(\boldsymbol{\theta} | \mathbf{x}_t, z_{t-1})) \quad (4.4)$$

with

$$z_t^2(\boldsymbol{\theta} | \mathbf{x}_t, z_{t-1}) = (a_1 x_{t-1} - \cdots - a_p x_{t-p} - b_1 z_{t-1} - \cdots - b_q z_{t-q})^2.$$

The residuals in the likelihood have to be calculated iteratively as functions of the parameters and the data, therefore, by fixing $z^* := (z_1, z_2, \ldots, z_{\max p,q})$ at their respective expected values, namely at $E(z) = 0$, we get the conditional likelihood

$$L(\boldsymbol{\theta} | \mathbf{x}, z^*) = -(2\pi)^{-n/2}(\sigma_Z)^{-n} \exp(-\frac{1}{2\sigma_Z^2} S(\boldsymbol{\theta} | \mathbf{x}, z^*)),$$

where

$$S(\boldsymbol{\theta} | \mathbf{x}, z^*) = \sum_{t=\max(p,q)+1}^{n} z_t^2(\boldsymbol{\theta} | \mathbf{x}, z^*),$$

is the conditional sums of squares.

The conditional sum of squares (equivalently the conditional likelihood) can then be minimized by using standard numerical methods to obtain the estimates for $\boldsymbol{\theta}$. The innovation variance σ_Z^2 is then estimated by

$$\hat{\sigma}_Z^2 = \frac{1}{n - 2p - q} S(\hat{\boldsymbol{\theta}} | \mathbf{x}, z^*).$$

Here $n - 2p - q$ represents the degrees of freedom resulting from fitting $p + q$ number of parameters, using $n - p$ observations.

Asymptotically, the conditional likelihood estimators have the same properties of the estimators obtained from the full likelihood. However, estimators can be obtained by maximizing the full (unconditional) likelihood. Assume that (X_t) is a Gaussian time series, not necessarily stationary, with mean 0 and ACF $r(i,j) := E(X_i X_j)$. Let $\mathbf{X}_n := (X_1, \ldots, X_n)$ and set $\hat{X}_n := E(X_n|\mathbf{X}_{n-1})$ and $\hat{\mathbf{X}}_n := (\hat{X}_1, \ldots, \hat{X}_n)$ (set $\hat{X}_1 \equiv X_1$). Denote by $\upsilon_n := E(X_{n+1} - \hat{X}_{n+1})^2$, the variance of the one-step ahead prediction errors. Since for linear models, one-step ahead prediction errors U_n are the innovations (Z_t), we have $\upsilon_n = \sigma_Z^2$. Therefore, we can write the likelihood in (4.4) as

$$L(\boldsymbol{\theta}|\mathbf{x}) = \frac{1}{(2\pi)^{n/2}}(\upsilon_0 \times \upsilon_1 \times \cdots \times \upsilon_{n-1})^{-1/2} \exp(-\frac{1}{2}\sum_{j=1}^{n}(X_j - \hat{X}_j)^2/\upsilon_{j-1}).$$

This intuitive argument can be shown rigorously by starting from the density function of $(X_1, \ldots, X_n)$. Let $\boldsymbol{\gamma} := E(\mathbf{X}'\mathbf{X})$, be the (non-singular) covariance matrix of $\mathbf{X}$. The likelihood is then given by

$$L(\boldsymbol{\gamma}|\mathbf{x}) = \frac{1}{(2\pi)^{n/2}}(\det\boldsymbol{\gamma})^{-1/2} \exp(-\frac{1}{2}\mathbf{X}'\boldsymbol{\gamma}^{-1}\mathbf{X}). \tag{4.5}$$

Since for each linear ARMA(p,q) model, we know the exact form of the covariance matrix $\boldsymbol{\gamma}$ in terms of the model parameters (direct maximization of the likelihood can be done by numerical methods). However, writing $\boldsymbol{\gamma}$ and its inverse in terms of the model parameters is not practical. But the likelihood can be written in terms of the one-step prediction errors $X_t - \hat{X}_t$. Since the time series X_t is linear, the mean square predictor $\hat{X}_n$ is a linear function of $\mathbf{X}_{n-1}$; therefore the one-step prediction errors $\mathbf{U}_n = (U_1, \ldots, U_n)$ are linear functions of $\mathbf{X}_n$. Let $\mathbf{U}_n = \mathbf{A}_n\mathbf{X}_n$. Here, $\mathbf{A}_n$ is of the lower triangular form

$$\mathbf{A}_n = \begin{pmatrix} 1 & 0 & 0 & \cdots 0 \\ a_{11} & 1 & 0 & \cdots 0 \\ a_{22} & 1 & 0 & \cdots 0 \\ a_{n-1,n-1} & a_{n-1,n-2} & a_{n-1,n-3} & \cdots 1 \end{pmatrix}.$$

Note that $\mathbf{A}_n$ is non-singular with

$$\mathbf{C}_n \equiv \mathbf{A}_n^{-1} = \begin{pmatrix} 1 & 0 & 0 & \cdots 0 \\ \theta_{11} & 1 & 0 & \cdots 0 \\ \theta_{22} & 1 & 0 & \cdots 0 \\ \theta_{n-1,n-1} & \theta_{n-1,n-2} & \theta_{n-1,n-3} & \cdots 1 \end{pmatrix}.$$

Hence

$$\hat{\mathbf{X}}_n = \mathbf{X}_n - \mathbf{U}_n = \mathbf{C}_n\mathbf{U}_n - \mathbf{U}_n,$$

so that for $n = 1, 2, \ldots,$

$$\hat{X}_{n+1} = \sum_{j=1}^{n} \theta_{nj} (X_{n+1-j} - \hat{X}_{n+1-j}).$$

Since

$$\mathbf{A}_n \mathbf{X}_n = \mathbf{U}_n = \mathbf{X}_n - \hat{\mathbf{X}}_n,$$

we see that

$$\mathbf{X}_n = \mathbf{C}_n (\mathbf{X}_n - \hat{\mathbf{X}}_n).$$

Note that $\mathbf{U}_n$ are one-step ahead prediction errors, which under linearity, are the uncorrelated innovations $\boldsymbol{Z}_n = (Z_1, \ldots, Z_n)$, therefore the covariance matrix $\mathbf{D}_n$ of $\mathbf{U}_n$ is a diagonal matrix $\mathrm{diag}(\upsilon_0, \ldots, \upsilon_{n-1})$. Since $\mathbf{X}_n = \mathbf{C}_n \mathbf{U}_n$, the covariance function $\boldsymbol{\gamma}$ of $\mathbf{X}_n$ can be written as

$$\boldsymbol{\gamma}_n = \mathbf{C}_n' \mathbf{D}_n \mathbf{C}_n,$$

with

$$\det \boldsymbol{\gamma}_n = (\det \mathbf{C}_n)^2 (\det \mathbf{D}_n) = \prod_{j=0}^{n-1} \upsilon_j$$

and since $\mathbf{C}_n$ is lower triangular,

$$\begin{aligned} \mathbf{X}_n' \boldsymbol{\gamma}^{-1} \mathbf{X}_n &= (\mathbf{X}_n - \hat{\mathbf{X}}_n)' \mathbf{D}_n^{-1} (\mathbf{X}_n - \hat{\mathbf{X}}_n) \\ &= \sum_{j=1}^{n} (X_j - \hat{X}_j)^2 / \upsilon_{j-1}. \end{aligned}$$

Hence, the likelihood in (4.5) can be written as

$$L(\boldsymbol{\gamma}|\mathbf{x}) = \frac{1}{\sqrt{(2\pi)^n \upsilon_0 \cdots \upsilon_{n-1}}} \exp(-\frac{1}{2} \sum_{j=1}^{n} (X_j - \hat{X}_j)^2 / \upsilon_{j-1}). \qquad (4.6)$$

One-step predictors $\hat{X}_{i+1}$ and their corresponding mean square errors υ_i can be calculated recursively by using the innovation algorithm (see e.g., Brockwell and Davis 1996) in terms of the model parameters through the covariance function. The likelihood can then be maximized numerically. If X_t is not Gaussian, maximizing the Gaussian likelihood (4.6) with respect to the model parameters to obtain estimators still makes sense. Estimating functions give justification for obtaining estimators using this Gaussian likelihood for non-Gaussian data.

4.4.2 Estimating Functions

In point estimation, the focus is on a function of the data, i.e., statistics, called estimators, having desirable properties such as unbiasedness, consistency, sufficiency and we search for the optimal estimator among these subclasses. Optimality is often defined in terms of maximization or minimization of some objective function. Minimum variance unbiased estimators which are built on the concept of sufficiency and obtained from Rao-Blackwell or Lehmann-Scheffé theorems have the most desired properties, but they only exist for certain families of distributions. Two methods of estimation, namely the methods of least squares and maximum likelihood (ML), are the most widely used methods of point estimation. The least squares method is widely applicable under the assumption of existence of the first two moments, whereas the maximum likelihood method needs the full specification of joint distributions, although there are variants such as composite likelihood methods which do not need the full specification of the joint distributions. ML estimators are generally optimal or asymptotically optimal, whereas least squares method is only optimal when the underlying distribution is Normal. Estimating functions or the method of quasi-likelihood permit the unification of these methods of estimation under a very general setting. Estimating functions are functions of data as well as the model parameters, and desirable characteristics are given in terms of these functions, rather than on the estimators. Typically, the focus is on functions $h(\mathbf{x}, \boldsymbol{\theta})$ of the data $\mathbf{x}$ and the model parameters $\boldsymbol{\theta}$, and the study is then centered in finding estimators as solutions of

$$h(\mathbf{x}, \boldsymbol{\theta}) = 0.$$

Estimating functions already play a very strong role in least squares and ML methods; for example the maximization of the likelihood passes through finding the roots of the score function, which is a function of the data and the parameters, and hence is an estimating function. The least squares estimators are obtained by finding the solutions of a system of equations which are obtained by differentiating the sum of squares of errors with respect to the model parameters, which again is an estimating function. For example, if we have a time series of size n from the AR(1) model

$$X_t = aX_{t-1} + Z_t, \; Z_t \sim N(0, \sigma^2),$$

one can obtain the least squares estimate for a by minimizing

$$\sum_{t=2}^{n}(x_t - ax_{t-1})^2,$$

for all possible a. Under the normality assumption is equivalent to maximizing the (conditional) likelihood function. The estimator is found by solving the equation

$$h(\mathbf{x}_t, a) = \sum_{t=2}^{n} x_t x_{t-1} - a\sum_{t=2}^{n} x_{t-1}^2 = 0,$$

which results in the estimate

$$\hat{a} = \sum_{t=2}^{n} x_t x_{t-1} / \sum_{t=2}^{n} x_{t-1}^2.$$

Most of the effort is then given to studying the properties of the estimator $\hat{\boldsymbol{\theta}}$ of $\boldsymbol{\theta}$. The method of estimating functions concentrates on studying the properties of the function $h(\mathbf{x}, \boldsymbol{\theta})$ rather than the estimator $\hat{\boldsymbol{\theta}}$ of $\boldsymbol{\theta}$. At first sight, by looking at this example, the benefits of such inferential strategy may not be clear. However, when the full likelihood for the model parameters is not known and due to the deviations from normality, the quality of least squares estimators is in doubt, then as we will see the method of estimating functions appears as the right approach. This is particulary relevant when dealing with different types of nonlinear time series models for which the likelihood functions cannot be fully specified. In this section, we give a brief description the theory of estimating functions. For the general treatment of estimation functions, see Godambe (1991), Heyde (1997) and Bera et al. (2006).

We start by the scalar parameter case. Assume that $(X_1, \ldots, X_n)$ is a time series with joint distribution function $F(\mathbf{x}, \theta)$, $\theta \in \Theta \subset \mathbb{R}$ and let $h(\mathbf{x}, \theta)$ be a measurable real valued function of the sample and the model parameter.

Definition 4.4.1. $h(\mathbf{x}, \theta)$ is called a regular unbiased estimating function if

1. $E(h(\mathbf{x}, \theta)) = \int h(\mathbf{x}, \theta) f(\mathbf{x}, \theta) d\mathbf{x} = 0$, where f is the density function of $\mathbf{X} := (X_1, \ldots, X_n)$;
2. $\frac{\partial h(\mathbf{x},\theta)}{\partial \theta}$ exists for all $\theta \in \boldsymbol{\Theta}$;
3. $\frac{\partial \int h(\mathbf{x},\theta) f(\mathbf{x},\theta) d\mathbf{x}}{\partial \theta} = \int \frac{\partial h(\mathbf{x},\theta) f(\mathbf{x},\theta)}{\partial \theta} d\mathbf{x}$;
4. $E[(\frac{\partial h(\mathbf{x},\theta)}{\partial \theta})^2] > 0$;
5. $V(h(\mathbf{x}, \theta)) < \infty$.

Let $\mathcal{H}$ be the class of all unbiased estimating functions. Note that when θ is the true value, $h(\mathbf{x}, \theta)$ should be as near as possible to zero and this requires $V(h(\mathbf{x}, \theta)) = E[(h^2(\mathbf{x}, y)]$ to be as small as possible. Hence the optimal estimating function $h^*(\mathbf{x}, \theta)$ should satisfy

$$V(h^*(\mathbf{x},\theta)) \leq V(h(\mathbf{x},\theta)),$$

for any $h \in \mathcal{H}$. Note also that, h^* should be sensitive to deviations from the true value of the parameter. Namely, $h^*(\mathbf{x},\theta+\delta)$ should differ from its expected value $E(h(\mathbf{x},\theta)) = 0$ as much as possible whenever $|\delta| > 0$. That is, $[E(\partial h^*/\partial\theta)]^2$ should be as large as possible. Hence, for any $h \in \mathcal{H}$,

$$[E(\partial h^*(\mathbf{x},\theta)/\partial\theta)]^2 \geq [E(\partial h(\mathbf{x},\theta)/\partial\theta)]^2.$$

These two arguments can be joined together to give the following definition of optimality.

Definition 4.4.2. $h^* \in \mathcal{H}$ is said to be optimal if

$$\frac{V(h^*(\mathbf{x},\theta))}{[E(\partial h^*(\mathbf{x},\theta)/\partial\theta)]^2} \leq \frac{V(h(\mathbf{x},\theta))}{[E(\partial h(\mathbf{x},\theta)/\partial\theta)]^2},$$

for all $h \in \mathcal{H}$ and $\theta \in \boldsymbol{\Theta}$.

The corresponding optimal estimator $\hat{\theta}$ based on the time series $\mathbf{x}_n := (x_1,\ldots,x_n)$ is then found by solving for θ the equation

$$h^*(\mathbf{x}_n,\theta) = 0.$$

Let $h_s(\mathbf{x},\theta) = h(\mathbf{x},\theta)/E(\partial h(\mathbf{x},\theta)/\partial\theta)$ be the normalized estimating function. Then the unbiased estimating function h^* is optimum if

$$V(h_s^*(\mathbf{x},\theta)) \leq V(h_s(\mathbf{x},\theta)),$$

for all $h \in \mathcal{H}$. There is an alternative intuitive argument to justify the definition of optimality in terms of the normalized estimating function rather than the optimizing function itself: for any constant $k \neq 0$, two unbiased estimating functions h and $g = kh$ will result in the same estimator. However, they have variances $V(h)$ and $k^2V(h)$, respectively and the latter can be made arbitrarily small (or large). Therefore the comparison of two non-normalized estimating functions based on their variances is not meaningful. On the other hand, with the proposed standardization we have $V(h_s) = V(g_s)$.

Let

$$S := \frac{\partial \log f(\mathbf{x},\theta)}{\partial\theta},$$

be the score function. The following theorem due to Godambe (1960) says that the score function is the optimal estimating function.

Theorem 4.4.1. *For all $h \in \mathcal{H}$,*

$$V(h_s(\mathbf{x}, \theta)) \geq \frac{1}{[E(\frac{\partial \log f(\mathbf{x},\theta)}{\partial \theta})]^2},$$

and the equality is attained by the estimating function

$$h^*(\mathbf{x}, \theta) = \frac{\partial \log f(\mathbf{x}, \theta)}{\partial \theta}.$$

The proof follows the same line of arguments for getting the Cramér-Rao lower bound. For simplicity, dropping the arguments in the respective functions and replacing the partial differential operator ∂ by d, $E(h) = 0$ implies that

$$\frac{d \int hf d\mathbf{x}}{d\theta} = \int \frac{dh}{d\theta} f d\mathbf{x} + \int h \frac{d \log f}{d\theta} f d\mathbf{x} = 0,$$

so that

$$Cov(h, S) = -E[\frac{dh}{d\theta}]. \tag{4.7}$$

Hence

$$\begin{aligned}\left[E(\frac{dh}{d\theta})\right]^2 &= \left[-Cov\left(h, \frac{d \log f}{d\theta}\right)\right]^2 \\ &\leq \text{Var}(h)\text{Var}\left(\frac{d \log f}{d\theta}\right),\end{aligned}$$

so that

$$V(h_s) = \frac{V(h)}{[E(\frac{dh}{d\theta})]^2} \geq \frac{1}{(E(\frac{d \log f}{d\theta}))^2}.$$

This result has several consequences. When the score function is known, it is the optimal estimating function; therefore the estimators obtained from the optimal estimating function and the ML estimators coincide. When the score function is not known, choosing the optimal estimating function by minimizing the variance of the normalized unbiased estimating function corresponds to choosing the estimating function which has the highest correlation with the score function. Let S be the score function. It follows from (4.7) that

$$\begin{aligned} Corr(h,S) &= \frac{[E(h,S)]^2}{E(h^2)E(S^2)} \\ &= \frac{(E(\frac{dh}{d\theta}))^2}{E(h^2)E(S^2)} \\ &= \frac{1}{V(h_s)V(s)}. \end{aligned}$$

Therefore choosing the normalized estimating function with the minimum variance is equivalent to choosing the estimating function which maximizes the correlation with the score function. Equivalently, we choose h in such a manner that the normalized estimating function minimizes the distance from the score function

$$h_s := \arg\inf_h E[(S-h)^2].$$

A useful interpretation of optimal estimating functions is given in terms of the Hilbert space setting: Consider again $\mathcal{L}^2$ norm and the Hilbert space of $\mathcal{L}^2(\Omega, \mathcal{F}, P)$ of r.v's with finite second-order moments, and let $\mathcal{H}$ be a closed subspace of this space. Let $h(X|\mathcal{H})$ be the orthogonal projection of X onto $\mathcal{H}$, so that

$$||X - h(X|\mathcal{H})|| = \inf_{h\in\mathcal{H}} ||X-h|| = \inf_{h\in\mathcal{H}} E(X-h)^2.$$

Specifically, if X is the score function, then the unbiased optimal estimating function, which is the orthogonal projection of the score function onto the chosen space of estimating functions, is the optimal estimating function. For these reasons, inference based on optimal estimating functions are called quasi-score or quasi-likelihood methods.

Now assume that X_t is a time series having finite dimensional distributions $F(\mathbf{x}, \boldsymbol{\theta})$, where $\boldsymbol{\theta} := (\theta_1, \ldots, \theta_p) \in \boldsymbol{\Theta} \subset \mathbb{R}^p$. Let

$$\mathbf{h}(\mathbf{x}, \boldsymbol{\theta}) := (h_1(\mathbf{x}, \boldsymbol{\theta}), \ldots, h_p(\mathbf{x}, \boldsymbol{\theta}))',$$

where each $h_1(\mathbf{x}, \boldsymbol{\theta})$ satisfies the conditions of regularity in Definition 4.4.1 for each θ_i. Let Σ_h be the $p \times p$ covariance matrix

$$\Sigma_h = V(\mathbf{h}(\mathbf{x}, \boldsymbol{\theta})\mathbf{h}(\mathbf{x}, \boldsymbol{\theta})') = E(\mathbf{h}(\mathbf{x}, \boldsymbol{\theta})\mathbf{h}'(\mathbf{x}, \boldsymbol{\theta}))$$

and let

$$H_h = E\left(\frac{\partial h(\mathbf{x}, \boldsymbol{\theta})}{\partial \boldsymbol{\theta}}\right),$$

where

$$\frac{\partial \mathbf{h}(\mathbf{x}, \boldsymbol{\theta})}{\partial \boldsymbol{\theta}} = (\frac{\partial h_j(\mathbf{x}, \boldsymbol{\theta})}{\partial \theta_i}, j = 1, \ldots, p, i = 1, \ldots, p),$$

is the $p \times p$ matrix of partial derivatives of jth estimating function with respect to the ith parameter θ_i, so that H_h is a $p \times p$ non-singular matrix. The normalized estimating function is then written as a $p \times 1$ vector

$$\mathbf{h}_s := H_h^{-1}\mathbf{h}(\mathbf{x}, \boldsymbol{\theta}),$$

with covariance matrix $\Sigma_s := V(\mathbf{h}_s)$. The optimal estimating function then can be defined as

Definition 4.4.3. $\mathbf{h}^*$ is said to be optimal if for any other unbiased regular estimating function vector $\mathbf{h}$, $V(\mathbf{h}_s^*) \leq V(\mathbf{h}_s)$.

Interpretation of this definition is not as straightforward as its scalar counterpart since $V(\mathbf{h}_s) = \Sigma_s$ is a $p \times p$ matrix and there are many ways one can compare matrices. For example, we may say that $V(\mathbf{h}_s^*) \leq V(\mathbf{h}_s)$ if for all $\mathbf{h}$,

- $V(\mathbf{h}_s^*) - V(\mathbf{h}_s)$ is nonnegative definite;
- $\text{Trace}(V(\mathbf{h}_s^*)) \leq \text{Trace}(V(\mathbf{h}_s))$;
- $|V(\mathbf{h}_s^*)| \leq V(\mathbf{h}_s)$.

However, all these optimality criteria are equivalent provided that an optimal estimating function exists; see Bera et al. (2006) and Heyde (1997) for details. The existence of an optimal estimating function is not guaranteed and the choice of optimal estimating functions in general is not straightforward except for certain classes of estimating functions. One such class of estimating functions, which is relatively easy to verify, is the class of estimating functions with orthogonal differences; see Heyde (1997) for details. This observation is the basis of the class of linear estimating functions given by Godambe (1985).

How can we apply the notion of estimating functions to find optimal estimators for nonlinear time series models? Unfortunately, for a broad class of unbiased estimating functions, it is not possible to find a unique optimal estimating function, but the general theory highlighted above can be used, provided we choose a good but restricted class of unbiased estimating functions from which we choose the optimal estimating function. We now define one such class. Assume that $(x_1, \ldots, x_n)$ are observations from a time series X_t with finite-dimensional distribution functions $F(\mathbf{x}, \boldsymbol{\theta})$. Godambe (1985) suggests starting from the elementary estimating functions $h_t(\mathbf{x}, \boldsymbol{\theta})$ which are martingale differences, that is,

$$E(h_t|\mathcal{F}_{t-1}) = 0, \ t = 1, \ldots, n,$$

where $\mathcal{F}_{t-1}$ is the σ-algebra generated by $(x_s, s \leq t-1)$. Such estimating functions are orthogonal in the sense that $E(h_t h_s) = 0$, for $t \neq s$. Clearly, one-step ahead prediction errors

$$X_t - E(X_t|x_1, \ldots, x_{t-1}),\ t = 1, \ldots, n,$$

have the properties of desired elementary estimating functions. Godambe (1985) suggests restricting the class of unbiased regular estimating functions to the linear combination of these elementary estimating functions. Let

$$\mathcal{H} := \{h : h(\mathbf{x}, \boldsymbol{\theta}) = \sum_{t=1}^{n} c_{t-1} h_t(\mathbf{x}, \boldsymbol{\theta})\}. \tag{4.8}$$

Here, the coefficients c_{t-1} are functions of $x_1, \ldots, x_{t-1}$ and $\boldsymbol{\theta}$. The objective is then to find the optimal estimating function and the corresponding estimator $\hat{\boldsymbol{\theta}}$. For this linear class of estimating functions, this optimization reduces to the optimal choice of the coefficients c_{t-1}. The following theorem gives us the optimal estimating function for the class given by (4.8).

Theorem 4.4.2. *Let $\mathcal{H}$ be the class of estimating functions given in (4.8). Then the optimal estimating function h^* within this class is given by*

$$h^*(\mathbf{x}, \boldsymbol{\theta}) = \sum_{t=1}^{n} c^*_{t-1} h_t(\mathbf{x}, \boldsymbol{\theta}),$$

where

$$\begin{aligned} c^*_{t-1} \equiv c^*_{t-1}(\mathbf{x}, \boldsymbol{\theta}) &= \frac{E(\frac{\partial h_t(\mathbf{x},\boldsymbol{\theta})}{\partial \boldsymbol{\theta}}|\mathcal{F}_{t-1})}{E(h_t^2|\mathcal{F}_{t-1})} \\ &= \left(\frac{E(\frac{\partial h_t(\mathbf{x},\boldsymbol{\theta})}{\partial \theta_j}|\mathcal{F}_{t-1})}{E(h_t^2|\mathcal{F}_{t-1})}, j = 1, \ldots, p \right). \end{aligned}$$

Note that the construction depends on the martingale differences $X_t - E(X_t|\mathcal{F}_{t-1})$ and one-step ahead prediction errors $E((X_t - E(X_t|\mathcal{F}_{t-1}))^2|\mathcal{F}_{t-1})$. We have seen in (4.6) that the Gaussian likelihood is also written in a similar fashion in terms of martingale differences and the one-step ahead prediction errors and, hence, when the time series is Gaussian, likelihood and quasi-likelihood inference based on optimal estimating functions coincide. This is one justification for the commonly used strategy in large sample estimation problems to make the Gaussian assumption and maximize the corresponding likelihood, even when the Gaussian assumption is not valid for the data set; see Brockwell and Davis (1991, 1996), Hannan (1970) and Heyde (1997).

Example 4.4.1. Consider the AR(1) process

$$X_t = aX_{t-1} + Z_t,$$

with $Z_t \sim WN(0, \sigma^2)$. Then,

$$h_t \equiv h(x_t, a) = x_t - a x_{t-1},\ t = 2, \ldots, n,$$

are $n - 1$ estimating functions with $E(h_t^2 | \mathcal{F}_{t-1}) = \sigma^2$. Let

$$h^* = \sum_{t=2}^{n} c_{t-1}^* h_t,$$

be the optimal estimating function. Then

$$\begin{aligned} c_{t-1}^* &= \frac{E(\frac{\partial h_t(\mathbf{x},a)}{\partial a} | \mathcal{F}_{t-1})}{E(h_t^2 | \mathcal{F}_{t-1})} \\ &= -x_{t-1}/\sigma^2. \end{aligned}$$

Solving for a in

$$h_t^*(\mathbf{x}, a) = \sum_{t=2}^{n} -x_{t-1}/\sigma^2 (x_t - a x_{t-1}) = 0,$$

gives

$$\hat{a} = \sum_{t=2}^{n} x_{t-1} x_t \Big/ \sum_{t=2}^{n} x_t^2.$$

If we consider the two parameter AR(1) with non-zero mean

$$X_t - \mu = a(X_t - \mu) + Z_t,$$

then we use the $n - 1$ basic unbiased estimating functions

$$h_t(\mathbf{x}, \mu, a) = (x_t - \mu) - a(x_{t-1} - \mu),\ \ t = 2, \ldots, n.$$

Note that again $E(h_t^2 | \mathcal{F}_{t-1}) = \sigma^2$, and the optimal coefficients are given by

$$\begin{aligned} c_{t-1}^*(\mathbf{x}, \mu, a) &= (c_{t-1}^*(\mathbf{x}, \mu), c_{t-1}^*(\mathbf{x}, a)) \\ &= \left(\frac{E(\frac{\partial h_t(\mathbf{x},\mu,a)}{\partial \mu} | \mathcal{F}_{t-1})}{E(h_t^2 | \mathcal{F}_{t-1})}, \frac{E(\frac{\partial h_t(\mathbf{x},\mu,a)}{\partial a} | \mathcal{F}_{t-1})}{E(h_t^2 | \mathcal{F}_{t-1})} \right) \\ &= \left(-x_{t-1}/\sigma^2, -(1-a)/\sigma^2 \right). \end{aligned}$$

Now solving the set of equations

$$-1/\sigma^2 \sum_{t=2}^{n} (x_{t-1} - \mu)(x_t - \mu) - a(x_{t-1} - \mu) = 0,$$

$$-(1-a)/\sigma^2 \sum_{t=2}^{n} (x_t - \mu) - a(x_{t-1} - \mu) = 0,$$

we get

$$\hat{\mu} = \frac{\sum_{t=2}^{n} x_t \sum_{t=2}^{n} x_{t-1}^2 - \sum_{t=2}^{n} x_{t-1} \sum_{t=2}^{n} x_t x_{t-1}}{(n-1)\sum_{t=2}^{n} x_{t-1}^2 - (\sum_{t=2}^{n} x_{t-1})^2(1-\hat{a})},$$

$$\hat{a} = \frac{\sum_{t=2}^{n} x_t \sum_{t=2}^{n} x_{t-1} - \sum_{t=2}^{n}(n-1)\sum_{t=2}^{n} x_t x_{t-1}}{(\sum_{t=2}^{n} x_{t-1})^2 - (n-1)\sum_{t=2}^{n} x_{t-1}^2}.$$

Note that these estimates coincide with the ML estimates conditional on $X_1 = x_1$, when Z_t is Gaussian noise.

Example 4.4.2. Consider the AR(p) model

$$X_t = a_1 X_{t-1} + \cdots + a_p X_{t-p} + Z_t$$

and the unbiased basic estimating functions

$$h_t(\mathbf{x}, \mathbf{a}) = x_t - a_1 x_{t-1} - \cdots - a_p x_{t-p}, \; t = p+1, \ldots, n.$$

One-step ahead prediction errors are given by σ^2 (see Chaps. 1 and 2), so that for each t,

$$E(h_t^2 | \mathcal{F}_{t-1}) = \sigma^2.$$

The optimal estimating function is given by

$$h_t^*(\mathbf{x}, \mathbf{a}) = \sum_{t=p+1}^{n} c_{t-1}^* h_t,$$

where

$$c_{t-1}^*(\mathbf{x}, a_j) = E(\frac{\partial h_t}{\partial a_j})/\sigma^2 = -x_{t-j}/\sigma^2.$$

Then the optimal estimates for a_j are obtained by solving the system of p equations

$$-1/\sigma^2 \sum_{t=p+1}^{n} x_{t-j}(x_t - a_1 x_{t-1} - \cdots - a_p x_{t-p}) = 0.$$

Again, these estimators are equivalent to ML estimators conditional on $(X_1, \ldots, X_p) = (x_1, \ldots, x_p)$ under the assumption of Gaussian error structure.

In these examples, due to the linearity of the models, the system of equations that we need to solve are simple, having analytical solutions. First, the one-step ahead prediction errors are constant being equal to the error variance. Second, the derivatives are nice linear functions of the data, again facilitating the estimating equations. In general, for nonlinear models, no such easy construction exists. Often solutions have to be found numerically and/or asymptotic arguments have to be brought in for approximations. Still, the estimating functions facilitate the construction of good estimators in the absence of a likelihood function, which is often the case for many nonlinear processes. We now give some examples on how estimating functions can be used in obtaining estimators for nonlinear time series. We follow Thavaneswaran and Abraham (1988) and Heyde (1997).

Example 4.4.3 (Random coefficient autoregressive model). Consider the AR(p) process

$$X_t = \sum_{i=1}^{p} a_{t,i} X_{t-i} + Z_t, \tag{4.9}$$

where Z_t are i.i.d. zero-mean r.v's having variance σ_Z^2. The random parameters $a_{t,i}$ are given independent random shocks at each time point, so that

$$a_{t,i} = a_i + W_{t-i}.$$

Here, the random shocks W_{t-i} are zero-mean independent random sequences, independent of each other and independent of Z_t, having common variance (for convenience in calculations) σ_W^2. Consider the estimating functions

$$\sum_{t=1}^{n} c_{i,t-1} h_t,$$

where $h_t := X_t - E(X_t|\mathcal{F}_{t-1})$. Then by Theorem 4.4.2, the optimal estimating function for a_i is given by

$$h_i^* = \sum_{t=1}^{n} c_{i,t-1}^* h_t,$$

where

$$c_{i,t-1}^{*} = \frac{E(\frac{\partial h_t}{\partial a_i}|\mathcal{F}_{t-1})}{E(h_t^2|\mathcal{F}_{t-1})}.$$

It follows from independence assumptions that

$$\begin{aligned} h_t &= \sum_{i=1}^{p} [a_i X_{t-i} + E(W_{t-i}|\mathcal{F}_{t-1})X_{t-i}] + E(Z_t|\mathcal{F}_{t-1}) \\ &= X_t - \sum_{i=1}^{p} a_i X_{t-i} \end{aligned}$$

and

$$\begin{aligned} E(h_t^2|\mathcal{F}_{t-1}) &= E([X_t - \sum_{i=1}^{p} a_i X_{t-i}]^2|\mathcal{F}_{t-1}) \\ &= E([Z_t + \sum_{i=1}^{p} W_{t-i} X_{t-1}]^2|\mathcal{F}_{t-1}) \\ &= \sigma_Z^2 + \sum_{i=1}^{p} X_{t-1}^2 \sigma_W^2, \end{aligned}$$

so that

$$c_{i,t-1}^{*} = \frac{-X_{t-i}}{\sigma_Z^2 + \sum_{i=1}^{p} X_{t-1}^2 \sigma_W^2}.$$

The optimal estimate for $\boldsymbol{\theta}_n := (a_i;\, i = 1, \ldots, p)$ is then obtained by solving i equations simultaneously

$$h_i^* = \sum_{t=1}^{n} c_{i,t-1}^{*} h_t = 0,$$

giving

$$\hat{\boldsymbol{\theta}}_n = \frac{\sum_{t=p}^{n} \mathbf{x}_{t-1}\mathbf{x}_t'/v_t}{\sum_{t=p}^{n} \mathbf{x}_{t-1}\mathbf{x}_{t-1}'/v_t},$$

where

$$\mathbf{x}_{t-1} := (x_{t-1}, x_{t-2}, \ldots, x_{t-p}),$$

and $v_t = \sigma_Z^2 + \sum_{i=1}^{p} x_{t-1}^2 \sigma_W^2$ are the one-step ahead prediction errors. The solution for the special AR(1) case

$$X_t = (a + W_t)X_{t-1} + Z_t,$$

is given by

$$\hat{\theta}_n = \frac{\sum_{t=2}^{n} x_{t-1}x_t/(\sigma_Z^2 + x_{t-1}^2\sigma_W^2)}{\sum_{t=2}^{n} x_{t-1}^2/(\sigma_Z^2 + x_{t-1}^2\sigma_W^2)}.$$

For the AR(1) case, Nicholls and Quinn (1980) (see also Tjøstheim 1986) give the estimator for $\tilde{\theta}_n$

$$\tilde{\theta}_n = \frac{\sum_{t=2}^{n} x_{t-1}x_t}{\sum_{t=2}^{n} x_{t-1}^2}.$$

The estimator $\hat{\theta}_n$ is a weighted version of $\tilde{\theta}_n$ and has the same properties of consistency and asymptotic normality. However, it is more efficient. Note that σ_Z^2 and σ_W^2 are nuisance parameters, which are generally unknown. Yet, initial estimates can be given; see Nicholls and Quinn (1980) for details.

Example 4.4.4 (First-order random coefficient autoregressive model). Consider the AR(1) version of (4.9) given by Heyde (1997) with different error assumptions. Let

$$X_t = (\theta + W_t)X_{t-1} + Z_t,$$

where $E(W_t|\mathcal{F}_{t-1}) = 0$, $E(Z_t|\mathcal{F}_{t-1}) = 0$ and $E(Z_t^2|\mathcal{F}_{t-1}) = \sigma^2 X_{t-1}$. The alternative representation for X_t is given in the form

$$X_t = \theta X_{t-1} + u_t,$$

$$u_t = W_t X_{t-1} + Z_t.$$

The optimal estimating function based on n observations is given as

$$h_n^* = \sum_{t=1}^{n} c_{t-1}^*(X_t - \theta X_{t-1}),$$

where

$$c_{t-1}^* = \frac{X_{t-1}}{E(u_t^2|\mathcal{F}_{t-1})}.$$

Here

$$E(u_t^2|\mathcal{F}_{t-1}) = X_{t-1}^2 E(W_t^2|\mathcal{F}_{t-1}) + 2X_{t-1}E(W_t Z_t|\mathcal{F}_{t-1}) + E(Z_t^2|\mathcal{F}_{t-1}),$$

may not be tractable without assuming additional structure for W_t and Z_t. For example, if we further assume that W_t and Z_t are independent sequences and $E(W_t|\mathcal{F}_{t-1}) = \sigma_W^2$, then

$$E(u_t^2|\mathcal{F}_{t-1}) = \sigma_W^2 X_{t-1}^2 + \sigma_Z^2 X_{t-1},$$

and the optimal estimating function is given by

$$h_n^* = \sum_{t=2}^{n} \frac{X_{t-1}}{\sigma_W^2 X_{t-1}^2 + \sigma_Z^2 X_{t-1}} (X_t - \theta X_{t-1}),$$

so that

$$\hat{\theta}_n = \sum_{t=2}^{n} \frac{x_t x_{t-1}}{\sigma_W^2 x_{t-1}^2 + \sigma_Z^2 x_{t-1}} \Big/ \sum_{t=1}^{n} \frac{x_{t-1}^2}{\sigma_W^2 x_{t-1}^2 + \sigma_Z^2 x_{t-1}}.$$

This expression depends on the nuisance parameters σ_W^2 and σ_Z^2, but if $X_t \to \infty$, then $\hat{\theta}_n$ has the asymptotic expression

$$\hat{\theta}_n = \frac{1}{n} \sum_{t=2}^{n} \frac{x_t}{x_{t-1}}.$$

See Heyde (1997) for the concept of asymptotic quasi-likelihood methods.

Example 4.4.5 (Doubly stochastic time series). Consider the process

$$X_t = a_t f(t, \mathcal{F}_{t-1}) + Z_t,$$

where a_t is a stochastic sequence, $f(t, \mathcal{F}_{t-1})$ is a measurable function with respect to the σ-field $\mathcal{F}_{t-1} := (X_{t-1}, X_{t-2}, \dots)$, and Z_t is an i.i.d., zero-mean random sequence with variance σ_Z^2. In order to be more precise, let us assume a specific form for the random sequence a_t by assuming that

$$a_t = a + W_t + W_{t-1},$$

where (W_t) is a i.i.d. zero-mean sequence, independent of Z_t, having variance σ_W^2. Hence,

$$X_t = af(t, \mathcal{F}_{t-1}) + W_t f(t, \mathcal{F}_{t-1}) + W_{t-1} f(t, \mathcal{F}_{t-1}) + Z_t,$$

so that from the independence assumptions

$$\begin{aligned} h_t &= X_t - E(X_t|\mathcal{F}_{t-1}) \\ &= af(t,\mathcal{F}_{t-1}) + E(W_t|\mathcal{F}_{t-1})f(t,\mathcal{F}_{t-1}) \\ &\quad + E(W_{t-1}|\mathcal{F}_{t-1})f(t,\mathcal{F}_{t-1}) + E(Z_t|\mathcal{F}_{t-1}) \\ &= (a + m_{t-1})f(t,\mathcal{F}_{t-1}), \end{aligned}$$

where $m_{t-1} = E(W_{t-1}|\mathcal{F}_{t-1})$. Also

$$\begin{aligned} E(h_t^2|\mathcal{F}_{t-1}) &= E((X_t - E(X_t|\mathcal{F}_{t-1})^2|\mathcal{F}_{t-1})) \\ &= E((Z_t + (W_t + W_{t-1} - m_{t-1})f(t,\mathcal{F}_{t-1}))^2|\mathcal{F}_{t-1}) \\ &= \sigma_Z^2 + f^2(t,\mathcal{F}_{t-1})E(W_t + W_{t-1} - m_{t-1}|\mathcal{F}_{t-1}) \\ &= \sigma_Z^2 + f^2(t,\mathcal{F}_{t-1})[\sigma_W^2 - \gamma_{t-1}], \end{aligned}$$

where

$$\begin{aligned} \gamma_{t-1} &= E((W_{t-1} - m_{t-1})^2|\mathcal{F}_{t-1}) \\ &= V(Z_{t-1}|\mathcal{F}_{t-1}). \end{aligned}$$

Hence the optimal estimate for a is given as a solution to

$$h_t^* = \sum_{t=2}^{n} c_{t-1}^* h_t = 0,$$

where

$$\begin{aligned} c_{t-1}^* &= \frac{E(\frac{\partial h_t}{\partial a})}{E(h_t^2|\mathcal{F}_{t-1})} \\ &= \frac{f(t,\mathcal{F}_{t-1})(1 + \frac{\partial m_{t-1}}{\partial a})}{\sigma_Z^2 + f^2(t,\mathcal{F}_{t-1})[\sigma_W^2 - \gamma_{t-1}]}, \end{aligned}$$

from which we obtain

$$\hat{a} = \frac{\sum_{t=2}^{n} c_{t-1}^* x_t}{\sum_{t=2}^{n} c_{t-1}^* f(t,\mathcal{F}_{t-1})}.$$

Here, m_{t-1}, $\partial m_{t-1}/\partial a$ and γ_{t-1} need to be calculated iteratively. If $x_0 = 0$, then $m_0 = 0$, $\gamma_0 = \sigma_W^2$ and m_t, γ_t satisfy the following recursive relations (see Thavaneswaran and Abraham 1988):

$$m_t = \frac{\sigma_W^2 f(t, \mathcal{F}_{t-1})[x_t - (a + m_{t-1}) f(t, \mathcal{F}_{t-1})]}{\sigma_Z^2 + f^2(t, \mathcal{F}_{t-1})[\sigma_W^2 - \gamma_{t-1}]},$$

$$\gamma_t = \sigma_W^2 + \frac{f^2(t, \mathcal{F}_{t-1})\sigma_W^4}{\sigma_Z^2 + f^2(t, \mathcal{F}_{t-1})[\sigma_W^2 - \gamma_{t-1}]}.$$

Note that γ_t does not depend on the parameter a, therefore

$$\frac{\partial m_t}{\partial \theta} = \frac{-[\sigma_W^2 f^2(t, \mathcal{F}_{t-1})(1 - \frac{\partial m_{t-1}}{\partial a})]}{\sigma_Z^2 + f^2(t, \mathcal{F}_{t-1})[\sigma_W^2 - \gamma_{t-1}]},$$

and consequently $\partial m_t / \partial a$ can be calculated recursively along with m_t and γ_t. This estimator depends on the nuisance parameters σ_W^2 and σ_Z^2, and these parameters will have to be estimated using a different set of data. Conditional least squares estimators are given by (see, Tjøstheim 1986)

$$\tilde{a} = \frac{\sum_{t=2}^{n} f(t, \mathcal{F}_{t-1})(1 + \frac{\partial m_{t-1}}{\partial a}) X_t}{\sum_{t=2}^{n} [f(t, \mathcal{F}_{t-1})(1 + \frac{\partial m_{t-1}}{\partial a})]}.$$

Again, $\hat{a}$ (given in page 149) is a weighted version of the least squares estimator $\tilde{a}$, using the information in the nuisance parameters σ_W^2 and σ_Z^2.

Example 4.4.6 (General recursive estimation). Consider the general nonlinear time series model given by (see Aase 1983)

$$X_t = a f(t-1, \mathcal{F}_{t-1}) + Z_t, \tag{4.10}$$

where f is a nonlinear measurable function of $\mathcal{F}_{t-1}$, which does not depend on the parameter a and (Z_t) are i.i.d., zero-mean r.v's with variance σ_Z^2. Then, based on estimating functions, it is possible to give a recursive scheme for estimating the parameter a; see Thavaneswaran and Abraham (1988) or Heyde (1997) for details. Set

$$h_t := X_t - a f(t-1, \mathcal{F}_{t-1}),$$

the optimal estimating function for n observations is given by

$$h_t^* = \sum_{t=1}^{n} c_{t-1}^* h_t,$$

where

$$c_{t-1}^{*} = \frac{E(\frac{\partial h_t}{\partial a}|\mathcal{F}_{t-1})}{E(h_t^2|\mathcal{F}_{t-1})}$$
$$= f(t-1,\mathcal{F}_{t-1})/\sigma_Z^2,$$

so that the optimal parameter estimator for a, based on the n observations, is given by

$$\hat{a}_n = \frac{\sum_{t=1}^{n} c_{t-1}^{*} X_{t-1}}{\sum_{t=1}^{n} c_{t-1}^{*} f(t-1,\mathcal{F}_{t-1})}.$$

Writing

$$K_n^{-1} = \sum_{t=1}^{n} c_{t-1}^{*} f(t-1,\mathcal{F}_{t-1}),$$

$$\hat{a}_n - \hat{a}_{n-1} = K_n\left(\sum_{t=1}^{n} c_{t-1}^{*} X_t - \hat{a}_{n-1} K_n^{-1}\right) \tag{4.11}$$

and after some calculations, (4.11) can be put in the form

$$\hat{a}_n = \hat{a}_{n-1} + \frac{K_{n-1} c_n^{*}}{1 + f(n-1,\mathcal{F}_{n-1}) c_n^{*} K_{n-1}} (X_n - \hat{a}_{n-1} f(n-1,\mathcal{F}_{n-1})).$$

Starting from initial values $\hat{a}_0$ and K_0, it is then possible to obtain the optimal estimator $\hat{a}_n$ recursively. The above recursive method of estimation can be generalized for a class of general nonlinear models given by

$$X_t = g(t-1,\mathcal{F}_{t-1}) + a f(t-1,\mathcal{F}_{t-1}) + \sigma(t-1,\mathcal{F}_{t-1}) Z_t.$$

See Aase (1983) and Thavaneswaran and Abraham (1988) for further details.

Example 4.4.7 (Threshold autoregressive model). Consider the process

$$X_t = \sum_{j=1}^{p} a_j I_j(X_{t-1}) X_{t-1} + Z_t,$$

where (Z_t) are zero-mean uncorrelated r.v's with variance σ_Z^2 and $I_j(\cdot)$ are indicator functions given by

$$I_j(X_{t-1}) = \begin{cases} 1, & X_{t-1} \in A_j \\ 0, & \text{otherwise} \end{cases},$$

for some disjoint regions of the sample space $\mathbb{R}$, such that $\cup_j A_j = \mathbb{R}$. Then

$$h_t = X_t - E(X_t|\mathcal{F}_{t-1})$$
$$= X_t - \sum_{j=1}^{p} a_j I_j(X_{t-1})X_{t-1},$$

and

$$E(h_t^2|\mathcal{F}_{t-1}) = \sigma_Z^2.$$

The optimal estimator for a_j is given by

$$\hat{a}_j = \sum_{t=2}^{n} x_t x_{t-1} I_j(x_{t-1}) / \sum_{t=2}^{n} x_{t-1}^2 I_j(x_{t-1}).$$

Note that these estimators are equal to the least squares estimators and ML estimators under normality assumptions (Tjøstheim 1986), since the weights $E(h_t^2|\mathcal{F}_{t-1})$ are constant.

Example 4.4.8 (First-order bilinear model). Consider the simple bilinear model

$$X_t = bX_{t-1}Z_{t-1} + Z_t,$$

where (Z_t) are i.i.d., zero-mean r.v's with finite variance σ_Z^2. We also assume that the process is invertible so that Z_{t-1} is measurable with respect to the σ-field $\mathcal{F}_{t-1} = (X_{t-1}, X_{t-2}, \ldots)$. Note that

$$E(X_t|\mathcal{F}_{t-1}) = bm_{t-1}x_{t-1},$$

where $m_{t-1} := E(Z_{t-1}|\mathcal{F}_{t-1})$, so that the one-step ahead prediction errors are given by $E(h_t^2|\mathcal{F}_{t-1}) = \sigma_Z^2$. The optimal estimating function for b is given by

$$h_t^* = \sum_{t=2}^{n} c_{t-1}^* [X_t - bX_{t-1}m_{t-1}], \tag{4.12}$$

where

$$c_{t-1}^* = \frac{-x_{t-1}m_{t-1} - b\frac{\partial m_{t-1}}{\partial b}}{\sigma_Z^2}.$$

Hence,

$$\hat{b} = \frac{\sum_{t=2}^{n} c_{t-1}^* x_t}{\sum_{t=2}^{n} c_{t-1}^* x_{t-1} m_{t-1}}.$$

Note that c_{t-1}^* depends on b and there is no easy solution for $\hat{b}_n$, even in recursive form. However,

$$m_{t-1} = b x_{t-2} m_{t-2},$$

$$\frac{\partial m_{t-1}}{\partial b} = -x_{t-2} m_{t-2} + b x_{t-2} \frac{\partial m_{t-2}}{\partial b},$$

therefore, starting from initial values b_0, $m_0 = Z_0$, and $\frac{\partial m_0}{\partial b} = 0$, m_n, $\frac{\partial m_n}{\partial b}$ and $\hat{b}$ can be calculated iteratively. Conditional least squares estimator for b is obtained by solving the equation

$$\sum_{t=2}^{n} (X_t - \theta X_{t-1} m_{t-1})(-X_{t-1} m_{t-1} - \theta X_{t-1} \frac{\partial m_{t-1}}{\partial b}) = 0. \tag{4.13}$$

By comparing (4.12) and (4.13) we see that quasi-likelihood estimator obtained from the optimal estimating function is equivalent to the conditional least squares estimator.

Example 4.4.9 (General bilinear model). Consider the bilinear model BL(p, q, m, l) defined in (2.25) where (Z_t) are i.i.d. zero-mean r.v's with variance σ_Z^2. We further assume that the model is invertible. We again consider optimal estimating functions based on elementary estimating functions

$$h_t = X_t - E(X_t|\mathcal{F}_{t-1}) = X_t - \sum_{i=1}^{p} a_t x_{t-i} + \sum_{j=1}^{q} c_j m_{t-j} + \sum_{i=1}^{m} \sum_{j=1}^{l} b_{ij} x_{t-i} m_{t-j}, \tag{4.14}$$

where $m_{t-j} = E(Z_{t-j}|\mathcal{F}_{t-1})$, $j = 1, 2, \ldots, q$. The optimal estimating functions are given by the following set of equations:

$$h_{t,a_i}^* = \sum_{t=k+1}^{n} c_{t-1,a_i}^* h_t,$$

$$h_{t,c_j}^* = \sum_{t=k+1}^{n} c_{t-1,c_j}^* h_t,$$

$$h_{t,b_{ij}}^* = \sum_{t=k+1}^{n} c_{t-1,b_{ij}}^* h_t, \tag{4.15}$$

where $k := \max(p, q, m, l)$. Note that the invertible process is of the form

$$X_t = f(\mathcal{F}_{t-1}) + Z_t,$$

where f is a $\mathcal{F}_{t-1}$-measurable function, hence one-step ahead prediction errors $E(h_t^2|\mathcal{F}_{t-1})$ are constant and equal to σ_Z^2. Then

$$\begin{aligned} c^*_{t-1,a_i} &= \frac{E(\frac{\partial h_t}{\partial a_j}|\mathcal{F}_{t-1})}{\sigma_Z^2} \\ &= \frac{-X_{t-i}}{\sigma_Z^2}, \end{aligned}$$

$$\begin{aligned} c^*_{t-1,c_j} &= \frac{E(\frac{\partial h_t}{\partial c_j}|\mathcal{F}_{t-1})}{\sigma_Z^2} \\ &= \frac{-m_{t-j} - c_j \frac{\partial m_{t-j}}{\partial c_j} - \sum_{s=1}^{l} b_{ij} x_{t-s} \frac{\partial m_{t-j}}{\partial c_j}}{\sigma_Z^2}, \end{aligned}$$

$$\begin{aligned} c^*_{t-1,b_{ij}} &= \frac{E(\frac{\partial h_t}{\partial b_{ij}}|\mathcal{F}_{t-1})}{\sigma_Z^2} \\ &= \frac{-x_{t-i} m_{t-j} - \theta_j \frac{\partial m_{t-j}}{\partial b_{ij}} - b_{l_1 l_2} x_{t-i} \frac{\partial m_{t-j}}{\partial b_{l_1 l_2}}}{\sigma_Z^2}. \end{aligned}$$

Optimal estimators of the parameters

$$\boldsymbol{\theta} := (a_1, \ldots, a_p, c_1, \ldots, c_q, b_{11}, \ldots, b_{ml}),$$

can be obtained iteratively by finding simultaneously the zeroes of the optimal estimating functions given in (4.15) using numerical methods. Let $\mathbf{G}(\boldsymbol{\theta})$ be the set of all estimating functions given in (4.15). Then, the solution for

$$\mathbf{G}(\boldsymbol{\theta}) = 0,$$

can be found iteratively using the Newton-Raphson method by

$$\hat{\boldsymbol{\theta}}^{(r+1)} = \hat{\boldsymbol{\theta}}^{(r)} - \left(\frac{\partial \mathbf{G}(\boldsymbol{\theta}^{(r)})}{\partial \boldsymbol{\theta}}\right)^{-1} \mathbf{G}(\boldsymbol{\theta}^{(r)}),$$

where $\frac{\partial \mathbf{G}(\boldsymbol{\theta}^{(r)})}{\partial \boldsymbol{\theta}}$ is the observed matrix of partial derivatives of the estimating functions with respect to the parameters. The success of the method depends heavily on the choice of the initial values. There may be multiple solutions for

$$\mathbf{G}(\boldsymbol{\theta}) = 0.$$

See Heyde (1997) for possible ways of discriminating between the multiple roots when they exist. Note that the estimators obtained from estimating functions are equivalent to the estimators obtained from the conditional least squares method. Similar results can be obtained for the extended bilinear model given in (2.26). Note that for these models $E(h_t^2|\mathcal{F}_{t-1})$ is no longer constant. For example, if

$$X_t = aX_{t-1} + bX_{t-1}Z_{t-1} + cX_{t-1}Z_t + Z_t,$$

then

$$E(h_t^2|\mathcal{F}_{t-1}) = (1 + c^2 x_{t-1}^2)\sigma^2$$

and the optimal estimating function and the corresponding estimators given in (4.12) can be adjusted accordingly.

4.4.3 Composite Likelihood Methods

We have seen in the previous section that when the likelihood does not have an analytical form or the likelihood requires prohibitive computations, it may be feasible to look for other pseudo-likelihood methods, such as the method of estimating functions for a specific choice of estimating functions. Composite likelihood is a pseudo-likelihood inference and is based on combinations of likelihoods for small subsets of the data or based on combinations of conditional likelihoods. Following Varin et al. (2011), let $\mathbf{X}$ be a n-dimensional vector of r.v's with density function $p(\mathbf{x}|\boldsymbol{\theta})$. Denote by $(\mathcal{A}_1, \ldots, \mathcal{A}_K)$ a set of marginal or conditional events with associated likelihoods $\mathcal{L}_k(\boldsymbol{\theta}|\mathbf{x}) \propto p(\mathbf{x} \in \mathcal{A}_k|\boldsymbol{\theta})$. A composite likelihood is the weighted product

$$\mathcal{L}_c(\boldsymbol{\theta}|\mathbf{x}) = \prod_{k=1}^{K} \mathcal{L}_k(\boldsymbol{\theta}|\mathbf{x})^{w_k},$$

where w_k are some nonnegative weights. Often these weights are taken to be equal, hence they can be ignored. The above definition is very flexible and allows for combinations of marginal and conditional densities. However, composite likelihood based on pairwise likelihoods, defined as the product of the bivariate likelihood of all possible pairs of observations, are often used due to their simplicity. Hence for the n observations $\mathbf{x} := (x_1, \ldots, x_n)$, the pairwise likelihood is defined as

$$\mathcal{L}_2(\boldsymbol{\theta}|\mathbf{x}) = \prod_{i<j}^{n} p(x_i, x_j|\boldsymbol{\theta}),$$

where $p(x_i, x_j | \boldsymbol{\theta})$ is the joint density of the r.v's X_i and X_j. Consequently, this pairwise likelihood can be seen as the likelihood based on $n(n-1)/2$ independent samples of bivariate observations. As such, composite likelihood is a pseudo-likelihood method based on a misspecified model. For example, the simplest composite likelihood is based on the one dimensional densities constructed under the full independence assumption

$$\mathcal{L}_1(\boldsymbol{\theta} | \mathbf{x}) = \prod_{i=1}^{n} p(x_i | \boldsymbol{\theta}).$$

In this case, composite likelihood permits inference on marginal parameters, but it contains no information on the parameters that govern the dependence structure. Thus, a compromise between loss of information and computational feasibility has to be made while choosing the right composite likelihood for the problem at hand, capturing some, but not, all features of the full likelihood. For time series models based on n observations $\mathbf{x}$, simplified versions of the pairwise likelihood of consecutive pairs of observations are often used, due to the fact that most of the dependence occurs in adjacent observations; therefore such pseudo-likelihoods have mathematical and computational tractability without loss of too much information on the dependence structure of the series.

Pairwise log-likelihood based on successive pairs of observations is given by

$$\log \mathcal{L}_c(\boldsymbol{\theta} | \mathbf{y}) = \sum_{j=1}^{n-1} \sum_{t=1}^{n-j} \log p(x_t, x_{t+j} | \boldsymbol{\theta}),$$

and the simplified kth order consecutive pairwise log-likelihood is given as

$$\log \mathcal{L}_k(\boldsymbol{\theta} | \mathbf{y}) = \sum_{j=1}^{k} \sum_{t=1}^{n-j} \log p(x_t, x_{t+j} | \boldsymbol{\theta}),$$

and in particular the first order consecutive pairwise log-likelihood is given as

$$\log \mathcal{L}_1(\boldsymbol{\theta} | \mathbf{y}) = \sum_{t=1}^{n-1} \log p(x_t, x_{t+1} | \boldsymbol{\theta}).$$

Note that the score function of the composite likelihood is the sum of the score functions based on the corresponding marginal densities. Hence, in the full likelihood, the estimators are found by solving the score function equation

$$S(\boldsymbol{\theta} | \mathbf{x}) = 0,$$

whereas in the composite likelihood case, the estimators are found by solving the composite score function

$$S_c(\boldsymbol{\theta}, \mathbf{x}) = \sum_k \frac{\partial \log \mathcal{L}_k(\boldsymbol{\theta}|\mathbf{x})}{\partial \boldsymbol{\theta}} = 0,$$

which is a linear combination of the scores associated with each of the likelihood term $\mathcal{L}_k(\boldsymbol{\theta}|\mathbf{x})$. Hence the composite likelihood method is a special case for estimating functions where the scores based on marginal likelihoods are taken as the unbiased estimating functions, as compared to the estimating functions based on one-step ahead prediction errors considered in the previous section. However, this does not guarantee satisfactory properties of the resulting estimator. An overview of the composite likelihood methods can be found in Varin et al. (2011); see Davis and Yau (2011) for the use of consecutive pairwise composite likelihood for linear time series models.

For general state space models, which are arguably most suitable representations for nonlinear time series, composite likelihood methods based on pairwise likelihood can result in robust estimation of the parameters. We have seen in Sect. 2.2.7 that a generalized state space model is specified by a system of equations given respectively by

$$\begin{aligned} Y_n &= f_1(X_n, Z_n) \\ X_n &= f_2(X_{n-1}, w_n) \end{aligned}$$

where X_0, (Z_n) and (w_n) are mutually independent r.v's and $f_1(\cdot)$ and $f_2(\cdot)$ are measurable functions. Often X_n is taken to be a first-order Markov process and the observations Y_n are assumed to be conditionally independent given the states X_n. Such state space models are often called latent Markov processes and the process (Y_n, X_n) is specified by

$$\{p(x_0|\boldsymbol{\theta}), p(x_n|x_{n-1}, \boldsymbol{\theta}), p(y_n|x_n, \boldsymbol{\theta})\},$$

where $\boldsymbol{\theta}$ is an unknown vector of parameters to be estimated from the data. As was explained in (2.47) and (2.48), the likelihood function depends on complicated recursive calculations based on n-dimensional integrals

$$\mathcal{L}(\boldsymbol{\theta}|\mathbf{y}) = \int \cdots \int p(x_0|\boldsymbol{\theta}) \prod_{i=1}^{n} p(x_i|x_{i-1}, \boldsymbol{\theta}) p(y_i|x_i, \boldsymbol{\theta}) dx_1 \cdots dx_n. \tag{4.16}$$

Expressions (2.47) and (2.48) explain how the recursive calculations may be done, although the solutions still require the calculation of n-dimensional integrals and except for special cases (for example when f_1 and f_2 are linear functions, in which case the Kalman filter provides an elegant solution) exact and efficient methods for calculating and maximizing the likelihood are not available. In the next section we give Bayesian Monte Carlo methods for obtaining numerical simulation based methods for calculating and maximizing the likelihood given in (4.16). Composite likelihood methods based on pairwise likelihood, and in particular pairwise likelihoods of consecutive observations, can reduce the computational burden significantly, replacing the calculation of n-dimensional integrals by the calculation of two-dimensional integrals. The composite likelihood $\mathcal{L}_C(\boldsymbol{\theta}|\mathbf{y})$ based on consecutive observations is given by

$$\mathcal{L}_C(\boldsymbol{\theta}|\mathbf{y}) = \prod_{i=1}^{n-1} p(y_i, y_{i+1}|\boldsymbol{\theta}),$$

with

$$\begin{aligned} p(y_i, y_{i+1}|\boldsymbol{\theta}) &= \int\int p(y_i, y_{i+1}, x_i, x_{i+1}|\boldsymbol{\theta})dx_i dx_{i+1} \\ &= \int\int p(x_i|\boldsymbol{\theta})p(x_{i+1}|x_i, \boldsymbol{\theta})p(y_{i+1}|y_i, x_{i+1}, \boldsymbol{\theta}) \\ &\quad \times p(y_i|x_i, \boldsymbol{\theta})dx_i dx_{i+1}. \end{aligned}$$

Although the integral still has to be calculated numerically and some of the conditional densities may not have explicit analytical expressions, this pseudo-likelihood is written in terms of bivariate integrals, resulting in more robust methods due to the lack of error propagation that is often present in sequential MCMC methods which we will study in next section. Hence such methods not only reduce the computational burden, but can result in more robust estimation. Note that this pseudo-likelihood method also avoids the miss-specification of higher-order dimensional distributions. However, we still need to know the properties of the estimators obtained when the full likelihood is substituted by a pairwise pseudo-likelihood based on a misspecified model.

Asymptotic Properties of Composite Likelihood Estimators

Here we give a very brief summary of the asymptotic properties of the composite likelihood estimators. Details can be found in Varin et al. (2011). If we have a composite likelihood based on m-dimensional marginal distributions and have n i.i.d. samples from this m-dimensional distributions, then the log composite likelihood can be written as

$$\begin{aligned} \log \mathcal{L}_C(\boldsymbol{\theta}|\mathbf{y}) &= \sum_{i=1}^{n} \log \mathcal{L}_m(\boldsymbol{\theta}|\mathbf{y}_i) \\ &= \sum_{i=1}^{n} \log p_m(\mathbf{y}_i|\boldsymbol{\theta}), \end{aligned}$$

and hence the score function is based on the marginal likelihoods. Under fairly reasonable regularity conditions on the component log densities, one may expect that, from the central limit theorem for the composite score statistics, the composite maximum likelihood estimator $\hat{\boldsymbol{\theta}}_C$ is unbiased and asymptotically normal with

$$\sqrt{n}(\hat{\boldsymbol{\theta}}_C - \theta) \to^D N(0, G_n^{-1}(\boldsymbol{\theta})),$$

where $G_n^{-1}(\boldsymbol{\theta})$ is called the Godambe information matrix and is given by

$$G_n^{-1}(\boldsymbol{\theta}) = H_n(\boldsymbol{\theta})^{-1} J_n(\boldsymbol{\theta}) H_n(\boldsymbol{\theta})^{-1}$$

with

$$J_n(\boldsymbol{\theta}) = V(\nabla \mathcal{L}_m(\boldsymbol{\theta}\,|\mathbf{y}))$$

and

$$H_n(\boldsymbol{\theta}) = E(\nabla^2 \mathcal{L}_m(\boldsymbol{\theta}\,|\mathbf{y})).$$

Here, $\nabla \mathcal{L}_m(\boldsymbol{\theta}\,|\mathbf{y})$ and $\nabla^2 \mathcal{L}_m(\boldsymbol{\theta}\,|\mathbf{y})$ are respectively the gradient vector and the Hessian matrix of the marginal score functions with respect to the model parameters. The ratio of the $G(\boldsymbol{\theta})$ to the Fisher's information matrix $I(\boldsymbol{\theta})$ may give an idea of the asymptotic efficiency of the composite likelihood estimators. This comparison ideally should be made by analytical calculation of $G(\boldsymbol{\theta})$ and $I(\boldsymbol{\theta})$, but these expressions are rarely available. Alternatively, the comparison can be made either by simulation-based estimates of $G(\boldsymbol{\theta})$ or $I(\boldsymbol{\theta})$. In exceptional cases pairwise likelihood estimators are efficient. In fact, Cox and Reid (2004) report that composite likelihood estimators may even be inconsistent. However, Mardia et al. (2009) show that composite likelihood estimators can give full efficiency in multivariate normal distribution. In fact composite likelihood estimators are fully efficient in exponential families that have certain closure properties. Davis and Yau (2011) look at the relative efficiency of the composite likelihood estimators in linear time series models and show that for AR(1) models, relative efficiency of the composite likelihood estimators based on pairwise likelihood, compared to full likelihood, is equal to one. In contrast, they show that the relative efficiency for MA(1) processes is disappointingly low, particularly considering that the correlations in MA(1) do not extend beyond lag 1. Thus, it is not very clear, in which time series models the composite likelihood based on pairwise densities performs well. Davis and Yau (2011) also report good performance of the pairwise likelihood applied to a nonlinear model for time series of counts and come to a conclusion that this good performance may suggest that composite likelihood methods may be quite promising for more complicated nonlinear models. Davison and Gholamrezaee (2012) also obtain similar encouraging results for composite likelihood methods based on pairwise likelihood applied to max-stable processes. Varin and Vidoni (2009) show that the composite likelihood estimators, based on pairwise likelihood for the generalized state space models, are consistent when the time series exhibit short range dependence, but if the series exhibits long-range dependence, then convergence (asymptotic normality) may be slow, or even the estimators may be inconsistent. These reported good results are possibly due to the robust, simple models and to lack of error propagation often seen in numerical methods in nonlinear processes based on full likelihood methods, which may compensate for the miss-specification that pairwise likelihoods bring into the estimation.

4.5 Bayesian Methods

We have seen in Sect. 2.2.7 that a generalized state space model is characterized by the conditional densities (2.47) and (2.48), which for non-Gaussian and nonlinear systems do not have closed forms. Therefore, inference based on likelihood is in general very difficult to carry out and one needs to resort to simulation-based methods. Bayesian hierarchical models and simulation-based inferential methods, known as Markov Chain Monte Carlo (MCMC) methods, are particularly suited for such complex systems. Here we give a very brief introduction to these methods. For complete and deeper treatment, see Congdon (2010), Prado and West (2010) and Liang et al. (2010). See also Andrieu et al. (2010) and Chopin et al. (2013) for a review of the recent advances in Sequential and Particle MCMC methods.

A generalized state space model is composed of three components, observations $\mathbf{y}_t$, the state or the hidden (latent) process $\mathbf{x}_t$ and the parameters $\boldsymbol{\theta}$. In the Bayesian context, the parameters are also considered as random and the process can totally be characterized by the joint density of $(\mathbf{y}_t, \mathbf{x}_t, \boldsymbol{\theta})$ which can be written componentwise as

$$p(\mathbf{y}_t, \mathbf{x}_t, \boldsymbol{\theta}) = p(\mathbf{y}_t|\mathbf{x}_t, \boldsymbol{\theta})p(\mathbf{x}_t|\boldsymbol{\theta})p(\boldsymbol{\theta}).$$

Here, $p(\mathbf{y}_t|\mathbf{x}_t, \boldsymbol{\theta})$ is the likelihood or the first stage of the hierarchical specification, $p(\mathbf{x}_t|\boldsymbol{\theta})$ is the conditional specification of the latent process or the state process and finally $p(\boldsymbol{\theta})$ is the third stage or the prior specification for the model parameters. Often there is a fourth stage for this hierarchical specification, namely the hyperparameters that may be needed while specifying the prior distribution for the parameters. Target quantities which we are of interested in this hierarchical specification are:

1. The joint posterior density of the parameters and the state

$$p(\boldsymbol{\theta}, \mathbf{x}_t|\mathbf{y}_t) = \frac{p(\mathbf{y}_t|\boldsymbol{\theta}, \mathbf{x}_t)p(\mathbf{x}_t|\boldsymbol{\theta})p(\boldsymbol{\theta})}{p(\mathbf{y}_t)}.$$

Here,

$$p(\mathbf{y}_t) = \int_{\mathbf{x}_t, \boldsymbol{\theta}} p(\mathbf{y}_t|\boldsymbol{\theta}, \mathbf{x}_t)p(\boldsymbol{\theta}, \mathbf{x}_t)d\boldsymbol{\theta}\, d\mathbf{x}_t, \tag{4.17}$$

does not depend on $(\boldsymbol{\theta}, \mathbf{x}_t)$ and hence is a constant. Therefore the joint posterior is proportional to the product of the likelihood and the prior

$$p(\boldsymbol{\theta}, \mathbf{x}_t | \mathbf{y}_t) \propto p(\mathbf{y}_t | \boldsymbol{\theta}, \mathbf{x}_t) p(\mathbf{x}_t | \boldsymbol{\theta}) p(\boldsymbol{\theta}).$$

This formulation is useful since the integral in (4.17), called the constant of proportionality, is in general difficult to calculate. Most simulation-based inferential techniques avoid this constant in calculations.

2. The marginal posterior densities $p(\boldsymbol{\theta} | \mathbf{y}_t)$ and $p(\mathbf{x}_t | \mathbf{y}_t)$, can be obtained from the joint posterior density by integration. Alternatively, the marginal posterior density $p(\boldsymbol{\theta} | \mathbf{y}_t)$ can be obtained from the marginal likelihood

$$p(\mathbf{y}_t | \boldsymbol{\theta}) = \int p(\mathbf{x}_t | \boldsymbol{\theta}) p(\mathbf{y}_t | \mathbf{x}_t, \boldsymbol{\theta}) d\mathbf{x}_t.$$

3. The joint predictive density of future observations is given by

$$p(\mathbf{y}_{t+1}, \mathbf{x}_{t+1} | \mathbf{y}_t) = \int p(\mathbf{y}_{t+1} | \mathbf{y}_t, \boldsymbol{\theta}, \mathbf{x}_{t+1}) p(\boldsymbol{\theta}, \mathbf{x}_{t+1} | \mathbf{y}_t) d\boldsymbol{\theta}, \quad (4.18)$$

from which marginal predictive distributions $p(\mathbf{y}_{t+1} | \mathbf{y}_t)$ and $p(\mathbf{x}_{t+1} | \mathbf{y}_t)$ can be obtained by integration. The future predicted values $\mathbf{y}_{t+1}$ and $\mathbf{x}_{t+1}$ then can be calculated as posterior means $E(\mathbf{y}_{t+1} | \mathbf{y}_t)$ and $E(\mathbf{x}_{t+1} | \mathbf{y}_t)$, respectively.

Often, we write $\boldsymbol{\theta} := (\boldsymbol{\theta}_y, \boldsymbol{\theta}_x)$ to highlight the parameters that are relevant to the observation equation and the state equation, respectively. Then it is generally assumed that the observations are conditionally independent of $\boldsymbol{\theta}_x$ given $\mathbf{x}_t$ so that

$$p(\mathbf{y}_t, \mathbf{x}_t, \boldsymbol{\theta}) = p(\mathbf{y}_t | \mathbf{x}_t, \boldsymbol{\theta}_y) p(\mathbf{x}_t | \boldsymbol{\theta}_x) p(\boldsymbol{\theta}_y | \boldsymbol{\theta}_x) p(\boldsymbol{\theta}_x),$$

which simplifies the hierarchical specification.

4.5.1 Simulation-Based Methods

Simulation-based inferential methods are iterative sampling methods, that generate samples from the target densities, namely the joint posterior distribution $p(\mathbf{x}_t, \boldsymbol{\theta} | \mathbf{y}_t)$ and marginal posterior densities $p(\boldsymbol{\theta} | \mathbf{y}_t)$ and $p(\mathbf{x}_t | \mathbf{y}_t)$. Once sufficient values from the target densities are simulated, posterior summaries such as posterior means and variances of the target densities, as well as credible intervals, can be obtained from these simulations. Simulated samples from the target joint posterior density $p(\mathbf{x}_t, \boldsymbol{\theta} | \mathbf{y}_t)$ will also permit obtaining samples from the joint predictive distribution through the relation (4.18), which will give us predictions for future values, as well as their credible intervals.

For nonlinear and non-Gaussian processes, the joint posterior $p(\boldsymbol{\theta}, \mathbf{x}_t | \mathbf{y}_t)$ as well as marginal posteriors $p(\mathbf{x}_t | \mathbf{y}_t)$ and $p(\boldsymbol{\theta} | \mathbf{y}_t)$ do not have closed form expressions, and sampling directly from them is impossible. It is therefore necessary to approximate target densities by simpler forms, often called proposal densities, which are relatively easy to sample from and yet sufficiently structured to capture the scale, as well as complex dependence structure that exists within the process. Successful MCMC sampling schemes depend in finding such approximate target densities. In small dimensions, this is relatively feasible, but for nonlinear time series it is a challenge. For example, the joint posterior density of $(\mathbf{x}_t, \boldsymbol{\theta})$ has dimension $n + p$, where n is the sample size and p is the dimension of the parameter space. Efficient approximate sampling schemes for such high dimensional densities is now a very active field of research. Sequential and particle MCMC methods are currently being used with success for these highly complex models; see Andrieu et al. (2010) and Chopin et al. (2013) for an overview of the field.

Before we give an overview of these sequential methods for state space models, consider the simpler case when we have a dependent observation $\mathbf{y} := (y_1, \ldots, y_n)$ coming from the model $p(\mathbf{y}|\boldsymbol{\theta})$, where $\boldsymbol{\theta}$ has prior density $p(\boldsymbol{\theta})$. If $p(\boldsymbol{\theta})$ is a conjugate prior, then it is possible to obtain the analytical expression for the marginal posterior density $p(\boldsymbol{\theta}|\mathbf{y}_t)$ and samples can be generated from this target density. Often due to high dimension it may not be feasible to sample directly from this target density and componentwise methods are used to sample from a portion of the parameter space. Assume that the parameter space can be written in components $\boldsymbol{\theta} := (\boldsymbol{\theta}_1, \boldsymbol{\theta}_2, \ldots, \boldsymbol{\theta}_q)$ and let

$$p(\boldsymbol{\theta}_i|\boldsymbol{\theta}_{[-i]}, \mathbf{y}) \propto p(\mathbf{y}|\boldsymbol{\theta}_i)p(\boldsymbol{\theta}_i),$$

be the density of the component $\boldsymbol{\theta}_i$ conditional on the rest of the parameters which we denote by $\boldsymbol{\theta}_{[-i]}$. These densities are called full conditionals. The Gibbs sampling scheme (Gelfand and Smith 1990) draws samples from $\boldsymbol{\theta}$ in a componentwise manner, using the full conditionals as the sampling densities. Thus samples are drawn sequentially in such a way that at iteration $i + 1$,

- $\boldsymbol{\theta}_1^{(i+1)} \sim p(\boldsymbol{\theta}_1|\boldsymbol{\theta}_{[-1]}^{(i)}, \mathbf{y}_t)$
- $\boldsymbol{\theta}_2^{(i+1)} \sim p(\boldsymbol{\theta}_2|\boldsymbol{\theta}_{[-2]}^{(i)}, \mathbf{y}_t)$

  ⋮
- $\boldsymbol{\theta}_q^{(i+1)} \sim p(\boldsymbol{\theta}_q|\boldsymbol{\theta}_{[-q]}^{(i)}, \mathbf{y}_t)$.

Here,

$$\boldsymbol{\theta}_{[-k]}^{(i+1)} := (\boldsymbol{\theta}_1^{(i+1)}, \ldots, \boldsymbol{\theta}_{k-1}^{(i+1)}, \boldsymbol{\theta}_{k+1}^{(i)}, \ldots, \boldsymbol{\theta}_q^{(i)}).$$

There are other variations on this general scheme of Gibbs sampler; see Congdon (2010), for details. Gibbs sampler is a very powerful tool applicable to many statistical models, but it is not particularly useful if we does not have expressions for the full conditionals. In this case, Metropolis-Hasting algorithm is used. Metropolis-Hasting (M-H) algorithm is based on two basic sampling methods:

1. *Acceptance-Rejection methods*: Acceptance-rejection methods are proposed for simulating random numbers from a d-dimensional random variable whose probability structure does not permit direct or efficient simulation of samples from its probability distribution. If $\mathbf{X}$ has a density $p(\mathbf{x})$ and U, conditional on $\mathbf{x}$, is uniform in $(0, p(\mathbf{x}))$, then the surface under the density curve

$$\mathcal{S}_p := \{(\mathbf{x}, u), 0 \leq u \leq p(\mathbf{x})\} \subset \mathbb{R}^{d+1},$$

has a unit volume. Hence, if the random pair $(\mathbf{X}, U)$ is uniformly distributed in this region then $\mathbf{X}$ will have density $p(\mathbf{x})$. If we rescale $p(\mathbf{x})$ by $f(\mathbf{x})$ such that

$$\sup_{\mathbf{x}} p(\mathbf{x}) \leq f(\mathbf{x}) \leq k < \infty, \tag{4.19}$$

and if $(\mathbf{X}, U)$ is uniform in

$$\mathcal{S}_f := \{(\mathbf{x}, u), 0 \leq u \leq f(\mathbf{x})\} \subset \mathbb{R}^{d+1},$$

then $\mathbf{X}$ will still have the marginal density $p(\mathbf{x})$.

If it is difficult to sample from $p(\mathbf{x})$ and if it is possible to find another density $f(\mathbf{x})$ satisfying (4.19), which is easier to sample from, then the above remarks suggest a way of sampling from $p(\mathbf{x})$. Note that $\mathcal{S}_p \subset \mathcal{S}_f$, hence we can sample from $p(\mathbf{x})$ by first getting samples from $\mathcal{S}_f$ and then by considering only those points which fall within $\mathcal{S}_p$. Here $p(\mathbf{x})$ is called the target density, whereas $f(\mathbf{x})$ is called the envelope or proposal density. Thus, based on these arguments, the following algorithm called the acceptance-rejection algorithm can be implemented to sample from $p(\mathbf{x})$:

 (a) Simulate $\mathbf{x}$ from the proposal density $f(\mathbf{x})$, and simulate $U \sim U(0, 1)$ independent of $\mathbf{x}$.
 (b) If

$$U \leq \frac{p(\mathbf{x})}{k f(\mathbf{x})},$$

 then accept $\mathbf{x}$ as a sample from $p(\mathbf{x})$. Otherwise reject $\mathbf{x}$.

 The acceptance rate is the ratio of the volume of the target region $\mathcal{S}_p$ to the volume of the proposal region $\mathcal{S}_f$.

2. *Sampling from a Markov process*
 When it is not feasible to simulate i.i.d. samples from a target distribution $P(d\mathbf{x})$ (here we use the notation $P(d\mathbf{x})$ for the distribution function whereas $p(\mathbf{x})$ for the density, which we assume that exists), it may be possible to simulate dependent samples with relative ease by loosing some efficiency in the simulations. This can be done by using discrete time homogeneous Markov

processes. A Markov process is defined by its initial distribution $P_0(d\mathbf{x})$ and the one-step transition kernel $P_{t+1}(\mathbf{x}, A)$, which represents the probability $P(\mathbf{X}_{t+1} \in A|\mathbf{X}_t = \mathbf{x})$. When the process is homogeneous, the transition kernel does not depend on time, so that $P_{t+1}(\mathbf{x}, A) = P(\mathbf{x}, A)$. In this case, at any time point $t+1$,

$$P_{t+1}(d\mathbf{y}) = P(\mathbf{y} \leq \mathbf{X}_{t+1} \leq \mathbf{y} + d\mathbf{y})$$
$$= \int_{\mathbf{x}} p_t(\mathbf{x}) P_{t+1}(\mathbf{x}, d\mathbf{y}) d\mathbf{x}.$$

Thus, at any time point t, $P_t(\mathbf{y})$ is uniquely determined by the initial distribution $P_0(d\mathbf{x})$ and the transitional kernel $P(\mathbf{x}, d\mathbf{y})$. Under fairly mild conditions (when the Markov process is Harris positive recurrent and aperiodic, that is when it is ergodic; see Meyn and Tweedie 2009)

$$\lim_{t\to\infty} P_t(d\mathbf{y}) = P(d\mathbf{y})$$

and $P(d\mathbf{y})$ is called the limiting or equilibrium distribution of the process which is independent of the initial distribution and satisfies the equality

$$P(d\mathbf{y}) = \int_{\mathbf{x}} p(\mathbf{x}) P(\mathbf{x}, d\mathbf{y}) d\mathbf{x}. \tag{4.20}$$

If $p(\mathbf{y})$ and $p(\mathbf{y}|\mathbf{x})$ are the corresponding densities for $P(d\mathbf{y})$ and $P(\mathbf{x}, d\mathbf{y})$ respectively, then (4.20) can be written as

$$p(\mathbf{y}) = \int_{\mathbf{x}} p(\mathbf{y}|\mathbf{x}) p(\mathbf{x}) d\mathbf{x}.$$

With a given initial distribution and a transition kernel, it is relatively easy to simulate dependent samples from a Markov process. This facility can be used to get independent samples from a target density $p(\mathbf{x})$ by constructing an ergodic Markov process with limiting distribution $p(\mathbf{x})$. If we can construct such a Markov process with known initial distribution and transition kernel, then a sample simulated from this Markov process, upon an initial period of convergence (burn-in) will be a sample from the limiting density $p(\mathbf{x})$, which by construction is our target density. However, for a given limiting density $p(\mathbf{x})$, there is no unique transition kernel, hence the construction of an ergodic Markov process for a given target density is not very simple. Imposing further the condition of reversibility on the Markov process facilitates this construction. Limiting distribution $p(\mathbf{x})$ (and the corresponding ergodic Markov process) is said to be reversible if

$$p(\mathbf{x}) p(\mathbf{y}|\mathbf{x}) = p(\mathbf{y}) p(\mathbf{x}|\mathbf{y}).$$

Kelly (1979) showed that if a Markov process is reversible, then it is also ergodic. It is relatively easy to take samples from a reversible Markov chain by

(a) Specifying a proposal density $p(\mathbf{y}|\mathbf{x})$ which is symmetrical so that $p(\mathbf{y}|\mathbf{x}) = p(\mathbf{x}|\mathbf{y})$.
(b) Carrying out an acceptance-rejection sampling from this proposal density in such a way that the resulting Markov process is reversible and thus ergodic with target stationary distribution $p(\mathbf{x})$.

The Metropolis algorithm is based on this principle:

(a) At any time point t, with $\mathbf{X}_t = \mathbf{x}_t$, sample a candidate value $\mathbf{y}$ from the symmetrical proposal density $p(\mathbf{y}|\mathbf{x}_t)$;
(b) Compute the acceptance ratio

$$\alpha(\mathbf{y}, \mathbf{x}_t) := \min\left\{1, \frac{p(\mathbf{y})}{p(\mathbf{x}_t)}\right\};$$

(c) Set $\mathbf{x}_{t+1} = \mathbf{y}$ with probability $\alpha(\mathbf{y}, \mathbf{x}_t)$ otherwise, set $\mathbf{x}_{t+1} = \mathbf{x}_t$.

This method samples from the ergodic Markov chain with stationary distribution $p(\mathbf{x})$; thus sampled values, after an initial burn-in, can be considered as observations from the stationary distribution. The Metropolis-Hastings (M-H) algorithm generalizes the Metropolis algorithm in a way that permits sampling from non-symmetrical densities, yet still sampling from an ergodic Markov chain with the desired stationary distribution.

The Metropolis-Hastings Algorithm

(a) At any time point t with $\mathbf{X}_t = \mathbf{x}_t$, draw a candidate sample $\mathbf{y}$ from $p(\mathbf{y}|\mathbf{x}_t)$, where $p(\mathbf{y}|\mathbf{x}_t)$ need not be equal to $p(\mathbf{x}_t|\mathbf{y})$;
(b) Compute the ration

$$\alpha(\mathbf{y}, \mathbf{x}_t) = \min\left\{1, \frac{p(\mathbf{y})p(\mathbf{x}_t|\mathbf{y})}{p(\mathbf{x}_t)p(\mathbf{y}|\mathbf{x}_t)}\right\};$$

(c) Set $\mathbf{x}_{t+1} = \mathbf{y}$ with probability $\alpha(\mathbf{y}, \mathbf{x}_t)$, otherwise set $\mathbf{x}_{t+1} = \mathbf{x}_t$.

See Congdon (2010) and the references therein, and in particular Roberts and Rosenthal (2004), for the construction of such Markov process and the resulting acceptance-rejection schemes for sampling.

In the simple time series case when we have dependent observations $\mathbf{y}_t$ from the model $p(\mathbf{y}_t|\boldsymbol{\theta})$, the target distribution from which we want to sample is the posterior distribution $p(\boldsymbol{\theta}|\mathbf{y}_t)$. Except for special cases, it is not possible to sample directly from this target density, therefore the M-H algorithm can be adopted.

By adopting the M-H algorithm, at any iteration $i+1$, a candidate value $\boldsymbol{\theta}^{(c)}$ is sampled from the proposal density $q(\boldsymbol{\theta}^{(c)}|\boldsymbol{\theta}^{(i)})$ and is accepted as the next generation sample $\boldsymbol{\theta}^{(i+1)}$ with probability $\alpha(\boldsymbol{\theta}^{(c)}|\boldsymbol{\theta}^{(i)})$ so that

$$\boldsymbol{\theta}^{(i+1)} = \begin{cases} \boldsymbol{\theta}^{(c)}, \text{ with probability } \alpha(\boldsymbol{\theta}^{(c)}|\boldsymbol{\theta}^{(i)}); \\ \boldsymbol{\theta}^{(i)}, \text{ with probability } 1-\alpha(\boldsymbol{\theta}^{(c)}|\boldsymbol{\theta}^{(i)}), \end{cases}$$

where

$$\alpha(\boldsymbol{\theta}^{(c)}|\boldsymbol{\theta}^{(i)}) = \min\left\{1, \frac{p(\boldsymbol{\theta}^{(c)}|\mathbf{y}_t)q(\boldsymbol{\theta}^{(i)}|\boldsymbol{\theta}^{(c)})}{p(\boldsymbol{\theta}^{(i)}|\mathbf{y}_t)q(\boldsymbol{\theta}^{(c)}|\boldsymbol{\theta}^{(i)})}\right\}.$$

With such a sampling scheme, the transition kernel of the Markov process is given by

$$p(\boldsymbol{\theta}^{(i+1)}|\boldsymbol{\theta}^{(i)}) = \alpha(\boldsymbol{\theta}^{(i+1)}|\boldsymbol{\theta}^{(i)})q(\boldsymbol{\theta}^{(i+1)}|\boldsymbol{\theta}^{(i)}),$$

so that the probability of staying at the current state $\boldsymbol{\theta}^{(i)}$ after an iteration is given by

$$p(\boldsymbol{\theta}^{(i+1)} = \boldsymbol{\theta}^{(i)}|\boldsymbol{\theta}^{(i)}) = 1 - \int_{\boldsymbol{\theta}^{(c)}} \alpha(\boldsymbol{\theta}^{(c)}|\boldsymbol{\theta}^{(i)})q(\boldsymbol{\theta}^{(c)}|\boldsymbol{\theta}^{(i)})d\boldsymbol{\theta}^{(c)}.$$

Such a scheme guarantees sampling from the stationary distribution of the Markov chain, that is from the posterior distribution upon initial burn-in period; see Roberts and Rosenthal (2004) for theoretical justification. When the chosen proposal distribution $q(\boldsymbol{\theta}^{(c)}|\boldsymbol{\theta}^{(i)})$ is symmetric, that is $q(\boldsymbol{\theta}^{(c)}|\boldsymbol{\theta}^{(i)}) = q(\boldsymbol{\theta}^{(i)}|\boldsymbol{\theta}^{(c)})$ then the acceptance probability reduces to

$$\alpha(\boldsymbol{\theta}^{(c)}|\boldsymbol{\theta}^{(i)}) = \min\left\{1, \frac{p(\boldsymbol{\theta}^{(c)}|\mathbf{y}_t)}{p(\boldsymbol{\theta}^{(i)}|\mathbf{y}_t)}\right\},$$

thus we accept the candidate $\boldsymbol{\theta}^{(c)}$ as the next generation sample, if the posterior probability at $\boldsymbol{\theta}^{(c)}$ is larger than the posterior probability calculated at $\boldsymbol{\theta}^{(i)}$. Since

$$p(\boldsymbol{\theta}|\mathbf{y}_t) = \frac{p(\mathbf{y}_t|\boldsymbol{\theta})p(\boldsymbol{\theta})}{p(\mathbf{y}_t)},$$

acceptance of the candidate value $\boldsymbol{\theta}^{(c)}$ as next generation sample is done with probability

$$\min\left\{1, \frac{p(\mathbf{y}_t|\boldsymbol{\theta}^{(c)})p(\boldsymbol{\theta}^{(c)})}{p(\mathbf{y}_t|\boldsymbol{\theta}^{(i)})p(\boldsymbol{\theta}^{(i)})}\right\},$$

Note also that since we compare ratios, the difficult integral $p(\mathbf{y})$ disappears in the comparison, and the M-H is implemented by comparing functions which are non-normalized posterior distributions calculated at $\boldsymbol{\theta}^{(c)}$ and $\boldsymbol{\theta}^{(i)}$.

There are other variations on M-H sampling scheme such as hit-and-run, the Langavin and the multiple-try algorithms. M-H algorithms can also be combined with the Gibbs sampler to give algorithms such as the Metropolis within Gibbs sampler; see Liang et al. (2010) and Congdon (2010) for details.

In order for these algorithms to work efficiently, the choice of the proposal distribution is crucial. The acceptance probability α depends on the variance of the proposal density. If the acceptance rate is too high (consequently smaller proposal variance), then the scheme will be sampling in a restricted zone of the parameter space, which may result in non-convergence to the target distribution. If the acceptance probability is too low, then sampling might get stuck around a fixed value of $\boldsymbol{\theta}^{(i)}$, thus slowing convergence. The proposal distribution is often chosen with some knowledge on how the posterior distribution might look like. The normal distribution is often used as a proposal distribution, since asymptotically most posterior distributions are Normal. We refer to Congdon (2010), and to the references there in for a good choice of acceptance and proposal distributions.

Importance Sampling

Importance sampling is a standard technique for estimating integrals. Suppose that we want to calculate $I = \int f(\mathbf{x})d\mathbf{x}$. If it is not possible to calculate this integral directly, then it may be possible to approximate it by drawing samples from an easy-to-sample density $q(\mathbf{x})$ such that $q(\mathbf{x}) > 0$ whenever $f(\mathbf{x}) > 0$. In this case, the integral I can be written as an expectation with respect to q

$$I = \int \frac{f(\mathbf{x})}{q(\mathbf{x})} q(\mathbf{x})d\mathbf{x} = E_q[w(\mathbf{X})],$$

where

$$w(\mathbf{x}) = \frac{f(\mathbf{x})}{q(\mathbf{x})}.$$

The integral I then can be approximated by

1. Drawing N samples $\mathbf{x}^{(1)}, \ldots, \mathbf{x}^{(N)}$ from $q(\mathbf{x})$,
2. Calculating the empirical mean

$$\hat{I} = \frac{1}{N}\sum_{i=1}^{N} w(\mathbf{x}^{(i)}).$$

$\hat{I}$ is a consistent estimator of I, as $N \to \infty$, and assuming $E_q[w^2(\mathbf{X})] < \infty$,

$$V(\hat{I}) = V_q[w(\mathbf{X})].$$

For generalized state space models, one of our target integrals is the marginal likelihood (most often, an analytical expression does not exist)

$$p(\mathbf{y}_t|\boldsymbol{\theta}) = \int_{\mathbf{x}_t} p(\mathbf{y}_t|\mathbf{x}_t, \boldsymbol{\theta})p(\mathbf{x}_t|\boldsymbol{\theta})d\mathbf{x}_t,$$

calculated at a set of $\boldsymbol{\theta}$ values. Writing

$$w(\mathbf{x}_t) = \frac{p(\mathbf{x}_t|\boldsymbol{\theta})p(\mathbf{y}_t|\mathbf{x}_t, \boldsymbol{\theta})}{q(\mathbf{x}_t)},$$

we can approximate this marginal likelihood by Importance Sampling, provided we have the analytical expressions for $p(\mathbf{x}_t|\boldsymbol{\theta})$ and for the likelihood $p(\mathbf{y}_t|\boldsymbol{\theta}, \mathbf{x}_t)$. This approximation can be done along the following steps:

1. Draw N samples $\mathbf{x}_t^{(1)}, \ldots, \mathbf{x}_t^{(N)}$ from the proposal density $q(\mathbf{x}_t)$.
2. Calculate the weights

$$w(\mathbf{x}_t^{(i)}) = \frac{p(\mathbf{x}_t^{(i)}|\boldsymbol{\theta})p(\mathbf{y}_t|\mathbf{x}_t^{(i)}, \boldsymbol{\theta})}{q(\mathbf{x}_t^{(i)})}, i = 1, \ldots, N.$$

Then, for a fixed value of $\boldsymbol{\theta}$, approximate $p(\mathbf{y}_t|\boldsymbol{\theta})$ by

$$\hat{p}(\mathbf{y}_t|\boldsymbol{\theta}) = \frac{1}{N}\sum_{i=1}^{N} \frac{p(\mathbf{x}_t^{(i)}|\boldsymbol{\theta})p(\mathbf{y}_t|\mathbf{x}_t^{(i)}, \boldsymbol{\theta})}{q(\mathbf{x}_t^{(i)})}.$$

The quality of the approximation $\hat{I}$ for I depends on the choice of the proposal density $q(\mathbf{x})$. The optimal choice, in the sense of reduced variance $V(\hat{I})$, is given by

$$q_{opt}(\mathbf{x}) = p(\mathbf{x})/I,$$

but this is hardly a useful proposal. The design of a proposal density will depend on the process and it is by no means an easy task; see Fearnhead (2008) for ways of designing proposal densities and some examples.

Importance sampling is extensively used as part of sequential MCMC methods for approximating posterior densities for generalized state space models. We now give a brief introduction to these sequential methods.

4.5.2 Sequential MCMC Methods for Generalized State Space Models

Generalized state space models, as explained in Sect. 2.2.7, bring extra difficulties in adopting appropriate MCMC methods. First, the target posterior distribution $p(\boldsymbol{\theta}, \mathbf{x}_t|\mathbf{y}_t)$, or the marginal posterior distributions $p(\boldsymbol{\theta}|\mathbf{y}_t)$ and $p(\mathbf{x}_t|\mathbf{y}_t)$, as well as the likelihood $p(\mathbf{y}_t|\boldsymbol{\theta})$, may not have closed form expressions. Second, the dimension of the parameter space (including the unknown values of the latent state process $\mathbf{x}_t$) is $n + p$, which brings very difficult design as well as computational problems. In this context, efficient sequential and particle MCMC methods (see for example Fearnhead 2008 and Andrieu et al. 2010) can be constructed to sample from the joint posterior density. Standard Sequential MCMC methods (SMCMC) are constructed with the aim of sampling from the marginal posterior density $p(\mathbf{x}_t|\mathbf{y}_t)$ for a fixed value of the parameter $\boldsymbol{\theta}$, whereas particle MCMC (PMCMC) methods are more complicated as they aim to sample from the joint posterior $p(\boldsymbol{\theta}, \mathbf{x}_t|\mathbf{y}_t)$.

These methods are typically implemented for the so called hidden Markov state process, where the latent process $\mathbf{X}_t$ is assumed to be a homogeneous Markov process with transition density $p(x_{t+1}|\mathbf{x}_t, \boldsymbol{\theta}) = p(x_{t+1}|x_t, \boldsymbol{\theta})$, and the observations $\mathbf{Y}_t$ are assumed to be independent conditional on the realizations of the state process, having common marginal density $p(y_t|X_1 = x_1, \ldots, X_t = x_t, \boldsymbol{\theta}) = p(y_t|x_t, \boldsymbol{\theta})$. In this case (assuming that $\mathbf{y}_T := (y_1, y_2, \ldots, y_T)$ is the observed time series), for a fixed value of $\boldsymbol{\theta}^*$

$$\begin{aligned} p(\mathbf{x}_t|\mathbf{y}_t, \boldsymbol{\theta}^*) &\propto p(\mathbf{x}_t, \mathbf{y}_t|\boldsymbol{\theta}^*) \\ &\propto p(x_1|\boldsymbol{\theta}^*) \prod_{t=2}^{T} p(x_t|x_{t-1}, \boldsymbol{\theta}^*) \prod_{t=1}^{T} p(y_t|x_t, \boldsymbol{\theta}^*). \end{aligned}$$

In SMCMC methods, the marginal posterior density $p(\mathbf{x}_t|\mathbf{y}_t, \boldsymbol{\theta}^*)$ is approximated sequentially by first approximating the pair $p(x_1|y_1, \boldsymbol{\theta}^*)$ and $p(y_1|\boldsymbol{\theta}^*)$, then $p(x_1, x_2|y_1, y_2, \boldsymbol{\theta}^*)$ and $p(y_1, y_2|\boldsymbol{\theta}^*)$ and these iterations are continued until the approximate densities for $\hat{p}(\mathbf{x}_T|\boldsymbol{\theta}, \mathbf{y}_T)$ and $\hat{p}(\mathbf{y}_T|\boldsymbol{\theta})$ are obtained. For this sequential approximation, it is more convenient to write

$$\begin{aligned} p(x_1, x_2|y_1, y_2, \boldsymbol{\theta}^*) &\propto p((x_1, y_1), (x_2, y_2)|\boldsymbol{\theta}^*) \\ &\propto p(y_2|x_2, \boldsymbol{\theta}^*) p(x_2|x_1, \boldsymbol{\theta}^*) p(x_1|y_1, \boldsymbol{\theta}^*). \end{aligned}$$

Hence, for any $t \leq T$

$$p(\mathbf{x}_t|\mathbf{y}_t, \boldsymbol{\theta}^*) \propto p(y_t|x_t, \boldsymbol{\theta}^*) p(x_t|x_{t-1}, \boldsymbol{\theta}^*) p(\mathbf{x}_{t-1}|\mathbf{y}_{t-1}, \boldsymbol{\theta}^*). \quad (4.21)$$

The expression in (4.21) clearly suggests that the posterior density $p(\mathbf{x}_t|\mathbf{y}_t, \boldsymbol{\theta}^*), t = 1, \ldots, T$ can be approximated sequentially using the following general guideline:

1. For $t = 1$, let

$$P(dx_1|y_1, \boldsymbol{\theta}^*) = \int_{\{x_1 \leq z \leq x_1+dx_1\}} p(z|y_1, \boldsymbol{\theta}^*)dz,$$

be the posterior distribution corresponding to the posterior density $p(x_1|y_1, \boldsymbol{\theta}^*)$ in the interval $[x_1, x_1+dx_1)$. Approximate this integral using IS with an adequate importance (proposal) density $q(x_1|y_1, \boldsymbol{\theta}^*)$. Note that by definition, for small dx_1, this will give the corresponding approximation for the posterior density $p(x_1|y_1, \boldsymbol{\theta}^*)$, which we denote by $\hat{p}(x_1|y_1, \boldsymbol{\theta}^*)$.
2. IS will produce $x_1^{(k)}$, $k = 1, \ldots, N$ samples or particles with corresponding importance weights $w_1^{(k)}$.
3. The corresponding approximation for the posterior density is given by

$$\hat{p}(x_1|y_1) = P(dx_1|y_1) = \sum_{k=1}^{N} w_1^{(k)} \mathbf{1}_{\{X_1^{(k)} \in [x_1, x_1+dx_1)\}},$$

where,

$$\mathbf{1}_{\{X_1^{(k)} \in [x_1, x_1+dx_1)\}} = \begin{cases} 1, \text{ if } X_1^{(k)} \in [x_1, x_1 + dx_1); \\ 0, \text{ otherwise,} \end{cases}$$

is the indicator function.
4. Further re-sample $x_1^{(k)}$, $k = 1, \ldots, N$ from the approximate density $\hat{p}(x_1|y_1)$ to be used as part of the sampling from $p(x_1, x_2|y_1, y_2, \boldsymbol{\theta}^*)$ in next stage.
5. At stage $t = 2$, aim to approximate the posterior density $p(x_1, x_2|y_1, y_2, \boldsymbol{\theta}^*)$. Since

$$\begin{aligned} p(x_1, x_2|y_1, y_2, \boldsymbol{\theta}^*) &\propto p((x_1, y_1), (x_2, y_2)|\boldsymbol{\theta}^*) \\ &\propto p(x_1|y_1, \boldsymbol{\theta}^*)p(y_2|x_2, \boldsymbol{\theta}^*)p(x_2|x_1, \boldsymbol{\theta}^*), \end{aligned}$$

use the sample $x_1^{(k)}$, $k = 1, \ldots, N$ obtained in the previous stage by re-sampling from the approximate density $\hat{p}(x_1|y_1, \boldsymbol{\theta}^*)$ and extend this to $(x_1^{(k)}, x_2^{(k)})$, $k = 1, \ldots, N$, where $x_2^{(k)}$ is now sampled from an importance sampling proposal density $q(x_2|y_2, x_1)$. The extended sample $(x_1^{(k)}, x_2^{(k)})$, $k = 1, \ldots, N$ is approximately distributed according to $p(x_1|y_1, \boldsymbol{\theta}^*)q(x_2|y_2, x_1, \boldsymbol{\theta}^*)$. Weights $\mathbf{w}_2^{(k)} = (w_1^{(k)}, w_2^{(k)})$, corresponding to this extended sample, are calculated and the approximate posterior density

$$\hat{p}(x_1, x_2|y_1, y_1, \boldsymbol{\theta}^*) = \sum_{k=1}^{N} \mathbf{w}_2^{(k)} \mathbf{1}_{\{\mathbf{x}_2^{(k)} \in [\mathbf{x}_2, \mathbf{x}_2+d\mathbf{x}_2)\}},$$

is obtained. Additional re-sample $(x_1^{(k)}, x_2^{(k)})$, $k = 1, \ldots, N$ is taken from this approximate density to keep as the seed for the next step.

6. For each $2 < t \leq T$, the above sampling, approximation and re-sampling steps are repeated, augmenting the parent sample

$$(x_1^{(k)}, x_2^{(k)}, \ldots, x_{t-1}^{(k)})$$

to

$$(x_1^{(k)}, x_2^{(k)}, \ldots, x_{t-1}^{(k)}, x_t^{(k)}).$$

However, at each step $t > 2$, unpromising samples are eliminated and only promising samples are allowed to propagate to the next generation augmented sample. This elimination is done as follows: at step $t-1$, standard multinomial resampling procedure is carried out in order to retain particles $x_j^{(k)}$, $j = 1, \ldots, t-1$ with larger importance weights and discard those particles which have smaller importance weights. The discarded particles are substituted by others according to the relative sizes of the importance weights.

7. Once the samples $(x_1^{(k)}, x_2^{(k)}, \ldots, x_T^{(k)})$ are available, the approximate densities $\hat{p}(\mathbf{x}_t|\mathbf{y}_t, \theta^*)$ and $\hat{p}(\mathbf{y}_t|\boldsymbol{\theta}^*)$ are calculated as

$$\hat{p}(\mathbf{x}_t|\mathbf{y}_t, \theta^*) = \sum_{k=1}^{N} \mathbf{w}_t^{(k)} \mathbf{1}_{\{\mathbf{x}_t^{(k)} \in [\mathbf{x}_t, \mathbf{x}_t + d\mathbf{x}_t)\}},$$

$$\hat{p}(\mathbf{y}_t|\boldsymbol{\theta}^*) = \hat{p}(y_1|\boldsymbol{\theta}^*) \prod_{t=2}^{T} \hat{p}(y_t|\mathbf{y}_{t-1}, \boldsymbol{\theta}^*),$$

where

$$\hat{p}(y_t|\mathbf{y}_{t-1}, \boldsymbol{\theta}^*) = \frac{1}{N} \sum_{k=1}^{N} w_n(\mathbf{x}_t^{(k)}),$$

which estimates at time $t \leq T$ the conditional density

$$p(y_t|\mathbf{y}_{t-1}, \boldsymbol{\theta}^*) = \int_{\mathbf{x}_t} w_n(\mathbf{x}_t) q(x_t|y_t, x_{t-1}, \boldsymbol{\theta}^*) p(\mathbf{x}_{t-1}|\mathbf{y}_{t-1}, \boldsymbol{\theta}^*) d\mathbf{x}_t.$$

We refer the reader to Andrieu et al. (2010) for the pseudo-code for this sequential Monte Carlo (SMC) algorithm, as well as for an adjusted M-H sampler, called Particle Independent M-H sampler, in which the SMC approximation $\hat{p}(\mathbf{x}_t|\boldsymbol{\theta}, \mathbf{y}_t)$ is used as the proposal density; See also Andrieu et al. (2010) for specific algorithms and design issues.

The main drawback of SMC algorithms is that, when the size T of the vector $\mathbf{x}_t$ is large, approximations for $p(\mathbf{x}_t|\boldsymbol{\theta},\mathbf{y}_t)$ deteriorate very quickly and extension of such algorithms to approximate joint posteriors $p(\boldsymbol{\theta},\mathbf{x}_t|\mathbf{y}_t)$ do not work so well.

Particle MCMC methods are specifically designed to sample from the joint posterior $p(\mathbf{x}_t,\boldsymbol{\theta}_t|\mathbf{y}_t)$. Typically these algorithms use sequential MCMC to sample from the approximate density $\hat{p}(\mathbf{x}_t|\boldsymbol{\theta},\mathbf{y}_t)$ as surrogate proposal density for $p(\mathbf{x}_t|\boldsymbol{\theta},\mathbf{y}_t)$. The particle marginal M-H sampler, given below, is suggested by Andrieu et al. (2010) to combine sequential MCMC and M-H sampler for taking samples from the joint posterior density:

1. Set $i = 0$, sample $\theta^{(0)}$ from the prior density for $\boldsymbol{\theta}$ and run SMCMC to draw samples from $\hat{p}(\mathbf{x}_t|\boldsymbol{\theta}^{(0)},\mathbf{y}_t)$. Sample $\mathbf{x}_t^{(0)}$ from approximate density $\hat{p}(\mathbf{x}_t|\boldsymbol{\theta}^{(0)},\mathbf{y}_t)$. Let $p(\mathbf{y}_t|\boldsymbol{\theta}^{(0)})$ be the marginal likelihood.
2. For any $i \geq 1$, sample the candidate value $\boldsymbol{\theta}^{(c)}$ from the proposal density $q(\boldsymbol{\theta}^{(c)}|\boldsymbol{\theta}^{(i-1)})$ and repeat step (1) with $\boldsymbol{\theta}^{(c)}$ by running SMCMC targeting $p(\mathbf{x}_t|\boldsymbol{\theta}^{(c)},\mathbf{y}_t)$ and then sampling $\mathbf{x}_t^{(c)}$ from $\hat{p}(\mathbf{x}_t|\boldsymbol{\theta}^{(c)},\mathbf{y}_t)$ giving the marginal likelihood $p(\mathbf{y}_t|\boldsymbol{\theta}^{(c)})$.
3. With probability

$$\min\left\{1, \frac{\hat{p}(\mathbf{y}_t|\boldsymbol{\theta}^{(c)})p(\boldsymbol{\theta}^{(c)})}{\hat{p}(\mathbf{y}_t|\boldsymbol{\theta}^{(i-1)})p(\boldsymbol{\theta}^{(i-1)})}\frac{q(\boldsymbol{\theta}^{(i-1)}|\boldsymbol{\theta}^{(c)})}{q(\boldsymbol{\theta}^{(c)}|\boldsymbol{\theta}^{(i-1)})}\right\},$$

set $\boldsymbol{\theta}^{(i)} = \boldsymbol{\theta}^{(c)}$, $\mathbf{x}_t^{(i)} = \mathbf{x}_t^{(c)}$ and $p(\mathbf{y}_t|\boldsymbol{\theta}^{(i)}) = p(\mathbf{y}_t|\boldsymbol{\theta}^{(c)})$. Otherwise, set $\boldsymbol{\theta}^{(i)} = \boldsymbol{\theta}^{(i-1)}$, $\mathbf{x}_t^{(i)} = \mathbf{x}_t^{(i-1)}$ and $p(\mathbf{y}_t|\boldsymbol{\theta}^{(i)}) = p(\mathbf{y}_t|\boldsymbol{\theta}^{(i-1)})$.

It is shown in Andrieu et al. (2010) that, under fairly mild conditions, this sampling scheme is ergodic and, after a sufficient burn-in, samples from the posterior density $p(\mathbf{x}_t,\boldsymbol{\theta}|\mathbf{y}_t)$. See Andrieu et al. (2010) for other sampling schemes, such as Particle Gibbs sampler and improvements, as well as extensions. We note however that due to the high dimensions (for example for $\boldsymbol{\theta}$ we sample from a p dimensional density, whereas for $\mathbf{x}_t$, we sample from a T dimensional density) there are very important computational issues, and the success of the methods depends on very good choice of proposal densities.

SMCMC and PMCMC methods are applicable to inference for nonlinear models, provided that the likelihood functions of these models are tractable. In general, as was explained above, the samples are taken from the posterior density

$$p(\boldsymbol{\theta}|\mathbf{y}_t) = \frac{p(\mathbf{y}_t|\boldsymbol{\theta})p(\boldsymbol{\theta})}{p(\mathbf{y}_t)}.$$

Often the constant $p(\mathbf{y}_t)$ is intractable, but MCMC methods are developed to deal with this case. The SMC and PMCMC methods that we examined in this section fall into this category. However, many nonlinear processes are doubly intractable in the sense that the likelihood $p(\mathbf{y}_t|\boldsymbol{\theta})$ is not available or is very difficult to calculate. Standard MCMC methods cannot be applied to these cases. For example, bilinear models are examples of such situations. For such processes, although the likelihood is intractable, it is relatively easy to simulate sample paths of the process. In such cases, there is a class of inferential methods, called Approximate Bayesian Computation (ABC) methods, which are likelihood free algorithms. These algorithms use as basis of inference the simulated sample paths of the process. The basic form of the ABC algorithm is given as follows: (see for example Wilkinson 2008 or Plagnol and Tavaré 2003)

1. Draw a sample from the prior density: $\boldsymbol{\theta} \sim p(\boldsymbol{\theta})$;
2. Simulate the sample path $\mathbf{y}_t^*$ from the model $M(\boldsymbol{\theta})$;
3. Accept $\boldsymbol{\theta}$ if the simulated data $\mathbf{y}_t^*$ do not differ from the observed data $\mathbf{y}_t$, that is, if $D(\mathbf{y}_t^*, \mathbf{y}_t) \leq \delta$.

Here, $M(\boldsymbol{\theta})$ is the model, often expressed in terms of a random difference equation, and from which it is supposed to be relatively easy to simulate sample paths starting from a set of fixed initial values. $D(\cdot, \cdot)$ is a distance measure on the realizations of the process and δ is a tolerance limit that establishes how much the simulated sample path for a given $\boldsymbol{\theta}$ value is allowed to differ from the observed time series. Accepted $\boldsymbol{\theta}$ values are not sampled from the true posterior density but from an approximation, written as $p(\boldsymbol{\theta} | D(\mathbf{y}_t, \mathbf{y}_t^*) \leq \delta)$. When $\delta \to 0$, accepted $\boldsymbol{\theta}$'s are drawn from the true posterior $p(\boldsymbol{\theta} | \mathbf{y}_t)$ and when $\delta \to \infty$, they are drawn from the prior $p(\boldsymbol{\theta})$; see Plagnol and Tavaré (2003) for details. When the sample size of the time series is large, it is suggested that rather than comparing $\mathbf{y}_t$ and $\mathbf{y}_t^*$, appropriately chosen summary statistics, say $S(\mathbf{y}_t)$ and $S(\mathbf{y}_t^*)$, can be compared. Ideally, $S(\cdot)$ should be a sufficient statistic. However, if the likelihood is not known, sufficient statistics cannot be identified. Other reasonable summary statistics can be used, but this adds another layer of deviation from the true model.

Although this algorithm is, in principle, easy to implement when the underlying process is easy to simulate, and gives good results when the process under study has discrete sample space, it is not clear how well it works for nonlinear time series models such as bilinear processes. The choice of the distance measure, as well as the tolerance limit δ are arbitrary and have significant effect on the simulations. However, there are variations to this basic ABC algorithm. See Biau et al. (2012) for a variation which avoids fixing the awkward tolerance limit δ.

4.6 Parameter Estimation for GARCH-Type Processes

Likelihood based methods work quite well for some classes of nonlinear models. GARCH type models fall within this category. In the next section, we give a brief summary of inferential methods for this class of models on the account of their prominent role in nonlinear time series analysis.

Most of the work in parameter estimation for GARCH-type processes is focused in the time-domain approach. In particular, likelihood-based estimators have become most popular for estimating the parameters of GARCH-type processes.

The conditional maximum likelihood estimator (CMLE, in short) based on Gaussian innovations, is arguably the most frequently used estimator in practice. The Gaussianity of the innovations is, however, not essential for the asymptotic properties of the CMLE as we will see. Recently, Bayesian methods for estimating the parameters of GARCH models have been also proposed, and we will give a brief summary of these methods in Sect. 4.6.3. A different approach considers frequency-domain methods and in particular the *Whittle criterion*. This method was originally proposed to estimate the parameters of Gaussian ARMA processes (i.e., linear processes with finite variance). The applicability of the Whittle criterion for non-Gaussian and nonlinear processes was discussed in detail by Dzhaparidze and Yaglom (1983) who introduced the class of non-Gaussian mixing processes. Dzhaparidze and Yaglom (1983) showed that, for this class of processes, the Whittle estimator is weakly consistent and asymptotically normal.

4.6.1 Conditional Maximum Likelihood

Consider the GARCH(p,q) model

$$X_t = \sigma_t Z_t, \; t \in \mathbb{Z}$$

with

$$\sigma_t^2 = a_0 + \sum_{i=1}^{p} a_i X_{t-i}^2 + \sum_{j=1}^{q} b_j \sigma_{t-j}^2,$$

where (Z_t) forms a sequence of i.i.d. r.v's with zero-mean and unit variance, and $\boldsymbol{\theta} := (a_0, a_1, \ldots, a_p, b_1, \ldots, b_q)'$ represents the parameters of the model. For the ARCH(p) case, with $q = 0$ and p fixed and assuming temporally that (Z_t) are i.i.d standard Normal r.v's, it is possible to write down explicitly the joint density of the vector $(X_{p+1}, \ldots, X_n)$ conditional on $(X_1, \ldots, X_p)$, which turns out to be proportional to

$$\left(\prod_{t=\nu+1}^{n} \sigma_t^2\right)^{1/2} \exp\left\{-\frac{1}{2} \sum_{t=p+1}^{n} \frac{X_t^2}{\sigma_t^2}\right\},$$

where $\nu = p$. The latter expression remains valid for GARCH(p,q) processes with $\nu \geq \max(p,q) + 1$. Maximizing this conditional likelihood, as a function of the model parameters, leads to the CMLE estimator $\hat{\boldsymbol{\theta}}$ defined as

$$\hat{\boldsymbol{\theta}} = \operatorname{argmin}_{\boldsymbol{\theta} \in \Theta} \frac{1}{n} \sum_{t=\nu+1}^{n} \left[\frac{X_t^2}{\sigma_t^2} + \log(\sigma_t^2)\right]. \tag{4.22}$$

If the condition

$$E(Z_0^2)\sum_{i=1}^{p} a_i + \sum_{j=1}^{q} b_j < 1,$$

holds, then a more convenient expression for σ_t^2 is given by

$$\sigma_t^2 = \frac{a_0}{1-\sum_{j=1}^{q} b_j} + \sum_{i=1}^{p} a_i X_{t-i}^2 \sum_{i=1}^{p} a_i \sum_{k=1}^{\infty} \sum_{j_1,\ldots,j_k=1}^{q} b_{j_1}\cdots b_{j_k} X_{t-i-j_1-\cdots-j_k}^2, \tag{4.23}$$

where the multiple sum vanishes if $q = 0$ (see Hall and Yao 2003). Thus, in the maximization procedure, σ_t^2 is replaced by a truncated version of the expression in (4.23), defined as

$$\tilde{\sigma}_t^2 = \frac{a_0}{1-\sum_{j=1}^{q} b_j} + \sum_{i=1}^{\min(p,t-1)} a_i X_{t-i}^2 \times \tag{4.24}$$

$$\times \sum_{i=1}^{p} a_i \sum_{k=1}^{\infty} \sum_{j_1,\ldots,j_k=1}^{q} b_{j_1}\cdots b_{j_k} X_{t-i-j_1-\cdots-j_k}^2 I(t-i-j_1-\cdots-j_k \geq 1).$$

One of the obvious problems with this estimation procedure is that, in practice, the Z_t's are better modeled by a heavy-tailed distribution. However, under certain conditions, asymptotic properties of the estimator in (4.22), such as the $\sqrt{n}$-consistency of the CMLE, remain valid for large classes of noise distribution. For example, under condition $E(|Z_t|)^{4-\epsilon} < \infty$ for $\epsilon > 0$, $\hat{\boldsymbol{\theta}}$ is asymptotically normal. Furthermore, the convergence rate is the standard $\sqrt{n}$ rate, provided $E(Z_t)^4 < \infty$ (see Hall and Yao 2003). Within this context, the asymptotic properties of the CMLE have been discussed by Weiss (1986), who treated the ARCH(p) case and by Lee and Hansen (1994) and Lumsdaine (1996) for GARCH(1, 1) processes. In contrast, however, when the innovation distribution is heavy-tailed with an infinite fourth moment, the estimators may not be asymptotically normal, the range of possible limit distributions is quite large and the convergence rate is slower than the standard $\sqrt{n}$-rate. Studies for general GARCH(p, q) processes, without the condition of fourth finite moment, may be found in Hall and Yao (2003), Berkes et al. (2003), Straumann and Mikosch (2006) and Mikosch and Straumann (2006). In particular, Hall and Yao (2003) obtained the following result.

Theorem 4.6.1. *Assume that* $\mathbf{M} = E_0(\sigma_1^{-4}\mathbf{U}\mathbf{U}^T)$ *is a* $(p+q+1)\times(p+q+1)$ *matrix, where* E_0 *denotes expectation when the components of* $\boldsymbol{\theta}$ *take their true values, say* $\boldsymbol{\theta}^0$, *and* $\mathbf{U} = \mathbf{U}(\boldsymbol{\theta})$, *the* $(p+q+1)$*-vector of first derivatives of* $\sigma_1^2 = \sigma_1^2(\boldsymbol{\theta})$ *with respect to the components of* $\boldsymbol{\theta}$, *is non-singular. Furthermore, assume that* $p \geq 1$ *and that* $a_0 > 0, a_1, \ldots, a_p$ *are nonzero, that for* $q \geq 1$ *all the* $b_1, \ldots, b_q$

are nonzero and that there exists a local minimum $\hat{\boldsymbol{\theta}}$ within radius δ, strictly positive and small, of $\boldsymbol{\theta}^0$.

1. *If $E(Z_0^4) = \infty$ but the distribution of Z_0^2 is in the domain of attraction of the normal law, then*

$$n(\lambda_n^*)^{-1}(\hat{\boldsymbol{\theta}} - \boldsymbol{\theta}) \xrightarrow{d} N(\mathbf{0}_{p+q+1}, \mathbf{M}^{-1}),$$

where $\lambda_n = \inf\{\lambda_n^* > 0 : nH(\lambda_n^*) \leq (\lambda_n^*)^2\}$, with $H(\lambda_n^*) = E(Z_0^4 I(Z_0^2 \leq \lambda_n^*))$.

2. *If the distribution of Z_0^2 is in the domain of attraction of a stable law with exponent $\alpha \in (1, 2)$, then*

$$n(\lambda_n^*)^{-1}(\hat{\boldsymbol{\theta}} - \boldsymbol{\theta}) \xrightarrow{d} \sum_{k=1}^{\infty}\{Y_k V_k - E(Y_k)E(V_1)\}$$

where $V_1, V_2 \ldots$, are i.i.d. as $\sigma_1^{-2}\mathbf{M}^{-1}\mathbf{U}$, $Y_1, Y_2 \ldots$, are r.v's with

$$F_{Y_k}(y) = \exp(-y^{-\alpha}) \sum_{j=0}^{k-1} \frac{y^{-j\alpha}}{j}, \quad y > 0$$

and $\lambda_n^* = \inf\{\lambda_n^* > 0 : nP(Z_1^2 > \lambda_n^*) \leq 1\}$.

3. *If the distribution of Z_0^2 is in the domain of attraction of a stable law with exponent $\alpha = 1$ and if $n(\lambda_n^*)^{-1}E(Z_0^2 I(Z_0^2 > \lambda_n^*))^2 \to 0$, then*

$$\begin{aligned} &n(\lambda_n^*)^{-1}(\hat{\boldsymbol{\theta}} - \boldsymbol{\theta}) \\ &+ n(\lambda_n^*)^{-1}E(Z_0^2 I(Z_0^2 > \lambda_n^*))E(V_1) \\ &- \gamma E(V_1) \xrightarrow{d} Y_1 V_1 + \sum_{k=2}^{\infty}\{Y_k V_k - E(Y_k)E(V_1)\}, \end{aligned}$$

where γ denotes Euler's constant.

On the other hand, in order to overcome the drawbacks due to the possible slow convergence rates of the CMLE, as an alternative Peng and Yao (2003) introduced the least absolute deviations estimator (LADE, in short) which is robust with respect to the heavy tails of the innovation distribution. The LADE estimator of Peng and Yao (2003) is as follows: let $c > 0$ be a constant such that the median of $\epsilon_t^2 = cZ_t^2$ is equal to 1. Then, (2.31) and (2.34) can be rewritten as

$$X_t = \sigma_t \epsilon_t, \; s_t^2 = \alpha_0^* + \sum_{i=1}^{p} \alpha_i^* X_{t-i}^2 \sum_{j=1}^{q} b_j s_{t-j}^2,$$

where $s_t^2 = c^{-1}\sigma_t^2$, $\alpha_0^* = c^{-1}a_0$, $\alpha_i^* = c^{-1}a_i$. In this case, the vector of true parameters is given by $\boldsymbol{\theta}^* = (\alpha_0^*, \alpha_1^*, \ldots, \alpha_p^*, b_1, \ldots, b_q)'$. Thus

$$\log(X_t^2) = \log(s_t^2) + \log(\epsilon_t^2),$$

being the median of $\log(\epsilon_t^2)$, equals 0. Hence, the true value of $\boldsymbol{\theta}^*$ minimizes $E(|\log(X_t^2) - \log(s_t^2)|)$ which motivates the LADE

$$\hat{\boldsymbol{\theta}}^* = \text{argmin}_{\boldsymbol{\theta}\in\Theta} \sum_{t=\nu+1}^{n} |\log(X_t^2) - \log(\tilde{s}_t^2)|,$$

where $\tilde{s}_t^2$ is a truncated version of s_t^2 defined as

$$\tilde{s}_t^2 = \frac{\alpha_0^*}{1 - \sum_{j=1}^{q} b_j} + \sum_{i=1}^{\min(p,t-1)} \alpha_i^* X_{t-i}^2 \times$$
$$\times \sum_{i=1}^{p} \alpha_i^* \sum_{k=1}^{\infty} \sum_{j_1,\ldots,j_k=1}^{q} b_{j_1} \cdots b_{j_k} X_{t-i-j_1-\cdots-j_k}^2 I(t - i - j_1 - \cdots - j_k \geq 1),$$

which directly follows from (4.24). Note that the LADE may also be viewed as a maximum conditional likelihood estimator by assuming that the log-squared innovations $\log(\epsilon_t^2)$ follow a Laplace distribution. Peng and Yao (2003) proved that the LADE is asymptotically normal with standard convergence $\sqrt{n}$-rate under the assumption that the second moment of the innovation distribution is finite. Simulation studies, carried out by Peng and Yao (2003), also indicate that the finite sample performance of the LADE is better than that of the CMLE when the innovations exhibit a heavy-tailed distribution; see Huang et al. (2008) for further details.

For the ARCH(∞) class of models, Robinson and Zaffaroni (2006) established strong consistency and asymptotic normality for the CMLE estimator of the model parameters, under some general conditions which also cover, for example, the FIGARCH model for which strict stationarity and ergodicity are not yet established properties. Dahlhaus and Rao (2006) generalized the class of ARCH(∞) processes to the non-stationary class of ARCH(∞) processes with time-varying coefficients, establishing consistency and asymptotic normality for the segmented CMLE estimator. Furthermore, the asymptotic properties of the CMLE for non-stationary GARCH(p, q) models have been also addressed recently. Jensen and Rahbek (2004a,b) proved the asymptotic behavior of the CMLE for non-stationary GARCH(1, 1) and ARCH(1) models. For the ARCH(1) model, Jensen and Rahbek (2004a) proved the following result.

Theorem 4.6.2. *Assume that the ARCH*(1) *process in (2.31) and (2.32) does not allow a stationary version or equivalently* $E(\log(a_1 Z_t^2)) \geq 0$. *Assume further that* *the i.i.d. innovations* (Z_t) *have* $E(Z_t) = 0$, $V(Z_t) = 1$, *and* $V(Z_t^2) = E(Z_t^4) - 1 = \zeta < \infty$ *and* a_0 *is known. Then, as* $n \to \infty$, *the CMLE estimator for* a_1 *obtained from (4.22) as a special case, is consistent and asymptotically normal*

$$\sqrt{n}(\hat{\boldsymbol{\theta}} - \boldsymbol{\theta}) \xrightarrow{d} N(0, \zeta a_1^2).$$

In a companion paper, Jensen and Rahbek (2004b) derived a similar result for an estimator of (a_1, b_1) in the GARCH(1, 1) framework.

Based on the results given in Klüppelberg et al. (2004) and Francq and Zakoïan (2008) concluded that the result in Theorem 4.6.2 is valid only under the assumption $E[\log(a_1 Z_t^2)] > 0$. Moreover, these authors also argued that the estimator studied in Jensen and Rahbek's paper is not the usual CMLE but a *constrained* estimator of the ARCH parameters, where a_0 is known, defined as

$$\hat{a}_1 = \text{argmin}_{a_1 \in [0,\infty)} \frac{1}{n} \sum_{t=1}^{n} \left[\frac{X_t^2}{\sigma_t^2} + \log(\sigma_t^2) \right].$$

Instead, to overcome these drawbacks, Francq and Zakoïan (2008) proved the following results.

Lemma 4.6.1. *Let the ARCH*(1) *model defined in (2.31) and (2.32), with* $X_0^2 \geq 0$. *Then, if* $E[\log(a_1 Z_t^2)] > 0$ *holds*

$$\frac{1}{\sigma_n^2} = o(d^n) \text{ and } \frac{1}{X_n^2} = o(d^n),$$

almost surely as $n \to \infty$ *for any constant* d *such that*

$$\exp\{E[\log(Z_0^2)]\}/a_1 < d < 1.$$

The lemma presented above allows to obtain the strong consistency and asymptotic normality of the CMLE of a_1.

Theorem 4.6.3 (Francq and Zakoïan 2008). *Under the assumptions of Lemma 4.6.1 and if* $\boldsymbol{\theta}_0 = (a_0, a_1) \in \Theta$, *the CMLE*

$$(\hat{a}_0, \hat{a}_1) = argmin_{\boldsymbol{\theta}_0 \in \Theta} \frac{1}{n} \sum_{t=\nu+1}^{n} \left[\frac{X_t^2}{\sigma_t^2} + \log(\sigma_t^2) \right],$$

where Θ *is a compact subset of* $(0, \infty)^2$, *satisfies* $\hat{a}_1 \to a_1$ *a.s. and, if* $\boldsymbol{\theta}_0$ *belongs to the interior of* Θ

$$\sqrt{n}(\hat{a}_1 - a_1) \xrightarrow{d} N(0, \zeta a_1^2),$$

as $n \to \infty$.

Weak consistency and asymptotic normality of the CMLE for the parameters of non-stationary GARCH(p, q) general models have been recently established by Chan and Ng (2009).

4.6.2 Whittle Estimation

An alternative to the conditional maximum likelihood approach is to use the so-called Whittle estimator. It is one of the standard estimators for ARMA processes which is asymptotically equivalent to the Gaussian maximum likelihood estimator and the least squares estimator. The Whittle estimator works in the spectral domain of the process. The idea of such estimation procedure in the GARCH case was first pointed out by Bollerslev (1986) who noted that X_t^2 in (2.31) and (2.34) can be rewritten as an ARMA(k, q) model, where $k = \max(p, q)$ as follows:

$$X_t^2 = a_0 + \sum_{i=1}^{k} \psi_i X_{t-i}^2 + \sum_{j=1}^{q} \varphi_j \upsilon_{t-j} + \upsilon_t, \tag{4.25}$$

where $\psi_i = a_i - \varphi_i$ with the convention $\varphi_i = 0$ if $i \in (q, p]$ and $a_i = 0$ if $i \in (p, q]$, $\varphi_j = -b_j$, and

$$\upsilon_t = X_t^2 - \sigma_t^2 = \sigma_t^2 (Z_t^2 - 1). \tag{4.26}$$

Under the assumptions that (σ_t^2) is strictly stationary and $V(X_t^2) < \infty$, the sequence (υ_t) constitutes a white noise sequence.

For estimating the parameters of an ARMA process, Whittle suggested a procedure which is based on the periodogram. In his setup (X_t) is a causal and invertible ARMA(p, q) of the form $\phi(B)X_t = \psi(B)Z_t$, where (Z_t) is a sequence of i.i.d. r.v's with $E(Z_0) = 0$ and $V(Z_0) < \infty$. Denoting by $\boldsymbol{\theta}$ the vector of unknown parameters and Θ the set of admissible values of $\boldsymbol{\theta}$, the Whittle estimator, say $\hat{\boldsymbol{\theta}}$, is given by

$$\hat{\boldsymbol{\theta}} = \operatorname{argmin}_{\boldsymbol{\theta} \in \Theta} \frac{1}{n} \sum_{j=1}^{n-1} \frac{I_{n,X}(\omega_j)}{g(\omega_j; \boldsymbol{\theta})}, \tag{4.27}$$

where $I_{n,X}(\omega)$ represents the periodogram of the mean-corrected sample $X_1, \ldots, X_n$ defined as

$$I_{n,X}(\omega) = \frac{1}{n} \left| \sum_{t=1}^{n} (X_t - \bar{X}) e^{-i\omega t} \right|^2, \quad \omega \in (-\pi, \pi].$$

For simplicity in computation, the periodogram is evaluated at the Fourier frequencies $\omega_j = 2\pi j/n$, $j = -[(n-1)/2], \ldots, [n/2]$. On the other hand, the normalized spectral density $g(\omega; \boldsymbol{\theta})$ is often explicitly given. For example, for GARCH(p, q) models it equals to

$$g(\omega; \boldsymbol{\theta}) := \left| \frac{1 - b(e^{i\omega})}{1 - a(e^{i\omega}) - b(e^{i\omega})} \right|^2, \tag{4.28}$$

where $a(z) = \sum_{i=1}^{p} a_i z^i$ and $b(z) = \sum_{j=1}^{q} b_j z^i$. The Whittle estimator for ARMA processes also works well if the innovations have infinite variance and its rate of convergence compares favorably to $\sqrt{n}$-rates.

Mikosch and Straumann (2002) investigated the large sample properties of the Whittle estimator based on the squares of a GARCH(1, 1) process. These authors proved that under the assumption $E(X_t^8) = \infty$ the Whittle estimator for the squared GARCH process is unreliable, being its limiting distribution an unfamiliar non-Gaussian law. Furthermore, if $E(X_t^4) = \infty$ the Whittle estimator is inconsistent. Note that this is in contrast to the ARMA case with i.i.d. innovations where the rate of convergence improves when the tails become fatter. The results obtained by Mikosch and Straumann (2002) are summarized below.

Let us rewrite the GARCH(1, 1) as the ARMA model in (4.25) with $k = q = 1$, and assume that $\boldsymbol{\theta} = (\psi_1, \varphi_1)$. In proving consistency, the following assumptions are needed.

1. Z_1 has a positive density on $\mathbb{R}$.
2. $a_0 > 0$. In addition, condition $E \log(a_1 Z_0^2 + b_1) < 0$ holds.
3. There exists $h_0 \leq \infty$ such that $E(Z_0^h) < \infty$ for all $h < h_0$ and $EZ_0^{h_0} = \infty$.

The conditions above lead us to conclude that the equation $E(a_1 Z_0^2 + b1)^{\kappa/2} = 1$ has a unique positive solution κ.

Theorem 4.6.4. *Let (X_t) be a strictly stationary GARCH$(1, 1)$ process. Assume that the vector of true parameters $\boldsymbol{\theta}_0 \in C$ with*

$$C := \{\boldsymbol{\theta} \in \mathbb{R}^2 : -1 < \varphi_1 \leq 0,\ -\varphi_1 \leq \psi_1 \leq 1\}.$$

Then the following statements hold:

1. *If $\kappa < 4$ and $a_1, b_1 > 0$, then the Whittle estimator (4.27) is not consistent.*
2. *If $\kappa > 4$, the Whittle estimator is strongly consistent.*

Finally, one arrives at the following result.

Theorem 4.6.5. *In addition to the conditions of Theorem 4.6.4 assume that $\kappa > 4$ and $E(Z_0^8) < \infty$. Then the following limit relation holds:*

$$x_n(\hat{\boldsymbol{\theta}} - \boldsymbol{\theta}_0) \xrightarrow{d} [W(\boldsymbol{\theta}_0)]^{-1}\left(f_0(\boldsymbol{\theta}_0)V_0 + 2\sum_{k=1}^{\infty} f_k(\boldsymbol{\theta}_0)V_k\right), \quad n \to \infty, \quad (4.29)$$

where

$$x_n = \begin{cases} n^{1-4/\kappa} & \kappa < 8 \\ n^{1/2} & \kappa > 8 \end{cases},$$

(V_k) is a sequence of $\kappa/4$-stable r.v's for $\kappa \in (4, 8)$ and a sequence of centered Gaussian r.v's for $\kappa > 8$. The infinite series on the right-hand side of (4.29) is understood as the weak limit of its partial sums. Moreover, $[W(\boldsymbol{\theta}_0)]^{-1}$ is the inverse of the matrix

$$W(\boldsymbol{\theta}_0) = \frac{\upsilon_1}{2\pi}\int_{-\pi}^{\pi}\left[\frac{\partial \log g(\omega;\boldsymbol{\theta}_0)}{\partial \boldsymbol{\theta}}\right]\left[\frac{\partial \log(\omega;\boldsymbol{\theta}_0)}{\partial \boldsymbol{\theta}}\right]' d\omega$$

and

$$f_k(\boldsymbol{\theta}_0) = \frac{1}{2}\int_{-\pi}^{\pi}\frac{\partial \log(1/g(\omega;\boldsymbol{\theta}_0))}{\partial \boldsymbol{\theta}} e^{-ik\omega} d\omega, \quad k \in \mathbb{N}_0.$$

The parameter a_0 is estimated through the expression $\hat{a}_0 = [n^{-1}\sum_{t=1}^{n}(X_t - \bar{X})^2](1-\hat{\psi}_1)$, where $\hat{\psi}_1$ is the Whittle estimator of ψ. If $\kappa > 4$, then $n^{-1}\sum_{t=1}^{n}(X_t - \bar{X})^2 \to V(X_1)$ a.s. and hence $\hat{a}_0$ is strongly consistent under the assumptions of Theorem 4.6.4.

Giraitis and Robinson (2001) considered Whittle estimation for the class of ARCH(∞) models with $E(X_t^8) < \infty$. Again the idea is to embed the model in (2.31) with

$$\sigma_t^2 = a_0 + \sum_{i=1}^{\infty} a_i X_{t-i}^2, \quad t \in \mathbb{Z},$$

into an ARMA-type representation of the form

$$\sigma_t^2 = \alpha_0(\boldsymbol{\theta}) + \sum_{i=1}^{\infty} \alpha_i(\boldsymbol{\theta}) X_{t-i}^2 + \upsilon_t, \quad t \in \mathbb{Z},$$

where $\nu_t = X_t^2 - \sigma_t^2$ are martingale differences and $\alpha_i(\boldsymbol{\theta})$, for $i = 0, 1, \ldots$ represent the parameters depending on the corresponding $\boldsymbol{\theta}$. Giraitis and Robinson (2001) proved that the Whittle estimator is consistent if the following assumptions hold:

1. The fourth conditional moment of Z_t is finite.
2. Θ is compact.
3. $\boldsymbol{\theta}_0 \in \Theta$ and $\sigma^2 > 0$.

4. For all $\boldsymbol{\theta} \in \Theta$

$$\int_{-\pi}^{\pi} \log g(\omega; \boldsymbol{\theta}) d\omega = 0$$

with $g(\cdot, \cdot)$ defined as in (4.28).
5. $g(\omega; \boldsymbol{\theta})^{-1}$ is continuous in $(\omega, \boldsymbol{\theta}) \in [-\pi, \pi] \times \Theta$.
6. The set $\{\omega : g(\omega; \boldsymbol{\theta}) \neq g(\omega; \boldsymbol{\theta}_0)\}$ has positive Lebesgue measure, for all $\boldsymbol{\theta} \in \Theta/\{\boldsymbol{\theta}_0\}$.

Under some additional regularity and moment conditions Giraitis and Robinson (2001) proved the asymptotic normality of the Whittle estimator through the next result.

Theorem 4.6.6. *Assume that the eighth conditional moment of Z_t is finite and that assumptions 2–6 hold. Assume further that*

- *$\boldsymbol{\theta}_0$ in an interior point of Θ.*
- *In a neighborhood of $\boldsymbol{\theta}_0$, $(\partial/\partial\boldsymbol{\theta})g^{-1}(\omega; \boldsymbol{\theta})$ and $(\partial^2/\partial\boldsymbol{\theta}\,\partial\boldsymbol{\theta}')g^{-1}(\omega; \boldsymbol{\theta})$ exist and are continuous in ω and $\boldsymbol{\theta}$.*
- *$(\partial/\partial\boldsymbol{\theta})g^{-1}(\omega; \boldsymbol{\theta}_0) \in Lip(\eta)$, $\eta > 0.5$.*
- *The matrix*

$$W(\boldsymbol{\theta}_0) = \frac{1}{2\pi} \int_{-\pi}^{\pi} \left[\frac{\partial \log g(\omega; \boldsymbol{\theta}_0)}{\partial \boldsymbol{\theta}}\right] \left[\frac{\partial \log(\omega; \boldsymbol{\theta}_0)}{\partial \boldsymbol{\theta}}\right]' d\omega$$

is nonsingular.

Then

$$n^{1/2}(\hat{\boldsymbol{\theta}} - \boldsymbol{\theta}_0) \xrightarrow{d} N(\mathbf{0}, 2W^{-1} + W^{-1}VW^{-1}),\ n \to \infty,$$

where

$$V = \frac{2\pi}{E(\upsilon_t^2)} \int_{-\pi}^{\pi} \frac{\partial g(\omega; \boldsymbol{\theta}_0)^{-1}}{\partial \boldsymbol{\theta}} \left[\frac{\partial g(\omega; \boldsymbol{\theta}_0)^{-1}}{\partial \boldsymbol{\theta}}\right]' f(\omega, -\lambda, \lambda) d\omega d\lambda < \infty$$

with $f(\cdot, \cdot, \cdot)$ representing the fourth-order cumulant spectrum of X_t^2.

Finally, the strong consistency and asymptotic normality of the Whittle estimator for a class of exponential volatility model, which contains as special cases the EGARCH model of Nelson (1991) and the GJR-GARCH model of Glosten et al. (1993), have been recently established by Zaffaroni (2009).

4.6.3 Bayesian Approach

A common feature of the previous works is that all of them rely on a likelihood-based approach. Recently, Bayesian methods for the estimation of GARCH-type models driven by normal or Student-t innovations have also been proposed. The analysis of these models, from a Bayesian point of view, is a recent area of research and can be considered very promising due to the advantages of the Bayesian approach, in particular the possibility of obtaining small-sample results and integrating these results in a formal decision model. Bayesian inference on GARCH-type model has been implemented using importance sampling (e.g., Geweke 1989 and Kleibergen and van Dijk 1993) and, more recently, using Markov chain Monte Carlo (MCMC) including Bauwens and Lubrano (1998) and Bauwens and Rombouts (2007) (Gibbs sampler), Geweke (1994), Nakatsuma (2000), Vrontos et al. (2000) and Ardia (2008) (Metropolis-Hastings algorithm) and Mitsui and Watanabe (2003) (acceptance-rejection/Metropolis-Hastings). Excellent reviews of MCMC methods for estimating GARCH models are Asai (2006) and Miazhynskaia and Dorffner (2006). In this section we briefly explain the approaches suggested by Ardia (2008) and Bauwens and Lubrano (1998) for the GARCH$(1, 1)$ and the ARCH(1) model, respectively.

We first consider the GARCH$(1, 1)$ model in (2.31) and

$$\sigma_t^2 = a_0 + a_1 X_{t-1}^2 + b_1 \sigma_{t-1}^2,$$

where (Z_t) are i.i.d. standard normal r.v's and $a_0 > 0, a_1, b_1 \geq 0$. Let $\boldsymbol{\theta} = (a_0, a_1, b_1)$ be the vector of unknown parameters and $\boldsymbol{x} = (x_1, \ldots, x_n)'$ a sample generated by the GARCH$(1, 1)$ model. In addition, we define the $(n \times n)$ diagonal matrix $\Sigma := \Sigma(\boldsymbol{\theta}) = \text{diag}(\{\sigma_t^2(\boldsymbol{\theta})\}_{t=1}^n)$ with $\sigma_t^2(\boldsymbol{\theta}) = a_0 + a_1 X_{t-1}^2 + b_1 \sigma_{t-1}^2(\boldsymbol{\theta})$. Then the likelihood function of $\boldsymbol{\theta}$ takes the form

$$L(\boldsymbol{\theta}|\boldsymbol{x}) \propto (\det\Sigma)^{-1/2} \exp\left\{-\frac{1}{2}\boldsymbol{x}'\Sigma^{-1}\boldsymbol{x}\right\},$$

where, for convenience, we have considered the first observation as an initial condition and the initial variance fixed and equal to α_0. In order to obtain the posterior distribution we have to consider prior distributions for the parameters. Prior distributions are intended to represent beliefs about parameter values, prior to the availability of data. Ardia (2008) proposed the joint prior $p(\boldsymbol{\theta}) = p(a_0, a_1)p(b_1)$ with $p(a_0, a_1) \propto N_2(a_0, a_1|\boldsymbol{\mu}_{(a_0,a_1)}, \sum_{(a_0,a_1)})I(a_0, a_1 > 0)$ and $p(b_1) \propto N(b_1|\boldsymbol{\mu}_{b_1}, \sum_{b_1})I(b_1 > 0)$, where $\boldsymbol{\mu}$ and $\sum$ are the hyperparameters and $I(\cdot)$ is an indicator function. The posterior distribution is proportional to

$$p(\boldsymbol{\theta}|\boldsymbol{x}) \propto L(\boldsymbol{\theta}|\boldsymbol{x})p(\boldsymbol{\theta}). \tag{4.30}$$

The joint posterior distribution of the parameters is sampled using Nakatsuma's (1998, 2000) approach, which provides a MCMC method based on the Metropolis-Hastings algorithm for a linear regression model with an ARMA-GARCH error. The steps required to implement the algorithm are the following:

1. Draw a set of initial values $\boldsymbol{\theta}^{(0)} = (a_0^{(0)}, a_1^{(0)}, b_1^{(0)})$ from the joint prior $p(\boldsymbol{\theta})$;
2. Draw $(a_0^{(1)}, a_1^{(1)})$ from $p(a_0, a_1 | b_1^{(0)}, \boldsymbol{x})$;
3. Draw $b_1^{(1)}$ from $p(b_1 | a_0^{(1)}, a_1^{(1)}, \boldsymbol{x})$;
4. Generate J sets of random numbers $(a_0^{(j)}, a_1^{(j)}, b_1^{(j)})$ for $j = 2, \ldots, J$ by repeating stages 2–3, with

$$(a_0^{(j)}, a_1^{(j)}) \sim p(a_0, a_1 | b_1^{(j-1)}, \boldsymbol{x}), \text{ and } b_1^{(j)} \sim p(b_1 | a_0^{(j)}, a_1^{(j)}, \boldsymbol{x}).$$

Point estimates can then be easily obtained by taking the mean or the median of the posterior distribution. It is important to stress that, since none of the full conditional distributions are known analytically, we sample $(a_0^{(j)}, a_1^{(j)}, b_1^{(j)})_{j=1}^{J}$ from two proposal distributions. These distributions are obtained by considering the ARMA representation of the GARCH$(1,1)$ given in (4.25) and (4.26) with $k = q = 1$. Note that $\upsilon_t = (\chi_1^2 - 1)\sigma_t^2$ so that, by construction, (υ_t) is a martingale difference process with variance $2\sigma_t^4$. Nakatsuma (1998) pointed out, however, that in practice is difficult to generate the sets $(a_0^{(j)}, a_1^{(j)}, b_1^{(j)})$ directly from (4.25). To overcome this difficulty, the idea is to *approximate* υ_t by a variable $\epsilon_t \sim N(0, 2\sigma_t^4)$. This leads to the following auxiliary model:

$$X_t^2 = a_0 + \psi_1 X_{t-1}^2 + \varphi_1 \epsilon_{t-1} + \epsilon_t,$$

By noting that ϵ_t and σ_t^2 are both functions of $\boldsymbol{\theta}$, respectively given by

$$\epsilon_t(\boldsymbol{\theta}) = X_t^2 - a_0 - \psi_1 X_{t-1}^2 - \varphi_1 \epsilon_{t-1}(\boldsymbol{\theta}) \tag{4.31}$$

and

$$\sigma_t^2(\boldsymbol{\theta}) = a_0 + a_1 X_{t-1}^2 + b_1 \sigma_{t-1}^2(\boldsymbol{\theta}),$$

by defining the $(n \times n)$ diagonal matrix $\Lambda := \Lambda(\boldsymbol{\theta}) = \text{diag}(\{2\sigma_t^4(\boldsymbol{\theta})\}_{t=1}^n)$ and the vector $\boldsymbol{\epsilon} := (\epsilon_1, \ldots, \epsilon_n)'$ we can approximate the likelihood function of $\boldsymbol{\theta}$ from the auxiliary model as follows:

$$L(\boldsymbol{\theta} | \boldsymbol{x}) \propto (\det \Lambda)^{-1/2} \exp\left\{-\frac{1}{2}\boldsymbol{\epsilon}'\Lambda^{-1}\boldsymbol{\epsilon}\right\}.$$

The construction of the proposal distribution for (a_0, a_1) and b_1 is based on this likelihood function.

In order to generate (a_0, a_1) we use the following recursive transformations (e.g., Chib and Greenberg 1994)

$$\begin{cases} r_t = 1 + b_1 r_{t-1} \\ s_t^2 = x_{t-1}^2 + b_1 s_{t-1}^2 \end{cases},$$

which allow to express the function $\epsilon_t(\boldsymbol{\theta})$ in (4.31) as a linear function of (a_0, a_1). The initial values of r_0 and s_0^2 are set to zero. If we regroup the terms within vectors $\boldsymbol{x}^2 = (x_1^2 \cdots x_n^2)'$, $c_t = (r_t\ Y_t)$ and construct the $(n \times 2)$ matrix C where the tth element is c_t, then, it turns out that $\boldsymbol{\epsilon} = \boldsymbol{x}^2 - C(a_0\ a_1)'$ and that we can express the likelihood function of $(a_0, a_1)'$ as follows:

$$L(a_0, a_1|b_1, \boldsymbol{x}) \propto (\det\Lambda\ )^{-1/2} \exp\left\{-\frac{1}{2}\boldsymbol{\epsilon}'\Lambda^{-1}\boldsymbol{\epsilon}\right\}. \tag{4.32}$$

Furthermore, the proposal distribution to sample (a_0, a_1) is obtained by combining this likelihood function and the prior distribution by the usual Bayes update:

- $q(\hat{a}_0, \hat{a}_1, a_0, a_1) \propto N(a_0, a_1|\hat{\boldsymbol{\mu}}_{(a_0,a_1)}, \hat{\Sigma}_{(a_0,a_1)})I(a_0, a_1 > 0)$;
- $\hat{\Sigma}_{(a_0,a_1)}^{-1} = C'\tilde{\Lambda}^{-1}C + \Sigma_{(a_0,a_1)}^{-1}$;
- $\hat{\boldsymbol{\mu}}_{(a_0,a_1)} = \hat{\Sigma}_{(a_0,a_1)}(C'\tilde{\Lambda}^{-1}\boldsymbol{x}^2 + \Sigma_{(a_0,a_1)}^{-1}\boldsymbol{\mu}_{(a_0,a_1)})$,

where the $(n \times n)$ diagonal $\tilde{\Lambda} := \text{diag}(\{\tilde{\sigma}_t^2\}_{t=1}^n)$ is such that

$$\tilde{\sigma}_t^2 = \tilde{a}_0 + \tilde{a}_1 x_{t-1}^2 + b_1\tilde{\sigma}_{t-1}^2.$$

The value $(\tilde{a}_0, \tilde{a}_1)$ is the previous draw of (a_0, a_1) in the Metropolis-Hastings sampler. A candidate (a_0^*, a_1^*) is sampled from this proposal distribution and accepted with probability

$$\min\left\{\frac{p(a_0^*, a_1^*, b_1|\boldsymbol{x})}{p(\tilde{a}_0, \tilde{a}_1, b_1|\boldsymbol{x})}\frac{q(a_0^*, a_1^*, \tilde{a}_0, \tilde{a}_1)}{q(\tilde{a}_0, \tilde{a}_1, a_0^*, a_1^*)}, 1\right\}.$$

Now, in order to generate b_1 the idea is to express $\epsilon_t(\boldsymbol{\theta})$ in (4.31) as a linear function of b_1 and proceed similar as in the case of (a_0, a_1). However, this approach is not feasible since it is not possible to rewrite ϵ_t as a linear function of b_1. To overcome the problem, we linearize $\epsilon_t(b_1)$ by the first order Taylor expansion at point, say, $\tilde{b}_1$ as follows:

$$\epsilon_t(b_1) \simeq \epsilon_t(\tilde{b}_1) + \frac{d\epsilon_t}{db_1}\bigg|_{b_1=\tilde{b}_1}(b_1 - \tilde{b}_1)$$

with $\tilde{b}_1$ representing the previous draw of b_1 in the Metropolis-Hastings sampler. Hence, the proposal distribution to sample b_1 results from combining the approximated likelihood and the prior distribution utilizing a similar scheme as in the case of (a_0, a_1) and the prior distribution by the usual Bayes update:

- $q(\hat{b}_1, b_1) \propto N(b_1|\hat{\mu}_{b_1}, \hat{\Sigma}_{b_1})I(b_1 > 0)$;
- $\hat{\Sigma}_{b_1}^{-1} = \nabla'\tilde{\Lambda}^{-1}\nabla + \Sigma_{b_1}^{-1}$;
- $\hat{\mu}_{b_1} = \hat{\Sigma}_{b_1}(\nabla'\tilde{\Lambda}^{-1}\boldsymbol{r} + \Sigma_{b_1}^{-1}\mu_{b_1})$,

where $\boldsymbol{r} = (r_1 \cdots r_n)'$ and $\nabla = (\nabla_1 \cdots \nabla_n)'$ with

$$r_t = \epsilon_t(\tilde{b}_1) + \tilde{b}_1\nabla_t$$

and

$$\nabla_t = -\frac{d\epsilon_t}{db_1}\bigg|_{b_1=\tilde{b}_1}(b_1 - \tilde{b}_1)\,.$$

The terms ∇_t can be computed by the following recursion

$$\nabla_t = X_{t-1}^2 - \epsilon_{t-1}(\tilde{b}_1) + \tilde{b}_1\nabla_{t-1}$$

with the initial value $\nabla_0 = 0$. Thus, we can approximate the expression in (4.32) by

$$\exp\left\{-\frac{1}{2}(\boldsymbol{r} - b_1\nabla)'\Lambda^{-1}(\boldsymbol{r} - b_1\nabla)\right\}.$$

Finally, a candidate b_1^* is sampled from this proposal distribution and accepted with probability

$$\min\left\{\frac{p(b_1^*, a_0, a_1|\boldsymbol{x})}{p(\tilde{b}_1, a_0, a_1|\boldsymbol{x})}\frac{q(b_1^*, \tilde{b}_1)}{q(\tilde{b}_1, b_1^*)}, 1\right\}.$$

An important issue, when implementing MCMC methods, is the assessment of the convergence of the algorithm. Although there is a reassuring theoretical literature concerning the convergence of MCMC methods, results do not easily translate into clear guidelines for the practitioners. The most straightforward approach for assessing convergence is based on simply plotting and inspecting traces of the observed MCMC sample. If the trace of values, for each of the parameters, stays relatively constant over the last, say, m iterations, this may be satisfactory evidence for convergence. It is worthwhile to mention that such diagnosis should be carried out for each of the estimated parameters by the MCMC algorithm, because convergent behavior by one parameter does not imply evidence for convergence for other parameters in the analysis. The ad hoc techniques described above fail, however, to guarantee convergence of the observed MCMC sample in the presence of a phenomenon called *metastability*, i.e., the Markov chain appears to have converged to the true equilibrium value, but after some period of stability around this value, the Markov chain may suddenly move to another region of the parameter space. Unfortunately, there is no statistical technique available for detecting metastability. Finally, we also would like to refer the work of Yu and Mykland (1998) who propose to monitor the mixing or convergence behavior of a Markov sampler using the CUSUM path plot of a chosen one-dimensional summary statistic. The authors proved that the smoothness of the CUSUM path plot corresponds to the slowness of the mixing behavior of the summary statistic. Along with the ad hoc techniques described above, a number of more formal methods exist which are prevalent in the literature. For example, Gelman and Rubin (1992) introduce a diagnostic based on multiple chains (with very dispersed starting values), in order to check if the posterior results are not too sensitive to the starting values. This criterion is not very attractive as it requires to run many times the Gibbs sampler, which is computationally demanding. Convergence criteria based on the examination of a single long run are thus preferable. Zellner and Min (1995) put forward three simple criteria which are useful only when $\boldsymbol{\theta}$ is partitioned in two blocks of parameters. Geweke (1992) provides a statistic that compares the estimate of a posterior mean from the first say, n_1 draws to the estimate from the last say, n_2 draws of the chain. The statistic is normally distributed provided n is large and the chain has converged. For an overview of the early work in this area see Cowles and Carlin (1996), Robert and Casella (2004) and Carlin and Thomas (2009) for recent developments.

Next, we focus on the estimation of the non-zero mean ARCH(1) model with (conditionally) normal innovations

$$X_t = u + \sigma_t Z_t, \; \sigma_t^2 = 1 + a_1(X_{t-1} - u)^2.$$

In this case the vector of unknown parameters is $\boldsymbol{\theta} := (u, a_1)$. Bauwens and Lubrano (1998) applied the Griddy-Gibbs sampler (GGS) proposed by Ritter and Tanner (1992) in order to conduct Bayesian inference. By assuming a flat prior, the conditional posterior density $p(u|a_1, \boldsymbol{x})$ has a kernel given by

$$r(u|a_1, \boldsymbol{x}) = \prod_{t=1}^{n} \sigma_t^{-1}(\boldsymbol{\theta}) \exp\left\{-\frac{1}{2}\frac{(x_t - u)^2}{\sigma_t^2(\boldsymbol{\theta})}\right\}. \tag{4.33}$$

Likewise, the conditional posterior density $p(a_1|u, \boldsymbol{x})$ has the same expression as in (4.33) but for a fixed u. Since σ_t^2 is a function of both u and a_1, the conditional posterior density of u cannot be a normal or any other well-known density from which random numbers could be easily generated. Hence, there is no property of conjugacy. Note, however, that the kernel in (4.33), conditionally on a previous draw of the conditioning parameter, can be evaluated over a grid of points. One can then compute the corresponding distribution function using a deterministic integration rule. Afterwards, one can generate a draw of u (respectively a_1) by inversion of the distribution at a random value sampled uniformly in [0, 1].

An advantage of the GGS is that it is successful in dealing with the shape of the posterior, such as the skewness, by using smaller MCMC outputs compared to other methods. This is due to the fact that integration is done on a grid so that every direction can be explored in detail.

Bauwens and Lubrano (1998) state that the choice of the grid of points has to be made carefully and constitutes the main difficulty in applying the GGS. Even if the parameter space is bounded, the authors recommend restricting the integration to the subset of the parameter space where the value of the posterior density is large enough to contribute to the integrals.

Bauwens and Lubrano (1998) implemented the following algorithm in order to extract n draws of the posterior distribution $p(\boldsymbol{\theta}|\boldsymbol{x})$ in (4.30).

1. Set an initial value $a_1^{(0)}$ for a_1;
2. Loop starting at $j = 1$;
3. Compute $r(u|a_1^{(j-1)}, \boldsymbol{x})$ over the grid $(u_1, \ldots, u_M)$ to obtain the vector $M_r = (r_1, \ldots, r_M)$;
4. By a deterministic integration rule[1] using M points, compute the values $M_\Phi = (0, \Phi_2, \ldots, \Phi_M)$ where

$$\Phi_i = \int_{u_1}^{u_i} r(u|a_1^{(j-1)}, \boldsymbol{x})du, \ i = 2, \ldots, M.$$

   Compute (and cumulate for the marginal) the normalized pdf values $M_\varphi = M_r/\Phi_M$ of $\varphi(u|a_1^{(j-1)}, \boldsymbol{x})$. Compute $E(u|a_1^{(j-1)}, \boldsymbol{x})$ and $V(u|a_1^{(j-1)}, \boldsymbol{x})$ by the same type of integration rule and store them in a table;
5. Generate $y \sim U[0, \Phi_M]$ and invert $\Phi(u|a_1^{(j-1)}, \boldsymbol{x})$ by numerical interpolation to get a draw of $u|a_1^{(j-1)}, \boldsymbol{x}$, indexed $u^{(j)}$. Store this draw in a table;
6. Repeat steps 3–5 for $a_1^{(j)}|u^{(j)}, \boldsymbol{x}$;
7. Increment j by one and go to step 3 unless $j > n$;
8. Compute the posterior moments of u and a_1 from the tables where conditional moments are stored (by averaging). Likewise, plot the marginal densities (cumulated M_φ). With the table containing $(u^{(1)}, a_1^{(1)}, \ldots, u^{(n)}, a_1^{(n)})$ one can compute posterior moments and draw a histogram of any function of the parameters.

References

Aase KK (1983) Recursive estimation in non-linear time series models of autoregressive type. J R Stat Soc B 45:228–237

Andrieu C, Doucet A, Holenstein R (2010) Particle Markov chain Monte Carlo methods. J R Stat Soc B 72:269–342

Ardia D (2008) Financial risk management with Bayesian estimation of GARCH models: theory and applications. Lecture notes in economics and mathematical systems, vol 612. Springer, Berlin

[1] Bauwens and Lubrano (1998) use trapezoidal rule of integration.

Asai M (2006) Comparison of MCMC methods for estimating GARCH models. J Jpn Stat Soc 36:199–212

Bauwens L, Lubrano M (1998) Bayesian inference on GARCH models using the Gibbs sampler. Econom J 1:C23–C46

Bauwens L, Rombouts JVK (2007) Bayesian inference for the mixed conditional heteroskedasticity model. Econom J 10:408–425

Beirlant J, Goegebeur Y, Segers J, Teugels J (2004) Statistics of extremes: theory and applications. Wiley, Chichester

Bera AK, Higgins ML (1997) ARCH and bilinearity as competing models for nonlinear dependence. J Bus Econ Stat 15:43–50

Bera A, Bilias Y, Simlai P (2006) Estimating functions and equations: an essay on historical developments with applications to econometrics. In: Mills TC, Patterson K (eds) Palgrave handbook of econometrics. Econometric theory, Palgrave MacMillan, vol 1, pp 427–476

Berkes I, Horváth L, Kokoszka P (2003) GARCH processes: structure and estimation. Bernoulli 9:201–227

Biau G, Cérou F, Guyader A (2012) New insights into Approximate Bayesian Computation. Ann Inst Henri Poincaré B, (to appear)

Bollerslev T (1986) Generalized autoregressive conditional heteroskedasticity. J Econom 31:307–327

Brock WA, Dechert WD, Scheinkman JA, LeBaron B (1996) A test for independence based on the correlation dimension. Econom Rev 15:197–235

Brockwell PJ, Davis RA (1991) Time series: theory and methods. Springer, New York

Brockwell PJ, Davis RA (1996) Introduction to time series and forecasting. Springer, New York

Brockwell PJ, Davis RA (2006) Time series: theory and methods, 2nd edn. Springer, New York

Carlin BP, Thomas AL (2009) Bayesian methods for data analysis, 3rd edn. CRC, Boca Raton

Chan NH, Ng CT (2009) Statistical inference for non-stationary GARCH(p,q) models. Electron J Stat 3:956–992

Chib S, Greenberg E (1994) Bayes inference in regression models with ARMA(p,q) errors. J Econom 64:183–206

Chopin N, Jacob PE, Papaspiliopoulos O (2013) SMC^2: an efficient algorithm for sequential analysis of state space models. J R Stat Soc B 75:397–426

Coles SG (2001) An introduction to statistical modeling of extreme values. Springer, London

Congdon P, (2010) Applied bayesian hierarchical methods. Chapman & Hall/CRC

Cowles MK, Carlin BP (1996) Markov chain Monte Carlo convergence diagnostics: a comparative review. J Am Stat Assoc 91:883–904

Cox DR, Reid N (2004) A note on pseudo-likelihood constructed from marginal densities. Biometrika 91:729–737

Dahlhaus R, Rao SS (2006) Statistical inference for time-varying ARCH processes. Ann Stat 34:1075–1114

Davis RA, Yau CY (2011) Comments on pairwise likelihood in time series models. Stat Sin 21:255–277

Davison A, Gholamrezaee M (2012) Geostatistics of extremes. Proc R Soc Lond Ser A Math Phys Eng Sci 468:581–608

Dzhaparidze KO, Yaglom AM (1983) Spectrum parameter estimation in time series analysis. Developments in statistics, vol 4. Academic, New York

Embrechts P, Klüppelberg C, Mikosch, T (1997) Modelling Extremal Events for Insurance and Finance. Springer-Verlag, Heildelberg

Fearnhead P (2008) Computational methods for complex stochastic systems: a review of some alternatives to MCMC. Stat Comput 18:151–171

Francq C, Zakoïan J-M (2008) Can one really estimate nonstationary GARCH models? Working paper 2008-06, Centre de Recherche en Economie et Statistique

Gelfand E, Smith AFM (1990) Sampling-based approaches to calculating marginal densities. J Am Stat Assoc 85:398–409

Gelman A, Rubin DB (1992) A single series from the Gibbs sampler provides a false sense of security. In: Bernardo JM, Berger JO, Dawid AP, Smith AFM (eds) Bayesian statistics, vol 4. Oxford University Press, Oxford, pp 625–631

Geweke J (1989) Exact predictive densities for linear models with ARCH disturbances. J Econom 40:63–86

Geweke J (1992) Evaluating the accuracy of sampling-based approaches to the calculation of posterior moments. In: Bernardo JM, Berger JO, Dawid AP, Smith AFM (eds) Bayesian statistics, vol 4. Oxford University Press, Oxford, pp 169–194 (with discussion)

Geweke J (1994) Bayesian comparison of econometric models. Working paper 532, Research Department, Federal Reserve Bank of Minneapolis

Giraitis L, Robinson PM (2001) Whittle estimation of ARCH models. Econom Theory 17:608–631

Glosten L, Jagannathan R, Runkle D (1993) On the relation between the expected value and the volatility of the nominal excess return on stocks. J Finance 48:1779–1801

Godambe VP (1960) An optimum property of regular maximum likelihood estimation. Ann Math Stat 31:1208–1211

Godambe VP (1985) The foundations of finite sample estimation in stochastic processes. Biometrika 72:419–428

Godambe VP (1991) Estimating functions. Oxford University Press, New York

Granger CWJ, Andersen A (1978) On the invertability of time series models. Stoch Proc Appl 8:87–92

Hall P, Yao Q (2003) Inference in ARCH and GARCH models with heavy-tailed errors. Econometrica 71:285–317

Hannan EJ (1970) Multiple time series. Wiley, New York

Heyde CC (1997) Quasi-likelihood and its application: a general approach to optimal parameter estimation. Springer, New York

Hinich MJ (1982) Testing for Gaussianity and linearity of a stationary time series. J Time Ser Anal 3:169–176

Huang D, Wang H, Yao Q (2008) Estimating GARCH models: when to use what? Econom J 11:27–38

Jensen ST, Rahbek A (2004a) Asymptotic normality of the QMLE estimator of ARCH in the nonstationary case. Econometrica 72:641–646

Jensen ST, Rahbek A (2004b) Asymptotic inference for nonstationary GARCH. Econom Theory 20:1203–1226

Keenan DM (1985) A Tukey nonadditivity-type test for time series nonlinearity. Biometrika 72:39–44

Kelly FP (1979) Reversibility and stochastic networks. Wiley, Chichester

Kleibergen F, van Dijk HK (1993) Non-stationarity in GARCH models: a Bayesian analysis. J Appl Econom 8(Suppl):41–61

Klüppelberg C, Lindner A, Maller R (2004) A continuous-time GARCH process driven by a Lévy process: stationarity and second-order behaviour. J Appl Probab 41:601–622

Lee SW, Hansen BE (1994) Asymptotic theory for the GARCH(1, 1) quasi-maximum likelihood estimator. Econom Theory 10:29–52

Liang F, Liu C, Caroll R (2010) Advanced Markov chain Monte Carlo methods: learning from Past samples. Wiley, New York

Lumsdaine R (1996) Consistency and asymptotic normality of the quasi-maximum likelihood estimator in IGARCH(1, 1) and covariance stationary GARCH(1, 1) models. Econometrica 64:575–596

Luukkonen R, Saikkonen P, Teräsvirta T (1988) Testing linearity against smooth transition autoregressive models. Biometrika 75:491–499

Mardia KV, Kent JT, Hughes G, Taylor CC (2009) Maximum likelihood estimation using composite likelihoods for closed exponential families. Biometrika 96:975–982

Mauricio JA (2008) Computing and using residuals in time series models. Comput Stat Data Anal 52:1746–1763

McLeod AI, Li WK (1983) Diagnostic checking of ARMA time series models using squared residual autocorrelations. J Time Ser Anal 4:269–273

Meyn S, Tweedie RL (2009) Markov chains and stochastic stability. Cambridge University Press, Cambridge

Miazhynskaia T, Dorffner G (2006) A comparison of Bayesian model selection based on MCMC with an application to GARCH-type models. Stat Pap 47:525–549

Mikosch T, Straumann T (2002) Whittle estimation in a heavy-tailed GARCH(1, 1) model. Stoch Process Appl 100:187–222

Mikosch T, Straumann T (2006) Stable limits of martingale transforms with application to the estimation of GARCH parameters. Ann Stat 34:493–522

Mitsui H, Watanabe T (2003) Bayesian analysis of GARCH option pricing models. J Jpn Stat Soc 33:307–324

Nakatsuma T (1998) A Markov-chain sampling algorithm for GARCH models. Stud Nonlinear Dyn Econom 3:107–117

Nakatsuma T (2000) Bayesian analysis of ARMA-GARCH models: a Markov chain sampling approach. J Econom 95:57–69

Nelson DB (1991) Conditional heteroskedasticity in asset returns: a new approach. Econometrica 2:347–370

Nicholls DF, Quinn BG (1980) The estimation of autoregressive models with random coefficients. J Time Ser Anal 13:914–931

Nummelin E (1984) General irreducible Markov chains and non-negative operators. Cambridge University Press, Cambridge

Peat M, Stevenson M (1996) Asymmetry in the business cycle: Evidence from the Australian labour market. J Econ Behav Organ 30:353–368

Peng L, Yao Q (2003) Least absolute deviations estimation for ARCH and GARCH models. Biometrika 90:967–75

Petruccelli JD, Davies N (1986) A portmanteau test for self-exciting threshold autoregressive-type nonlinearity in time series. Biometrika 73:687–694

Pickands J (1975) Statistical inference using extreme order statistics. Ann Stat 3:119–131

Plagnol V, Tavaré S (2003) Approximate Bayesian computation and MCMC. In: Niederreiter H (ed) Monte Carlo and quasi-Monte Carlo methods. Springer, Heidelberg

Prado R, West M (2010) Time series: modeling, computation, and inference. Texts in statistical science. Chapman & Hall/CRC, Boca Raton

Priestley MB (1981) Spectral analysis and time series. Academic, London

Quinn BG (1982) Stationarity and invertibility of simple bilinear models. Stoch Process Appl 12:225–230

Ritter C, Tanner MA (1992) The Gibbs stopper and the Griddy Gibbs sampler. J Am Stat Assoc 87:861–868

Robert CP, Casella G (2004) Monte Carlo statistical methods, 2nd edn. Springer, New York

Roberts GO, Rosenthal JS (2004) General state space Markov chains and MCMC algorithms. Probab Surv 1:20–71

Robinson PM, Zaffaroni P (2006) Pseudo-maximum likelihood estimation of ARCH(∞) models. Ann Stat 34:1049–1074

Straumann D, Mikosch T (2006) Quasi-MLE in heteroscedastic times series: a stochastic recurrence equations approach. Ann Stat 34:2449–2495

Subba Rao T, Gabr MM (1980) A test for linearity of stationary time series. J Time Ser Anal 1:145–158

Subba Rao T, Gabr MM (1984) An introduction to bispectral analysis and bilinear time series models. Springer, New York

Thavaneswaran A, Abraham B (1988) Estimation for non-linear time series models using estimating equations. J Time Ser Anal 9:99–108

Tjøstheim D (1986) Some doubly stochastic time series models. J Time Ser Anal 7:51–72

Tjøstheim D (1990) Non-linear time series and Markov chains. Adv Appl Probab 22:587–611

Tong H (1990) Non-linear time series. Oxford Science Publications, Oxford

Tsay RS (1989) Testing and modeling threshold autoregressive processes. J Am Stat Assoc 84:231–240
Tsay RS (1991) Detecting and modelling nonlinearity in univariate time series analysis. Stat Sin 1:431–451
Varin C, Vidoni P (2009) Pairwise likelihood inference for general state space models. Econom Rev 28:170–185
Varin C, Reid N, Firth D (2011) An overview of composite likelihood methods. Stat Sin 21:5–43
Vrontos ID, Dellaportas P, Politis D (2000) Full Bayesian inference for GARCH and EGARCH models. J Bus Econom Stat 18:187–198
Weiss A (1986) Asymptotic theory for ARCH models: estimation and testing. Econom Theory 2:107–131
Wilkinson RD (2008) Approximate Bayesian computation (ABC) gives exact results under the assumption of model error. Technical report, Sheffield University
Yu B, Mykland P (1998) Looking at Markov samplers through CUMSUM path plots: a simple diagnostic idea. Stat Comput 8:275–286
Zaffaroni P (2009) Whittle estimation of EGARCH and other exponential volatility models. J Econom 151:190–200
Zellner A, Min C (1995) Gibbs sampler convergence criteria. J Am Stat Assoc 90:921–927

Chapter 5
Models for Integer-Valued Time Series

5.1 Introduction

The analysis of time series of (small) counts has become an important area of research in the last two decades partially because of its wide applicability to

- Social science (McCabe and Martin 2005);
- Queueing systems (Ahn et al. 2000);
- Experimental biology (Zhou and Basawa 2005);
- Public health and medicine (Yu et al. 2013; Weiß 2013; Moriña et al. 2011; Andersson and Karlis 2010; Alosh 2009);
- Environmental processes (Scotto et al. 2014; Villarini et al. 2010; Cui and Lund 2009; Thyregod et al. 1999);
- Economy (Jung and Tremayne 2011; Fokianos et al. 2009; Quoreshi 2006; Blundell et al. 2002);
- International tourism demand (Brännäs and Nordström 2006; Brännäs et al. 2002; Garcia-Ferrer and Queralt 1997; Nordström 1996);
- Statistical control processes (Weiß 2007, 2009; Lambert and Liu 2006; Ye et al. 2001);
- Telecommunications (Weiß 2008a);
- Alarm systems (Monteiro et al. 2008).

We refer to McKenzie (2003) and Kedem and Fokianos (2002) for an overview of the early work in this area and to Tjøstheim (2012), Fokianos (2011), Jung and Tremayne (2006, 2011), and Weiß (2008b) for recent developments.

As an example, Fig. 5.1 displays a time series containing the monthly number of fires in Faro district (Portugal) for the period January 2004–January 2011. One way to obtain models for integer-valued data is to replace multiplication in conventional ARMA models by an appropriate *thinning operator*[1] to ensure the

[1] Several other approaches have been proposed in the literature for the analysis of the time series of counts, including static regression models and autoregressive conditional mean models; see Jung and Tremayne (2011) for further details.

K.F. Turkman et al., *Non-Linear Time Series*, DOI 10.1007/978-3-319-07028-5_5,

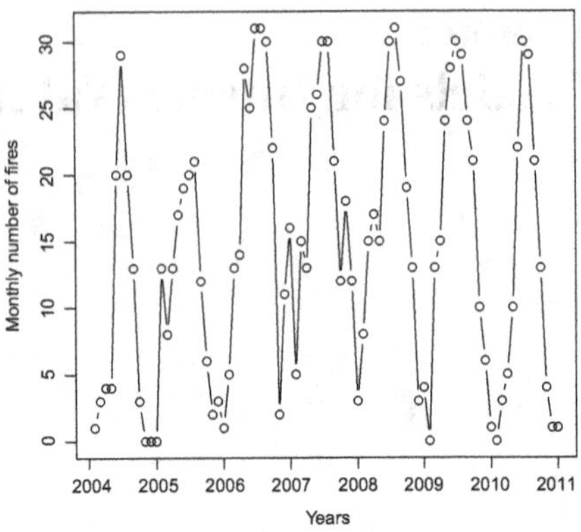

Fig. 5.1 Time series plot for the monthly number of fires in Faro district (Portugal) for the period January 2004–January 2011

integer discreteness of the process. The procedure of thinning is more common in the study of point processes but is also appropriate here as many discrete variate processes arise as aggregated point processes, i.e., counts of a point process in consecutive intervals of time. The most popular thinning operator is the *binomial thinning*, first introduced by Steutel and van Harn (1979), to adapt the terms of *self-decomposability* and *stability* for integer-valued time series.

Definition 5.1.1 (Binomial thinning). Let (ξ_j) be a counting sequence of i.i.d. Bernoulli r.v's with mean $a \in [0, 1]$ and Z a non-negative integer-valued random variable with range $\{0, 1, \dots, n\}$ or $\mathbb{N}_0$, independent of the counting sequence. The binomial thinning operator $a\circ$ is defined by

$$a \circ Z := \begin{cases} \sum_{j=1}^{Z} \xi_j(a) & Z > 0 \\ 0 & Z = 0 \end{cases}.$$

Some important properties of the binomial thinning operator are stated in the following lemma.

Lemma 5.1.1. *Let Z and V two r.v's with support in $\mathbb{N}_0$. For the binomial thinning operator it holds that, for $a_1, a_2 \in [0, 1]$*

1. $0 \circ Z = 0$;
2. $1 \circ Z = Z$;
3. $a_1 \circ a_2 \circ Z \overset{d}{=} a_2 \circ a_1 \circ Z$;
4. $a_1 \circ (Z + V) \overset{d}{=} a_1 \circ Z + a_1 \circ V$;
5. *Let* $a_1 \overset{j_1+Y}{\overset{j_1+1}{\circ}} Y := \sum_{j=j_1+1}^{j_1+Y} \xi_j(a_1)$. *Then,* $a_1 \circ (Z + V) \overset{d}{=} a_1 \overset{Z}{\overset{1}{\circ}} Z + a_1 \overset{Z+V}{\overset{Z+1}{\circ}} V$;
6. $a_1 \circ (a_2 \circ Z) \overset{d}{=} (a_1 a_2) \circ Z$;
7. $a_1 \circ Z + a_2 \circ Z \overset{d}{\neq} (a_1 + a_2) \circ Z$;

8. *Let* $X := a_1 \circ Z$. *The mean value and the variance of* X *are given by* $E(X) = a_1 E(Z)$ *and* $V(X) = a_1^2 V(Z) + a_1(1 - a_1)E(Z)$, *respectively.*

Property 4 in the previous lemma holds true provided that the counting sequences involved in defining $a_1 \circ Z$ and $a_1 \circ V$ are independent.

Various modifications of the binomial thinning operator have been proposed to make the integer-valued models based on thinning more flexible for practical purposes. Brännäs and Hellström (2001), for instance, suggested allowing for dependence between the indicators of the counting series (ξ_j). A different generalization was proposed by Latour (1998). Latour's operator is defined as follows:

$$a \bullet_\beta Z := \begin{cases} \sum_{j=1}^{Z} \xi_j(a) & Z > 0 \\ 0 & Z = 0 \end{cases},$$

where the r.v's ξ_j's are i.i.d. and independent of Z. However, in contrast to binomial thinning in Definition 5.1.1, the variables ξ_j's are now allowed to have the full range $\mathbb{N}_0$, with mean a and variance β. Latour's operator is referred to as *generalized thinning*. Note that this operator includes the binomial thinning as a special case where $\beta = a(1 - a)$. Another special case of the generalized thinning operator is the *extended thinning* operator proposed by Zhu and Joe (2003) which takes the form

$$a \circledast Z := \begin{cases} \sum_{j=1}^{Z} \xi_j(a) & Z > 0 \\ 0 & Z = 0 \end{cases},$$

where the ξ_j's are i.i.d. r.v's independent of Z, with the same distribution as a random variable $\xi(a)$ with generating probability function

$$P_{\xi(a)}(s; a) := E[s^{\xi(a)}] = \frac{(1 - a) + (a - \gamma)s}{(1 - a\gamma) - (1 - a)\gamma s}, \quad \gamma \in (0, 1] \tag{5.1}$$

with mean $E(\xi(a)) = a$ and variance $V(\xi(a)) = a(1 - a)\frac{1+\gamma}{1-\gamma}$. Clearly, the binomial thinning operator corresponds to $\gamma = 0$ in (5.1). Further extensions of the binomial thinning operator have been recently proposed by Zhu and Joe (2010) who introduced the *expectation thinning* operator based on a family of self-generalized[2] r.v's, as follows: let $\{\xi(a) : 0 \le a \le 1\}$ be a family of self-generalized r.v's with support on $\mathbb{N}_0$ and finite mean. The expectation thinning operator between a and Z is defined as

$$a \otimes Z := \begin{cases} \sum_{j=1}^{Z} \xi_j(a) & Z > 0 \\ 0 & Z = 0 \end{cases},$$

[2] A r.v. $Y(a)$ is said to be *self-generalized* with respect to parameter a, if $P_{Y(a)}(P_{Y(a)}(s; a); a') = P_{Y(a)}(s; aa')$, for all $a, a' \in [0, 1]$.

where the ξ_j's are i.i.d. r.v's independent of Z, with the same distribution as a r.v. $\xi(a)$ with $E(\xi(a)) < 1$, for all $a \in (0, 1)$.

Another important modification of the binomial thinning operator was proposed by Kim and Park (2008) who introduced the *signed binomial thinning* operator for handling over-dispersed and non-stationary integer-valued time series. One advantage of Kim and Park's operator is that it can handle negative integer-valued time series. This is in contrast with the binomial and the generalized operator which are only applicable to non-negative integer-valued time series. The signed binomial thinning is as follows:

$$a \odot Z := \text{sgn}(a) \cdot \text{sgn}(Z) \cdot \sum_{j=1}^{|Z|} \xi_j(a), \; a \in (-1, 1), \tag{5.2}$$

where

$$\text{sgn}(x) := \begin{cases} 1 & x \geq 0 \\ -1 & x < 0 \end{cases}$$

and (ξ_j) are a counting sequence of i.i.d. Bernoulli r.v's with mean $|a|$.

Kim and Park's operator has been generalized by Zhang et al. (2010) who introduced the *signed generalized power series thinning operator* $\circledast$ defined as

$$a \circledast Z := \text{sgn}(a) \cdot \text{sgn}(Z) \cdot \sum_{j=1}^{|Z|} \xi_j(a), \tag{5.3}$$

where (ξ_j) are i.i.d. with generalized power series distribution of the form

$$P(\xi_1 = n) = \frac{c(n)[g(\omega)]^n}{h(\omega)}, \; \omega > 0 \; n \in S,$$

where S is any non-empty enumerable set of non-negative integers and $c(\cdot)$, $g(\cdot)$, and $h(\cdot)$ are positive, finite, and differentiable functions.

Note that a common feature of all these thinning operators is the assumption of *independence* of the counting variables ξ_j's. Recently, Ristić et al. (2013) introduced a new binomial thinning operator based on a counting sequence of Bernoulli dependent r.v's. Ristić and co-authors' thinning operator is defined as

$$a \circ_\theta Z := \begin{cases} \sum_{j=1}^{Z} \xi_j(a) & Z > 0 \\ 0 & Z = 0 \end{cases}$$

with

$$\xi_j = (1 - V_j)W_j + V_j Y, \; j \in \mathbb{N}, \tag{5.4}$$

where (W_j) and (V_j) are sequences of i.i.d. Bernoulli r.v's with parameters $a \in [0,1]$ and $\theta \in [0,1]$, respectively, and Y is also a Bernoulli random variable with parameter $a \in [0,1]$. Furthermore, it is assumed that W_j, V_i and Y are independent for all $i, j \in \mathbb{N}$. The representation in (5.4) implies that (ξ_j) forms a sequence of dependent Bernoulli r.v's with parameter $a \in [0,1]$, since $Corr(\xi_i, \xi_j) = \theta^2 \neq 0$ for $\theta \neq 0$ and $i \neq j$. The case $\theta = 0$ corresponds to the binomial thinning operator.

Joe (1996) and Zheng et al. (2007) suggested extending the thinning concept by allowing a to be random itself. The resulting thinning operation is then called *random coefficient thinning.*

Definition 5.1.2 (Random coefficient thinning). Let Z be a random variable having the range $\mathbb{N}_0$. Let $\phi \in [0,1]$ be a random variable. Further, assume that ϕ is independent of Z. Then the random variable $\phi \circ Z$ is obtained from Z by random coefficient thinning if the operator $\circ$ is the binomial thinning operator, performed independently of Z and ϕ.

The concept of random coefficient thinning has been recently extended by Gomes and Canto e Castro (2009) by allowing a different discrete distribution associated to the thinning operator.

Definition 5.1.3 (Generalized random coefficient thinning). Let Z be a random variable having the range $\mathbb{N}_0$. Let φ be a random variable with support on $\mathbb{R}_+$. Then, $\varphi \circ^G Z$ is a random variable satisfying

$$(\varphi \circ^G Z|\varphi, Z) \sim G(\mu, \sigma^2),$$

where G is a given discrete-type distribution associated to the generalized thinning operation, with mean $\mu = \varphi Z$ and finite variance $\sigma^2 = \delta Z$ with δ possibly depending on φ and Z.

5.2 Integer-Valued ARMA Models

Among the most successful integer-valued time series models based on thinning operators are the INteger-valued AutoRegressive model of order p (INAR(p)) and the INteger-valued Moving Average model of order q (INMA(q)). The former was first introduced by McKenzie (1985) and Al-Osh and Alzaid (1987) for the case $p = 1$. It is worth noting that the INAR(1) model is a subcritical Galton-Watson process with immigration. Empirical relevant extensions for the INAR(1) model have been suggested by Brännäs (1995, explanatory variables), Brännäs et al. (2002, panel data), Brännäs and Hellström (2001, extended dependence structure), and Silva et al. (2005, replicated data). Further extensions and generalizations were proposed by Latour (1998), Du and Li (1991), Ispány et al. (2003), Zhu and Joe (2006), and Weiß (2008a). The INMA(q) model was proposed by Al-Osh and

Alzaid (1988) and McKenzie (1988) and subsequently studied by Brännäs and Hellström (2001) and Weiß (2008c). Related models were introduced by Aly and Bouzar (1994, 2005) and Zhu and Joe (2003). Moreover, there have been also some attempts to make the models based on thinning more attractive for economic applications by introducing covariates. Blundell et al. (2002), for example, studied the number of patents in firms, Rudholm (2001) analyzed competition in the generic pharmaceutical markets, Brännäs et al. (2002) estimated a nonlinear integer-valued model for tourism demand, and Brännäs and Nordström (2006) focused on the tourist accommodation impact of two large festivals in Sweden: the Water Festival Stockholm and the Gothenburg Party. Extensions for random coefficients integer-valued autoregressive models have been proposed by Zheng et al. (2006, 2007) who investigated basic probabilistic and statistical properties of these models. Zheng and co-authors illustrated their performance in the analysis of epileptic seizure counts (e.g., Latour 1998) and in the analysis of the monthly number cases of poliomyelitis in the US for the period 1970–1983.

This section summarizes some of the basic probabilistic properties of INAR and INMA models, including discussions on the existence of stationarity solutions and the tails of the marginal distribution. Furthermore some complementary extreme value results are also presented. We restrict our attention to time series with an *infinite* range of counts.[3]

5.2.1 INAR(1) Model

A discrete time non-negative, integer-valued process (X_t) is called an INAR(1) process, if it follows the recursion

$$X_t = M_t(X_{t-1}, a) + Z_t, \quad t \in \mathbb{Z}, \tag{5.5}$$

where $M_t(\cdot, \cdot)$ is a random operator which preserves the discreteness of the counts at time t and (Z_t) is an i.i.d. sequence of non-negative integer-valued r.v's, stochastically independent of X_{t-1} for all points in time, with finite mean $\mu_Z > 0$ and variance $\sigma_Z^2 \geq 0$. The most commonly used random operator is the binomial thinning operator introduced in Definition 5.1.1, i.e.,

$$M_t(X_{t-1}, a) = \sum_{j=1}^{X_{t-1}} \xi_j(a) \equiv a \circ X_{t-1},$$

[3]An excellent account for the theory of INAR(1) models with a finite range of counts can be found in Weiß and Kim (2013) and the references therein.

Table 5.1 Important properties of the INAR(1) based on the binomial thinning operator

1. $E(X_t) = \frac{\mu_Z}{1-a}$;
2. $V(X_t) = \frac{a\mu_Z + \sigma_Z^2}{1-a^2}$;
3. $E(X_t \vert X_{t-1}) = aX_{t-1} + \mu_Z$;
4. $V(X_t \vert X_{t-1}) = a(1-a)X_{t-1} + \sigma_Z^2$;
5. $Cov(X_t, X_{t+j}) = a^j V(X_t)$;
6. $Corr(X_t, X_{t+j}) = a^j$;
7. $P_{X_t}(s) = P_{Z_t}(s) P_{X_t}(1 - a + as)$, with $P_X(s) := E(s^X)$.

where (ξ_j) is a counting sequence of i.i.d. Bernoulli r.v's with mean $a \in [0, 1]$,[4] independent of X_{t-1}. It is important to stress the fact that the binomial thinning operator is performed at each time t. Hence, it would be more appropriate to write $\circ_t$ instead of $\circ$ to emphasized this fact. Nevertheless, since in the case of the INAR(1) model there is no risk of misinterpretation, the index t of the thinning operators is omitted since there is no advantage in retaining it. Note that X_t can be interpreted as being the sum of $a \circ X_{t-1}$, i.e., those who survive (with equal probability a) from time $(t-1)$ and t, and Z_t which represents those who arrive in the interval $(t-1, t)$ and survive until time t. It is worth noting that the INAR(1) process in (5.5) based on binomial thinning operator is stationary. Important properties of this model are given in Table 5.1. Note that the ACF at lag k is of the same form as in the case of the conventional AR(1) model.[5] Despite this fact, a remarkable difference between the conventional AR(1) and the INAR(1) process is that the latter is nonlinear which implies that the second-order structure will not necessarily show up accurately all features exhibited in the process.

Al-Osh and Alzaid (1987) proved that the stationary distribution of (X_t) is given by that of

$$X \stackrel{d}{=} \sum_{i=0}^{\infty} a^i \circ Z_{t-j}.$$

Furthermore, by the stationarity of (X_t) and property 7 in Table 5.1, it follows that the distribution of X_t can be determined through the relation

$$P_X(s) = P_Z(s) P_X(1 - a + as),$$

which allows us to conclude that the marginal distribution of the INAR(1) model must be *discrete self-decomposable*[6] (DSD). Important distributions belonging to

[4]The case $a \in (0, 1)$ is called the *stable* case. Ispány et al. (2003) introduced a *nearly-unstable* INAR(1) model with $a_n = 1 - \delta_n/n$, and $\delta_n \to \delta$ as $n \to \infty$.

[5]Although for the INAR(1) model the values of the ACF are always non-negative.

[6]A discrete distribution in $\mathbb{N}_0$ with probability generating function $P(z)$ is called DSD if $P(z) = P(1 - a + az)P_a(z)$, for $|z| < 1$ and $a \in (0, 1)$, with $P_a(\cdot)$ being some probability generating function. Alternatively, a non-negative integer-valued random variable X is DSD if for each $a \in$

the class of DSD distributions are the negative binomial, Poisson, generalized Poisson distribution and discrete stable distributions as a sub-class. This class, however, does not include distributions defined on bounded sets ruling out the possibility to consider the binomial distribution, for example, as a marginal distribution for X_t. Alzaid and Al-Osh (1988) proved that the distributions of Z_t and X_t belong to the same family of distributions iff one of them belongs to the class of discrete stable distributions. Within the context of the INAR(1) process the Poisson distribution plays a role similar to that of the Gaussian distribution in the conventional AR(1) process. Specifically, it is easy to check that the marginal distribution of (X_t) is $\text{Po}(\lambda/(1-a))$ iff the innovations constitutes an i.i.d. sequence of Poisson-distributed r.v's, $\text{Po}(\lambda)$ with $\mu_Z = \sigma_Z^2 = \lambda$, and if the initial distribution X_0 is $\text{Po}(\lambda/(1-a))$.

In order to obtain INAR(1) models with negative binomial (and in particular geometric) marginals, a number of approaches have been proposed in the literature. McKenzie (1986) first introduced the INAR(1) model based on the binomial thinning operator with binomial negative marginals. The author proved that $Z_t \overset{d}{=} \sum_{i=1}^{N} a^{U_i} \circ W_i$, where N is a Poisson-distributed random variable with parameter $-\eta \ln(a)$, (U_i) forms a sequence of i.i.d. uniformly distributed r.v's in the interval $(0, 1)$ and (W_i) is a sequence of i.i.d. binomial negative r.v's with parameters $\eta = 1$ and κ. The author, however, did not identify the distribution of Z_t.

Leonenko et al. (2007) extended McKenzie's results and proved that if X_t has a negative binomial distribution with parameters $\eta > 0$ and $\kappa := \eta/m$ with $m > 0$, that is

$$P(X_t = n) = \frac{\Gamma(\eta+n)}{n!\Gamma(\eta)}\left(1+\frac{m}{\eta}\right)^{-\eta}\left(\frac{m}{m+\eta}\right)^{n}, \quad n = 0, 1, \ldots,$$

then Z_t has a negative-binomial geometric distribution with parameters $(\eta, \kappa/(\kappa + a), a)$, that is

$$P(Z_t = n) = \sum_{i=0}^{\infty}\binom{i+n-1}{n}\left(\frac{\kappa}{\kappa+a}\right)^{i}\left(\frac{a}{\kappa+a}\right)^{n}\binom{\eta+i-1}{i}a^{\eta}(1-a)^{i}.$$

Ristić et al. (2009) showed that if ξ_j's are i.i.d. geometric r.v's with parameter $a/(1+a)$, being $a \in [0, 1)$, and (X_t) is a stationary process with marginal geometric with parameter $\mu/(1+\mu)$ where $\mu := E(X_t)$, then

$$P(Z_t = n) = \left(1-\frac{a\mu}{\mu-a}\right)\frac{\mu^n}{(1+\mu)^{n+1}} + \frac{a\mu}{\mu-a}\frac{a^n}{(1+a)^{n+1}}, \quad n = 0, 1, \ldots.$$

$(0, 1)$ there is a non-negative random variable X_a such that $X \overset{d}{=} a \circ X' + X_a$, where $a \circ X'$ and X_a independent, and X' is distributed as X.

Hence, Z_t is a mixture of two geometric r.v's with parameters $\mu/(1+\mu)$ and $a/(1+a)$, respectively.

A different approach for generating INAR(1) models with negative binomial marginals is due to Al-Osh and Aly (1992). These authors consider the INAR(1) in (5.5) with the random operator

$$M_t(X_{t-1}, a \times b) = \sum_{j=1}^{(a\times b)\circ X_{t-1}} \xi_j(b), \; b \in (0, 1),$$

provided that $X_{t-1} > 0$ and 0 otherwise. The ξ_j's are i.i.d. r.v's with range contained in $\mathbb{N}_0$ and distribution depending on the parameter b. In this case, it follows that if X_0 is $\mathrm{BN}(n, b(1-a)/[1+b(1-a)])$ and that Z_t has a $\mathrm{BN}(n, b/[1+b])$, then X_t is $\mathrm{BN}(n, b(1-a)/[1+b(1-a)])$.

The INAR(1) model in (5.5) based on the binomial thinning operator has been generalized in several ways. A possible extension is to consider an INAR(1) model with periodically varying parameter of the form

$$X_t = \phi_t \circ X_{t-1} + Z_t \tag{5.6}$$

with $\phi_t = a_j \in (0, 1)$ for $t = j + kT$, $(j = 1, \ldots, T,\ k \in \mathbb{N}_0)$, where the thinning operator $\circ$ is defined as

$$\phi_t \circ X_{t-1} \stackrel{d}{=} \sum_{i=1}^{X_{t-1}} \xi_{i,t}(\phi_t),$$

where $(\xi_{i,t}(\phi_t))$, for $i = 1, 2, \ldots$, is a periodic sequence of independent Bernoulli r.v's with success probability $P[\xi_{i,t}(\phi_t) = 1] = \phi_t$. The model in (5.6) was first introduced by Monteiro et al. (2010) and is referred to as Periodic INteger-valued AutoRegressive process of order one with period T (hereafter $\mathrm{PINAR}(1)_T$). Furthermore, Monteiro et al. (2010) assumed that (Z_t) constitutes a periodic sequence of independent, Poisson-distributed r.v's with mean $\upsilon_t = \lambda_j$ for $t = j + kT$, $(j = 1, \ldots, T,\ k \in \mathbb{N}_0)$, which are assumed to be independent of X_{t-1} and $\phi_t \circ X_{t-1}$. To avoid ambiguity, T is taken as the smallest positive integer satisfying (5.6). For this model Monteiro and co-workers proved the following result. For simplicity in notation we define

$$\beta_{t,i} := \begin{cases} \prod_{j=0}^{i-1} \phi_{t-j} & i > 0 \\ 1 & i = 0 \end{cases}.$$

Proposition 5.2.1. *For a fixed value of $j = 1, \ldots, T$, with $T \in \mathbb{N}$, the process (X_t) for $t = j + kT$ and $k \in \mathbb{N}_0$ is an irreducible, aperiodic and positive recurrent* *(and hence ergodic) Markov chain. Moreover, the stationary distribution of* (X_t) *is given by that of*

$$V_j := \sum_{m=1}^{+\infty}\sum_{s=0}^{T-1} (\beta_{j,j}\beta_{T,s}\beta_{T,T}^{m-1}) \circ Z_{T(m+1)-s} + \sum_{m=0}^{j-1} \beta_{j,m} \circ Z_{j-m},$$

where the series converges almost surely and also in L_2.

Now we are prepared to obtain the periodic mean and autocovariance function of (X_t).

Lemma 5.2.1. *For a fixed value of* $j = 1,\ldots,T$, *with* $T \in \mathbb{N}$, $t = j + kT$ *and* $k \in \mathbb{N}_0$

$$\mu_j = \mu_t = E(X_t) = V(X_t) = \frac{\sum_{k=0}^{j-1} \beta_{j,k}\lambda_{j-k} + \beta_{j,j}\sum_{i=0}^{T-j-1} \beta_{T,i}\lambda_{T-i}}{1-\beta_{T,T}}$$

with the convention $\sum_{i=0}^{-1} = 0$. *Moreover, for* $j = 1,\ldots,T$ *and* $h \geq 0$, $\gamma_{j+kT}(h) = \gamma_j(h) = \beta_{j+h,h}\mu_j$ *and* $\gamma_{j+kT}(-h) = \gamma_j(-h) = \beta_{j+kT,h}\mu_{j+kT-h}$.

Note that, in contrast to the autocovariance function of a stationary series, $\gamma_j(\cdot)$ is not symmetric in h; however $\gamma_t(-h) = \gamma_{t-h}(h)$ and $\gamma_t(h) = \gamma_{t+h}(-h)$. Furthermore, in view of the fact that h can be rewritten in the form $h = i + mT$, for some $i \in \{1,\ldots,T\}$ and $m \in \mathbb{N}$, the autocovariance function takes the form $\gamma_j(h) = \beta_{T,T}^m\beta_{j+i,i}\mu_j$ and $\gamma_j(-h) = \beta_{T,T}^m\beta_{j+T,i}\mu_{j+T-i}$.

Within the context of the PINAR$(1)_T$ process the Poisson distribution also plays a role similar to that of the Gaussian distribution in the conventional PAR(1) process, as the following result shows.

Theorem 5.2.1. *The marginal distribution of* (X_t), *with* $t = j + kT$ *for a fixed value of* $j = 1,\ldots,T$, *with* $T \in \mathbb{N}$, *and* $k \in \mathbb{N}_0$, *is Poisson with mean* μ_j *if and only if* (Z_t) *forms a sequence of independent Poisson-distributed r.v's with mean* λ_j.

In order to account for the so-called piecewise phenomena, Monteiro et al. (2012) introduced the class of self-exciting, integer-valued autoregressive model of order one with two regimes defined by the recursive equation

$$X_t = \begin{cases} a_1 \circ X_{t-1} + Z_t, & X_{t-1} \leq R \\ a_2 \circ X_{t-1} + Z_t, & X_{t-1} > R \end{cases}, \tag{5.7}$$

where R represents the threshold level and (Z_t) constitutes an i.i.d. sequence of Poisson-distributed r.v's with mean λ, for each t. The authors proved that (X_t) is an irreducible, aperiodic and positive recurrent (and hence ergodic) Markov chain, implying that there exists a strictly stationary process satisfying (5.7).

Another possible extension of the INAR(1) model is to replace the binomial thinning operator by the operator $\varphi \circ^G$ of Gomes and Canto e Castro (2009) which leads us to express X_t as

$$X_t = \varphi_t \circ^G X_{t-1} + Z_t. \tag{5.8}$$

Gomes and Canto e Castro (2009) obtained necessary and sufficient conditions to ensure the existence of a weak stationary process satisfying (5.8).

Theorem 5.2.2. *Let (φ_t) be a sequence of i.i.d. r.v's with $\varphi_t \in \mathbb{R}_+$ and (Z_t) a non-correlated sequence of non-negative integer-valued r.v's, independent of (φ_t). If, for all $t \in \mathbb{Z}$, $E(\varphi_t^2) < 1$, then there exists a unique non-negative integer-valued weakly stationary process (X_t) satisfying (5.8).*

An explicit form for the causal solution of (5.8) is given in the following result.

Proposition 5.2.2. *Under the conditions of the theorem above, there exists a sequence of non-correlated r.v's (ϵ_t), such that the process (X_t) in (5.8) admits as a stationary causal representation*

$$X_t = \sum_{i=0}^{\infty} \beta_{t,i}^* \epsilon_{t-i}, \tag{5.9}$$

where $\epsilon_t = \varphi_t \circ^G X_{t-1} - \varphi_t X_{t-1} + Z_t$ and

$$\beta_{t,i}^* := \begin{cases} \prod_{j=0}^{i-1} \varphi_{t-j} & i > 0 \\ 1 & i = 0 \end{cases}.$$

A major drawback with the representation (5.9) is that it is not canonical in the sense that the sequences (ϵ_t) and (X_t) are not independent. The expressions for the mean, variance and autocovariance function are the following:

$$E(X_t) = \frac{\mu_Z}{1 - E(\varphi_t)};$$

$$V(X_t) = \frac{\sigma_\epsilon^2}{1 - [V(\varphi_t) + E^2(\varphi_t)]} + V(\varphi_t)\frac{\mu_Z^2}{1 - [V(\varphi_t) + E^2(\varphi_t)]};$$

$$\gamma(k) = [E(\varphi_t)]^k V(X_t), \; k \geq 0.$$

Example 5.2.1. Assume that

$$X_t = \varphi_t \circ^P X_{t-1} + Z_t, \; t \in \mathbb{Z},$$

where P stands for a Poisson distribution with mean value and variance equal to $\varphi_t X_{t-1}$ and $Z_t \sim Po(\lambda)$. Then

$$E(X_t) = \frac{\lambda}{1 - E(\varphi_t)}, \quad V(X_t) = \frac{E(X_t)}{1 - E(\varphi_t^2)} + [E(X_t)]^2 \frac{V(\varphi_t)}{1 - E(\varphi_t^2)}.$$

As a third extension of the INAR(1) model in (5.5) we consider the zero-truncated Poisson INAR(1) (in short ZTPINAR(1)) model with parameter $\lambda > 0$ introduced by Bakouch and Ristić (2010), defined as

$$X_t = \begin{cases} Z_t & \text{with probability } e^{-\lambda} \\ a \circ X_{t-1} + Z_t & \text{with probability } 1 - e^{-\lambda} \end{cases},$$

where (Z_t) is a sequence of i.i.d. r.v's. Bakouch and Ristić (2010) proved that for this model

$$P(Z_t = n) = \frac{[(1-a)^n - (-a)^n]\lambda^n e^{\lambda a}}{n!(e^\lambda - 1)}. \tag{5.10}$$

Note that condition $a \in (0, \frac{1}{2})$ ensures the non-negativity of the probabilities in (5.10).

For the zero-truncated Poisson INAR(1) model the expressions for the mean, variance and autocovariance function are as follows:

$$E(X_t) = \frac{\lambda}{1 - a(1 - e^{-\lambda})};$$

$$V(X_t) = \frac{e^\lambda}{e^\lambda + a^2 - a^2 e^\lambda} \left(\frac{(e^\lambda - 1)a(1-a)}{e^\lambda + a - a e^\lambda} \mu_Z + \sigma_Z^2 \right);$$

$$\gamma(k) = a^k (1 - e^{-\lambda})^k, \ k \geq 0.$$

It is straightforward to show that the ZTPINAR(1) model can be rewritten as

$$X_t = a_t \circ X_{t-1} + Z_t, \tag{5.11}$$

where (a_t) is a sequence of i.i.d r.v's independent of (X_t) and (Z_t) with distribution $P(a_t = 0) = 1 - P(a_t = a) = e^{-\lambda}$. Bakouch and Ristić (2010) showed that the mean square solution of Eq. (5.11) is

$$X_t = \sum_{i=1}^{\infty} \left(\prod_{j=0}^{i-1} a_{t-j} \right) \circ Z_{t-i} + Z_t, \tag{5.12}$$

leading to the following result.

Theorem 5.2.3. *Equation (5.11) has the unique, strictly stationary and ergodic solution given by (5.12).*

5.2.2 *INAR(p) Model*

The counterpart of the conventional AR(p) model in the context of integer-valued time series is the so-called INAR(p) model which follows the recursion

$$X_t = M_t(X_{t-1}, \dots, X_{t-p}, a_1, \dots, a_p) + Z_t, \ t \in \mathbb{Z}, \tag{5.13}$$

where $M_t(\cdot)$ represents again a random operator which preserves the discreteness of the counts at time t with $0 \le a_i < 1$, $i = 1 \dots, p-1$, $0 < a_p < 1$, and (Z_t) an i.i.d. sequence of non-negative integer-valued r.v's, stochastically independent of X_{t-i} for all points in time, with finite mean $\mu_Z > 0$ and variance $\sigma_Z^2 \ge 0$. The binomial thinning operator is obtained by considering

$$\begin{aligned} M_t(X_{t-1}, \dots, X_{t-p}, a_1, \dots, a_p) &= \sum_{j=1}^{X_{t-1}} \xi_{j,1}^{(t)}(a_1) + \dots + \sum_{j=1}^{X_{t-p}} \xi_{j,p}^{(t)}(a_p) \\ &\equiv a_1 \circ_t X_{t-1} + \dots + a_p \circ_t X_{t-p}, \end{aligned}$$

the counting sequences $(\xi_{j,i}^{(t)})$ involved in $a_i \circ_t X_{t-i}$, at each time t are mutually independent and independent of (Z_t). Note that since the thinning operators are probabilistic, the joint distribution of

$$(a_1 \circ_{t+1} X_t, \dots, a_p \circ_{t+p} X_t),$$

has to be considered, leading to different types of INAR(p) models. Alzaid and Al-Osh (1990) assume a conditional multinomial distribution whereas Du and Li (1991) require conditional independence. We note that the statistical properties of the Alzaid and Al-Osh (1990) model are very different from the properties of the model suggested by of Du and Li (1991) which is less tractable. The stationarity condition for the INAR(p) process is that the roots of the polynomial

$$z^p - a_1 z^{p-1} - \dots - a_{p-1} z - a_p = 0,$$

are inside the unit circle, that is $|z| < 1$. This condition, however, is equivalent to $\sum_{i=1}^{p} a_i < 1$. These two different formulations of the INAR(p) model lead us to obtain different second-order structures for the processes. For the Alzaid and Al-Osh representation (INAR(p)-AA) the ACF resembles the one obtained for the

ARMA($p, p-1$) model whereas Du and Li formulation (INAR(p)-DL) implies that the ACF is the same as that of an AR(p) model.

The major drawback of the representations considered by Alzaid and Al-Osh and Du and Li is that the stationary marginal distribution of (5.13), for $p \geq 2$ is not necessarily in the DSD class. To circumvent this difficulty Zhu and Joe (2006) and Weiß (2008a) introduced the so-called *Combined* INAR(p) (hereafter CINAR(p)) model which is constructed so that the stationary marginal distribution is, indeed, part of the DSD family.

Definition 5.2.1. Let (Z_t) be an i.i.d. sequence with range $\mathbb{N}_0$ and $a \in (0,1)$. Let $(\mathbf{D}_t)$ be an i.i.d. process of r.v's $\mathbf{D}_t = (D_{t,1}, \ldots, D_{t,p}) \sim MULT(1; \phi_1, \ldots, \phi_p)$ independent of (Z_t). A process (X_t) which follows the recursion

$$X_t = D_{t,1}(a \circ_t X_{t-1}) + \cdots + D_{t,p}(a \circ_t X_{t-p}) + Z_t,\ t \in \mathbb{Z},$$

is called a CINAR(p) process if

1. The thinning operators at time n are performed independently of each other, of (Z_t) and $(\mathbf{D}_t)$. Additionally, the thinning operators of X_t are independent of those of $(X_s)_{s<t}$;
2. Z_t and $\mathbf{D}_t$ are independent of all X_s and $a \circ_{s+j} X_s$ with $s < t$, $j = 1, \ldots, p$;
3. The conditional probability $P(a \circ_{t+1} X_t, \ldots, a \circ_{t+p} X_t | X_t = x_t, \mathcal{H}_{t-1})$ equals $P(a \circ_{t+1} X_t, \ldots, a \circ_{t+p} X_t | X_t = x_t)$, where $\mathcal{H}_{t-1}$ abbreviates the process history of all X_s and $a \circ_{s+j} X_s$ for $s \leq t-1$, and $j = 1, \ldots, p$.

Note that the above definition states that X_t is equal to $a \circ_t X_{t-i} + Z_t$ with probability ϕ_i, $i = 1, \ldots, p$. The CINAR(p) model shares some properties with the INAR(1) model, namely that the expectation and the variance are given by the same form. The autocorrelation structure was first derived for the CINAR(2) model by Zhu and Joe (2006) and for the general case by Weiß (2008a).

Theorem 5.2.4 (ACF of the CINAR(p) model). *Let (X_t) be a stationary CINAR(p) process according to Definition 5.2.1. Let $\gamma(k)$ be the autocovariance function and*

$$\mu(i,j) := E((a \circ_t X_{t-i})X_{t-k}) - aE(X_{t-i}X_{t-k}),\ k \geq 1.$$

Then the autocovariances can be determined recursively from the equations

$$\gamma(k) = a\sum_{i=1}^{p} \phi_i \gamma(|k-i|) + \sum_{i=k+1}^{p} \phi_i \mu(i,k),$$

where $\mu(i,k) = 0$ *for* $i \leq k$*, and otherwise*

$$\mu(i,k) = \phi_{i-k}(Cov(a \circ_t X_{t-i}, a \circ_{t-k} X_{t-i}) - a^2\sigma_X^2) + a \sum_{r=k+1}^{i-1} \phi_{r-k}\mu(i,r).$$

In particular, $\mu(i, i-1) = \phi_1(Cov(a \circ_t X_{t-i}, a \circ_{t-i+1} X_{t-i}) - a^2\sigma_X^2)$.

Extensions of the INAR(p) model in (5.13) can be obtained by replacing the binomial thinning operator by the operators defined in (5.2) and (5.3). For example, the integer-valued autoregressive model of order p based on the signed binomial thinning (5.2) (INARS(p)) introduced by Kim and Park (2008) takes the form

$$X_t = a_1 \odot_t X_{t-1} + \cdots + a_p \odot_t X_{t-p} + Z_t, \tag{5.14}$$

where $|a_i| \leq 1$, $i = 1 \ldots, p$ and (Z_t) is an i.i.d. sequence of non-negative integer-valued r.v's, stochastically independent of X_{t-i} for all points in time, with finite mean $\mu_Z > 0$ and variance $\sigma_Z^2 \geq 0$. One advantage of the INARS(p) model is that it can handle *negative* integer-valued time series, whereas the previous integer-valued time series models are only applicable to non-negative integer-valued time series. Furthermore, the INARS(p) model also allows negative autocorrelations of the general pth-order to an integer-valued time series, whereas the previous integer-valued time series models can only work with positive autocorrelations. Kim and Park (2008) proved that the INARS(p) process is stationary and ergodic. The result states as follows:

Theorem 5.2.5. *Suppose that all roots of the polynomial* $z^p - a_1 z^{p-1} - \cdots - a_{p-1}z - a_p = 0$ *are inside the unit circle. Then, the process* X_t *in* L_2*-space uniquely satisfies (5.14) and* $Cov(X_s, Z_t) = 0$ *for* $s < t$*. Furthermore, the process is stationary and ergodic.*

To handle non-stationary integer-valued time series with a large dispersion Zhang et al. (2010) introduced the so-called GINARS(p) model based on the thinning operator defined in (5.3). The GINARS(p) model is defined as follows:

$$X_t = a_1 \circledast_t X_{t-1} + \cdots + a_p \circledast_t X_{t-p} + Z_t,\ t \in \mathbb{Z}, \tag{5.15}$$

where $a_i \circledast_t X_{t-i} := \text{sgn}(a_i) \cdot \text{sgn}(X_{t-i}) \cdot \sum_{j=1}^{|X_{t-i}|} \zeta_{j,i}$, with $\zeta_{j,i}$ being i.i.d. with generalized power-series distribution with finite mean $|a_i|$ and variance $\sigma_{\zeta_i}^2 < \infty$ for $i = 1, \ldots, p$. Zhang et al. (2010) give a sufficient condition to ensure that the GINARS(p) process has a unique strictly stationary and ergodic solution. The result is as follows. For simplicity in notation we define the $(p \times p)$ matrix

$$\mathbf{A} = \begin{bmatrix} |a_1| & |a_2| & \cdots & |a_{p-1}| & |a_p| \\ 1 & 0 & \cdots & 0 & 0 \\ 0 & 1 & \cdots & 0 & 0 \\ \vdots & \vdots & \ddots & \vdots & \vdots \\ 0 & 0 & \cdots & 1 & 0 \end{bmatrix}.$$

Theorem 5.2.6. *Suppose all the eigenvalues of* $\mathbf{A}$ *are inside the unit circle, then there exists a unique strictly stationary integer-valued random series* (X_t) *that satisfies (5.15) and* $Cov(X_s, Z_t) = 0$ *for* $s < t$. *Furthermore, the process is also ergodic.*

For the above GINARS(p) model, we have the following results.

Proposition 5.2.3. *Suppose* (X_t) *is a stationary process and* $\sum_{i=1}^{p} a_i \neq 1$. *Then*

$$E(X_t|X_{t-i}, 1 \le i \le p) = \sum_{i=1}^{p} a_i X_{t-i} + \mu_Z;$$

$$E(X_t) = \mu_Z/(1 - \sum_{i=1}^{p} a_i);$$

$$V(X_t|X_{t-i}, 1 \le i \le p) = \sum_{i=1}^{p} \sigma_{\xi_i}|X_{t-i}| + \sigma_Z^2;$$

$$\gamma(k) = \sum_{i=1}^{p} a_i \gamma(k-i).$$

5.2.3 INMA(q) Model

The discrete analogue of the conventional qth-order moving average process (in short INMA(q)) is defined by

$$X_t = b_0 \circ_t Z_t + b_1 \circ_t Z_{t-1} + \cdots + b_q \circ_t Z_{t-q}, \; t \in \mathbb{Z}, \tag{5.16}$$

with (Z_t) being an i.i.d. sequence of non-negative integer-valued r.v's with finite mean $\mu_Z > 0$ and variance $\sigma_Z^2 \ge 0$, and $b_0, \ldots, b_q \in [0, 1]$ with $b_q \neq 0$ and usually $b_0 = 1$. Note that a closer look to (5.16) reveals that each Z_t is involved in $q + 1$ thinning operators, at $q + 1$ different times $t, \ldots, t + q$. This implies, for example, that for the same set of parameters $(b_0, \ldots, b_q)$ a wide variety of models exhibiting very different autocovariance functions and joint distribution structure can be constructed by only changing the dependence structure of the thinning operators. Al-Osh and Alzaid (1988) suggest considering a dependence structure between the thinning operators of $b_i \circ_t Z_t$ and $b_j \circ_t Z_t$, involved in the variables X_{t+i} and X_{t+j} respectively, for all $i \neq j$ and t. One interpretation is the following: at each time t, Z_t individuals, say, enter into the system independent of previous inputs and have a maximum life-time of $q + 1$ time units. The presence of these individuals in the process (X_t) during each of their $q+1$ time units of total life-times may depend on their past and future presence along these units. Thus each of the individuals will be present in the system during its life-time according to some set of probabilities according with a vector of $q+1$ Bernoulli r.v's. Note that a dependence structure between the thinning operators $b_0 \circ_t Z_t, b_1 \circ_t Z_{t-1}, \ldots, b_q \circ_t Z_{t-q}$ is then generated for all t. We will further assume that all other thinning operators are independent. According with the dependence structure described above, several different INMA(q) models can be defined:

Model 1 (McKenzie 1988): in this case all thinning operators are independent, provided that, for finite q, all the Z_t individuals that come into the systems at time t are available until time $t + q$. Furthermore, it is also assumed that the thinning operators performed on them are independent. For this model the ACF is given by

$$\gamma(k) = \begin{cases} \sigma_Z^2 \sum_{i=0}^{q-k} b_i b_{i+k} & k = 1, \ldots, q \\ 0 & k \geq q+1 \end{cases}.$$

Model 2 (Brännäs and Hall 2001): suppose that Z_t represents the number of individuals that join the system at time t having a fixed maximum life-time. Furthermore, suppose that if an individual remains *alive* at time $t + j$, with $j = 0, \ldots, q$, then it is present in X_{t+j}. Let η_j, $j = 1, \ldots, q$, be the probability of surviving an extra time-unit conditional on survival for j time-units, then the probability of survival until time $i = 1, \ldots, q$, for each individual, is given by $b_i = \eta_i b_{i-1} = b_0 \prod_{j=1}^{i} \eta_j$. According with this definition, the INMA(q) model in (5.16) expresses the total number of individuals present in the system at time t, which is defined as the sum of the contributions of survivors of the present and past q time units. For this model the ACF is given by

$$\gamma(k) = \begin{cases} \sum_{i=0}^{q-k} \mu_Z (1 - b_i) b_{i+k} & \\ \quad + \sigma_Z^2 b_i b_{i+k} & k = 1, \ldots, q. \\ 0 & k \geq q+1 \end{cases}$$

Model 3 (Al-Osh and Alzaid 1988): Al-Osh and Alzaid's model resembles Model 2. However, this model allows migration of individuals and therefore, individuals can come and go several times during their fixed life-times. For this model the ACF is given by

$$\gamma(k) = \begin{cases} \sum_{i=0}^{q-k} \mu_Z b_i (b_k - b_{i+k}) & \\ \quad + \sigma_Z^2 b_i b_{i+k} & k = 1, \ldots, q. \\ 0 & k \geq q+1 \end{cases}$$

Model 4 (Brännäs and Hall 2001): this model is useful for describing time series representing volumes of sold goods of gradually deteriorating quality. Suppose now that Z_t represents the number of new items that enter the system (i.e. a store) at time t having a fixed maximum life $q + 1$ time units. Note that from a sales point of view, each item will either be sold at some instant of its life-time, or put away after the instant $t + q$. By assuming that b_i represents the probability of each item in Z_t being sold at time $t + i$, the joint distribution $(b_0 \circ_t Z_t, b_1 \circ_t Z_t, \dots, b_q \circ_t Z_t)$, conditioned on Z_t is multinomial and $\sum_{i=0}^{q} b_i \leq 1$. Within this framework the INMA(q) model in (5.16) expresses the total sales volume at time t. For this model the ACF is given by

$$\gamma(k) = \begin{cases} (\sigma_Z^2 - \mu_Z) \sum_{i=0}^{q-k} b_i b_{i+k} & k = 1, \dots, q \\ 0 & k \geq q+1 \end{cases}.$$

Note that for the four models the cutoff property of the ACF is equivalent to that of a conventional MA(q) process. Weiß (2008c) showed that these four models can be embedded into a single family of INMA(q) models.

5.3 Parameter Estimation

Parameter estimation can be grouped according to three broad categories: moment-based, regression-based, and likelihood-based estimators. We present the results only for models based on the binomial thinning operator.

5.3.1 Moment-Based Estimators

Estimation procedures based on sample moments are appealing because of their simplicity. Without employing any distributional assumption, the Yule-Walker (YW) estimators for the binomial thinning parameters $a_1, \dots, a_p$ of the INAR(p) model satisfy the following system of linear equations

$$\begin{bmatrix} \hat{\rho}(0) & \hat{\rho}(1) & \cdots & \hat{\rho}(p-1) \\ \hat{\rho}(1) & \hat{\rho}(0) & \cdots & \hat{\rho}(p-2) \\ \vdots & \vdots & \ddots & \vdots \\ \hat{\rho}(p-1) & \hat{\rho}(p-2) & \cdots & \hat{\rho}(0) \end{bmatrix} \begin{bmatrix} \hat{a}_1 \\ \hat{a}_2 \\ \vdots \\ \hat{a}_p \end{bmatrix} = \begin{bmatrix} \hat{\rho}(0) \\ \hat{\rho}(1) \\ \vdots \\ \hat{\rho}(p) \end{bmatrix}.$$

For the Poisson INAR (hereafter PoINAR) model of order one, the YW estimators for the thinning parameter a and the mean λ are given by

$$\hat{a}_{YW} = \frac{\sum_{i=1}^{n-1}(X_{i+1} - \bar{X})(X_i - \bar{X})}{\sum_{i=1}^{n}(X_i - \bar{X})^2}$$

and

$$\hat{\lambda}_{YW} = \bar{X}(1 - \hat{a}_{YW}),$$

respectively. Jung et al. (2005) proposed the following alternative estimator for the thinning parameter when dealing with small and medium size samples

$$\hat{a}_{YW} = \frac{n\sum_{i=1}^{n-1}(X_{i+1} - \bar{X})(X_i - \bar{X})}{(n-1)\sum_{i=1}^{n}(X_i - \bar{X})^2}.$$

Freeland and McCabe (2005) proved that the asymptotic distribution of the estimators based on YW equations is asymptotically equivalent to that of estimators based on the Conditional Least Squares (CLS) estimator.

For the DL and AA specifications of the INAR(2) model, YW estimators can be obtained from the first- and second-order sample autocorrelations. The ACF of the PoINAR(2)-AA model yields the following YW estimators for a_1 and a_2

$$\hat{a}_{1|YW-AA} = \hat{\rho}(1)$$

and

$$\hat{a}_{2|YW-AA} = \hat{\rho}(2) - \hat{\rho}(1)^2,$$

whereas the associated YW estimator for λ can be obtained from the unconditional mean of the PoINAR(2)-AA and is given by

$$\hat{\lambda}_{YW-AA} = \bar{X}(1 - \hat{a}_{1|YW-AA} - \hat{a}_{2|YW-AA}). \tag{5.17}$$

Moreover the corresponding YW estimators for the INAR(2)-DL model are obtained by recalling that the ACF is the same as that of an AR(2) model, yielding

$$\hat{a}_{1|YW-DL} = \hat{\rho}(1)\left(\frac{1 - \hat{\rho}(2)}{1 - \hat{\rho}(1)^2}\right)$$

and

$$\hat{a}_{2|YW-DL} = 1 - \frac{1 - \hat{\rho}(2)}{1 - \hat{\rho}(1)^2}.$$

The YW estimator for the parameter λ is as in (5.17), being $\hat{a}_{1|YW-AA}$ and $\hat{a}_{2|YW-AA}$ replaced by $\hat{a}_{1|YW-DL}$ and $\hat{a}_{2|YW-DL}$, respectively.

5.3.2 Conditional Least Squares

The CLS estimator is based on the minimization of the sum of squared deviations about the conditional mean $E(X_i|\mathcal{F}_{i-1})$

$$Q(\boldsymbol{\theta}) := \sum_{i=1}^{n}(X_i - E(X_i|\mathcal{F}_{i-1}))^2,$$

with $\boldsymbol{\theta}$ being the vector of unknown parameters and $\mathcal{F}_k$ the σ-algebra generated by the r.v's $X_1, \ldots, X_k$. By assuming that μ_Z and σ_Z^2 are known, $Q(\boldsymbol{\theta})$ has to be minimized with respect to the parameter a. The resulting CLS estimator $\hat{a}_{CLS}$, based on the observations $(X_1, \ldots, X_n)$, takes the form

$$\hat{a}_{CLS} = \frac{\sum_{i=2}^{n}(X_i - \mu_Z)X_{i-1}}{\sum_{i=1}^{n} X_{i-1}^2}.$$

Using the arguments given in Hall and Heyde (1980, §6.3), it follows easily that $\hat{a}_{CLS}$ is a strongly consistent estimator of a, as $n \to \infty$, for all $a \in [0, 1)$. Furthermore, under the assumption $E(Z_1^3) < \infty$ it follows that

$$\sqrt{n}(\hat{a}_{CLS} - a) \xrightarrow{d} N(0, \sigma_a^2), \; n \to \infty,$$

with

$$\sigma_a^2 = \frac{a(1-a)E(X^3) + \sigma_Z^2 E(X^2)}{[E(X^2)]^2},$$

where the distribution of X is the common distribution of the unique stationary solution of the INAR(1) model.

By considering now the case (a, μ_Z) unknown, it turns out that the CLS estimators $(\hat{a}_{CLS}, \hat{\mu}_{Z|CLS})$ of (a, μ_Z) are given by

$$\hat{a}_{CLS} = \frac{n\sum_{i=2}^{n} X_i X_{i-1} - \sum_{i=1}^{n} X_i \sum_{i=2}^{n} X_{i-1}}{n\sum_{i=2}^{n} X_{i-1}^2 - \left(\sum_{i=2}^{n} X_{i-1}\right)^2}$$

and

$$\hat{\mu}_{Z|CLS} = \frac{1}{n}\left(\sum_{i=1}^{n} X_i - \hat{a}_{CLS}\sum_{i=2}^{n} X_{i-1}\right).$$

Following again the arguments given in Hall and Heyde (1980, §6.3) it follows that the pair $(\hat{a}_{CLS}, \hat{\mu}_{Z|CLS})$ is a strongly consistent estimator of (a, μ_Z), as $n \to \infty$, for all $(a, \mu_Z) \in [0, 1) \times (0, \infty)$. Furthermore, as in the previous case under the assumption $E(Z_1^3) < \infty$, it follows that

$$\begin{bmatrix} \sqrt{n}(\hat{a}_{CLS} - a) \\ \sqrt{n}(\hat{\mu}_{Z|CLS} - \mu_Z) \end{bmatrix} \xrightarrow{d} N(\mathbf{0}_2, W),$$

with

$$W = \frac{1}{V(X)^2} \begin{bmatrix} 1 & -E(X) \\ -E(X) & E(X^2) \end{bmatrix} V \begin{bmatrix} 1 & -E(X) \\ -E(X) & E(X^2) \end{bmatrix}$$

and

$$V = a(1-a) \begin{bmatrix} E(X^3) & E(X^2) \\ E(X^2) & E(X) \end{bmatrix} + \sigma_Z^2 \begin{bmatrix} E(X^2) & E(X) \\ E(X) & 1 \end{bmatrix}.$$

As stated above, Freeland and McCabe (2005) proved that for the PoINAR(1) model the asymptotic distribution of the resulting CLS estimators are asymptotically equivalent to that of estimators based on the YW estimators as the next result shows.

Theorem 5.3.1. *In the PoINAR*(1) *model*

$$\begin{pmatrix} \hat{a}_{YW} - \hat{a}_{CLS} \\ \hat{\lambda}_{YW} - \hat{\lambda}_{CLS} \end{pmatrix} = o_p(n^{-1/2}),$$

and this is sufficient for the CLS and YW estimators to have the same asymptotic distribution.

The asymptotic distribution of the CLS estimators is given in the next result.

Theorem 5.3.2. *In the PoINAR*(1) *model the CLS estimators are asymptotically normal, i.e.*

$$\sqrt{n} \begin{pmatrix} \hat{a}_{CLS} - a \\ \hat{\lambda}_{CLS} - \lambda \end{pmatrix} \xrightarrow{d} N(\mathbf{0}_2, V^{-1}),$$

where

$$V^{-1} = \begin{bmatrix} \frac{a(1-a)^2}{\lambda} + (1+a)(1-a) & -(1+a)\lambda \\ -(1+a)\lambda & \lambda + \frac{1+a}{1-a}\lambda^2 \end{bmatrix}.$$

For the INAR(p)-DL model, Du and Li (1991) showed the strongly consistency and asymptotic normality of the CLS estimators. Later on Latour (1998) corrected some misprints and errors found in Du and Li's paper.

Theorem 5.3.3. *The CLS estimators* $\hat{\boldsymbol{\theta}} = (\hat{a}_{1|CLS-DL}, \ldots, \hat{a}_{p|CLS-DL}, \hat{\mu}_Z)$ *of* $\boldsymbol{\theta} = (a_1, \ldots, a_p, \mu_Z)$ *are strongly consistent. In addition, under the assumption* $E(Z_1^3) < \infty$

$$\sqrt{n}(\hat{\boldsymbol{\theta}} - \boldsymbol{\theta}) \xrightarrow{d} N(\mathbf{0}_{p+1}, V^{-1}WV^{-1}).$$

Here, the elements of the matrices V *and* W *are given by*

$$[W]_{i,j} = \sum_{k=1}^{p} a_k E(X_{p-i+1}X_{p-j+1}) + \sigma_Z^2 [V]_{i,j},\ 1 \le i, j \le p+1$$

and

$$[V]_{i,j} = E\left\{\frac{\partial}{\partial\theta_i} E(X_{p+1}|\mathcal{F}_p) \frac{\partial}{\partial\theta_j} E(X_{p+1}|\mathcal{F}_p)\right\},\ 1 \le i, j \le p+1,$$

where

$$\frac{\partial}{\partial\theta_i} E(X_{p+1}|\mathcal{F}_p) = \begin{cases} X_{p-i+1} & i = 1, \ldots, p \\ 1 & i = p+1 \end{cases}.$$

Furthermore, the inverse of matrix V *is*

$$V^{-1} = \begin{bmatrix} \boldsymbol{\Gamma}_p^{-1} & -\mu_X \boldsymbol{\Gamma}_p^{-1}\mathbf{1}_p \\ -\mu_X \boldsymbol{\Gamma}_p^{-1}\mathbf{1}_p & 1 + \mu_X^2 \mathbf{1}_p' \boldsymbol{\Gamma}_p^{-1}\mathbf{1}_p \end{bmatrix},$$

with $\boldsymbol{\Gamma}_p := [\gamma(i-j)]_{1 \le i,j \le p}$ *and* $\mathbf{1}_p := [1, \ldots, 1]'$.

5.3.3 Conditional Maximum Likelihood Estimation

Conditional Maximum Likelihood (CML) estimation for the INAR(p) model is based upon the fact that conditioning on the first p observations the conditional likelihood $L(\boldsymbol{\theta}|x_1, \ldots, x_p)$ can be expressed as

$$L(\boldsymbol{\theta}|x_1, \ldots, x_p) = \prod_{i=p+1}^{n} P(x_i|x_{i-1}, \ldots, x_{i-p}), \tag{5.18}$$

and so the knowledge of the transition probabilities is sufficient for its construction. For the PoINAR(p)-DL model, Bu et al. (2008) derived the following explicit expression for the conditional probabilities in (5.18).

Proposition 5.3.1. *For the PoINAR(p)-DL model, it follows that*

$$
\begin{aligned}
& P(x_i|x_{i-1},\ldots,x_{i-p}) \\
& = \sum_{j_1=0}^{\min(x_{i-1},x_i)} \binom{x_{i-1}}{j_1} a_1^{j_1}(1-a_1)^{x_{i-1}-j_1} \sum_{j_2=0}^{\min(x_{i-2},x_i-j_1)} \binom{x_{i-2}}{j_2} a_1^{j_2}(1-a_2)^{x_{i-1}-j_2} \\
& \cdots \sum_{i_p=0}^{\min(x_{i-p},x_i-(j_1+\cdots+j_{p-1}))} \binom{x_{i-p}}{j_p} a_p^{j_p}(1-a_p)^{x_{i-p}-j_p} \\
& \times \frac{e^{-\lambda}\lambda^{x_i-(j_1+\cdots+j_p)}}{[x_i-(j_1+\cdots+j_p)]!}.
\end{aligned}
$$

The score function for a_i, $i=1,\ldots,p$, is given by

$$
\ell'_{a_i} = \frac{1}{a_i(1-a_i)} \sum_{t=p+1}^{n} \{E_t(a_i \circ X_{t-i}) - E_{t-1}(a_i \circ X_{t-i})\},
$$

with

$$
E_t(a_i \circ X_{t-i}) = \frac{a_i X_{t-i} P(x_t-1|x_{t-1},\ldots,x_{t-i}-1,\ldots,x_{t-p})}{P(x_t|x_{t-1},\ldots,x_{t-p})},
$$

whereas the score for λ is

$$
\ell'_{\lambda} = \frac{1}{\lambda} \sum_{t=p+1}^{n} \{E_t(Z_t) - E_{t-1}(Z_t)\},
$$

where

$$
E_t(Z_t) = \frac{\lambda P(x_t-1|x_{t-1},\ldots,x_{t-p})}{P(x_t|x_{t-1},\ldots,x_{t-p})}.
$$

The asymptotic distribution of the CML estimators is given below.

Theorem 5.3.4. *In the PoINAR(p)-DL model the CML estimators $\hat{\boldsymbol{\theta}}$ are asymptotically normal, i.e.*

$$
\sqrt{n}(\hat{\boldsymbol{\theta}} - \boldsymbol{\theta}) \xrightarrow{d} N(\mathbf{0}_{p+1}, V^{-1}),
$$

where V denotes the Fisher information matrix whose elements are given by

$$\ell''_{a_i a_i} = \frac{1}{a_i^2(1-a_i)^2} \sum_{t=p+1}^{n} \{(2a_i - 1)E_t(a_i \circ X_{t-i})$$
$$+V_t(a_i \circ X_{t-i}) - a_i E_{t-1}(a_i \circ X_{t-i})\},$$

$$\ell''_{a_i a_j} = \frac{1}{a_i a_j (1-a_i)(1-a_j)} \sum_{t=p+1}^{n} Cov_t(a_i \circ X_{t-i}, a_j \circ X_{t-j}),$$

$$\ell''_{a_i \lambda} = \frac{1}{\lambda a_i (1-a_i)} \sum_{t=p+1}^{n} Cov_t(a_i \circ X_{t-i}, Z_t)$$

and

$$\ell''_{\lambda\lambda} = \frac{1}{\lambda^2} \sum_{t=p+1}^{n} \{V_t(Z_t) - E_t(Z_t)\}.$$

For the PoINAR(1)-DL model the derivatives of the log-conditional likelihood function with respect to λ and a are

$$\begin{cases} S_\lambda = \frac{\partial \log L(a,\lambda|X_1,\ldots,X_p)}{\partial \lambda} = 0 \Leftrightarrow \sum_{i=2}^{n} \frac{P_i(x_i-1)}{P_i(x_i)} = n-1 \\ S_a = \frac{\partial \log L(a,\lambda|X_1,\ldots,X_p)}{\partial a} = 0 \Leftrightarrow \sum_{i=2}^{n} \left[(x_i - ax_{i-1}) - \lambda \frac{P_i(x_i-1)}{P_i(x_i)}\right] / a(1-a) = 0 \end{cases},$$

where

$$P_i(x) = e^{-\lambda} \sum_{j_1=0}^{\min(x_{i-1},x_i)} \binom{x_{i-1}}{j_1} a_1^{j_1}(1-a_1)^{x_{i-1}-j_1} \frac{\lambda^{x_i - j_1}}{(x_i - j_1)!}, \quad i = 2, \ldots, n.$$

Note that

$$\sum_{i=2}^{n} x_i - a \sum_{i=2}^{n} x_{i-1} = (n-1)\lambda, \tag{5.19}$$

implying that Eq. (5.19) can be used to eliminate one of the parameters, say a, providing that S_λ can be written as a function of λ only, and hence the CML estimate can be found by iterating on S_λ.

5.3.4 Bayesian Approach

In this section, a Bayesian analysis is conducted in order to estimate the parameters of the integer-valued ARMA (in short INARMA (p, q)) model

$$X_t = \sum_{i=1}^{p} a_i \circ X_{t-i} + \sum_{j=1}^{q} b_i \circ Z_{t-j} + Z_t, \tag{5.20}$$

where (Z_t) is an i.i.d. sequence of non-negative, integer-valued r.v's with finite variance $\sigma_Z^2 \geq 0$. For conciseness, throughout this section we shall assume that Z_1 is Poisson-distributed with parameter λ. Furthermore, it is assumed that the thinning operators are all independent and that the orders p and q are fixed parameters. For simplicity in notation we define $\mathbf{a}^{(p)} := (a_1, \ldots, a_p)$ and $\mathbf{b}^{(q)} := (b_1, \ldots, b_q)$.

In order to facilitate inference for the INARMA(p, q) model it is necessary to augment the data. For this purpose, Neal and Subba Rao (2007) introduced the following procedure: for a fixed value of $t \in \mathbb{Z}$, let $Y_{t,i} := a_i \circ X_{t-i}, i = \ldots, p$, and $V_{t,j} := b_j \circ Z_{t-j}, j = 1 \ldots, q$. Thus,

$$Z_t = X_t - \sum_{i=1}^{p} Y_{n,i} - \sum_{j=1}^{q} V_{n,j}, \ t \in \mathbb{Z}.$$

In addition, let $\boldsymbol{Y}_t^{(p)} := (Y_{t,1}, \ldots, Y_{t,p})$, $\boldsymbol{Y} := (\boldsymbol{Y}_1^{(p)}, \ldots, \boldsymbol{Y}_n^{(p)})$, $\boldsymbol{V}_t^{(q)} := (V_{t,1}, \ldots, V_{t,q})$, $\boldsymbol{V} := (\boldsymbol{V}_1^{(q)}, \ldots, \boldsymbol{V}_n^{(q)})$, and let $\boldsymbol{Z}_t := (Z_1, \ldots, Z_t)$. For $t \geq 1$, let $\boldsymbol{y}_t := (y_{t,1}, \ldots, y_{t,p})$ and $\boldsymbol{v}_t := (v_{t,1}, \ldots, v_{t,q})$ with $\boldsymbol{y} := (\boldsymbol{y}_1, \ldots, \boldsymbol{y}_n)$ and $\boldsymbol{v} := (\boldsymbol{v}_1, \ldots, \boldsymbol{v}_n)$. For $q \geq 1$, let $\mathbf{z}_{IN} := (z_{1-q}, \ldots, z_0)$ representing the initial values of Z, and for $t \geq 1$, let $\mathbf{z}_t := (z_1, \ldots, z_t)$ with $\mathbf{z}_0$ corresponding to the empty set. Furthermore, assume that we have the observed time series $\boldsymbol{x} := (x_{1-\max\{p,q\}}, \ldots, x_n)$. Note that given $(\mathbf{z}_{IN}, \mathbf{z}_{t-1}, \boldsymbol{x}_{t-1}, \mathbf{a}^{(p)}, \mathbf{b}^{(q)}, \lambda, p, q)$, with $\boldsymbol{x}_s := (x_{1-\max\{p,q\}}, \ldots, x_s)$, $(s \geq 0)$, each of the components $(\boldsymbol{V}_t^{(q)}, \boldsymbol{Y}_t^{(p)}, \mathbf{Z}_t)$ is independent. Moreover, in a Bayesian framework it is also necessary to assign prior distributions to each parameter. Let us consider the independent priors

$$g(\mathbf{a}^{(p)}) \propto 1, \ 0 \leq a_i < 1, (1 \leq i \leq p), \sum_{i=1}^{p} a_i < 1;$$

$$g(\mathbf{b}^{(q)}) \propto 1, \ 0 \leq b_j < 1, (1 \leq j \leq q), \sum_{j=1}^{q} b_i < 1;$$

$$g(\lambda) = \Gamma(A_\lambda, B_\lambda), \ A_\lambda, B_\lambda > 0,$$

where Γ stands for the Gamma distribution. Therefore, it follows that

$$f(\mathbf{v},\mathbf{y},\mathbf{z}_n,\mathbf{z}_{IN},\mathbf{a}^{(p)},\mathbf{b}^{(q)},\lambda|\mathbf{x}) \propto \prod_{t=1}^{n}\left\{\frac{\lambda^{z_t}}{z_t!}e^{-\lambda}\prod_{i=1}^{p}\binom{x_{t-i}}{y_{t,i}}a_i^{y_{t,i}}(1-a_i)^{x_{t-i}-y_{t,i}}\times\right.$$
$$\left.\times\prod_{j=1}^{q}\binom{z_{t-j}}{v_{t,j}}b_j^{v_{t,j}}(1-b_i)^{z_{t-j}-v_{t,i}}\right\}\lambda^{A_\lambda}e^{-B_\lambda\lambda}\times$$
$$\times\prod_{i=1}^{\max\{p,q\}}\frac{\lambda^{z_{1-i}}}{z_{1-i}!}e^{-\lambda}, \tag{5.21}$$

subject to the constrains

$$\sum_{i=1}^{p}y_{t,i}+\sum_{j=1}^{q}v_{t,j}+z_t=x_t,\ v_{t,j}\le z_{t-j}\ (1\le j\le q),\ y_{t,i}\le x_{t-i}\ (1\le i\le p). \tag{5.22}$$

Furthermore, since the inference for the vector of unknown parameters $(\mathbf{a}^{(p)},\mathbf{b}^{(q)},\lambda)$ shall be done through the Gibbs/rejection sampling procedure, we have to derive the set of full conditional posterior densities. The results are summarized below.

Lemma 5.3.1. *Given (5.21), the full conditional posterior densities are*

$$\pi(a_i|\mathbf{v},\mathbf{y},\mathbf{z}_n,\mathbf{z}_{IN},\mathbf{a}_{i-}^{(p)},\mathbf{b}^{(q)},\lambda,\mathbf{x}) \sim Beta\left(1+\sum_{t=1}^{n}y_{t,i},1+\sum_{t=1}^{n}(x_{t-i}-y_{t,i})\right), \tag{5.23}$$

for $i=1,\dots,p$ *where* $\mathbf{a}_{i-}^{(p)}:=(a_1,\dots,a_{i-1},a_{i+1},\dots,a_p)$, *and subject to the constrain that* $\sum_{k=1}^{p}a_k<1$;

$$\pi(b_i|\mathbf{v},\mathbf{y},\mathbf{z}_n,\mathbf{z}_{IN},\mathbf{a}^{(p)},\mathbf{b}_{j-}^{(q)},\lambda,\mathbf{x}) \sim Beta\left(1+\sum_{t=1}^{n}v_{t,i},1+\sum_{t=1}^{n}(z_{t-i}-v_{t,i})\right), \tag{5.24}$$

for $j=1,\dots,q$ *where* $\mathbf{b}_{j-}^{(q)}:=(b_1,\dots,b_{j-1},b_{j+1},\dots,b_q)$, *and subject to the constrain that* $\sum_{l=1}^{q}b_l<1$; *and*

$$\pi(\lambda|\mathbf{v},\mathbf{y},\mathbf{z}_n,\mathbf{z}_{IN},\mathbf{a}^{(p)},\mathbf{b}^{(q)},\mathbf{x}) \sim \Gamma\left(A_\lambda+\sum_{t=1-q}^{n}z_t,B_\lambda+n+q\right). \tag{5.25}$$

The steps required to implement the MCMC algorithm are the following:

1. One component at a time, draw a proposed value for a_i from (5.23).
   (a) If a_i is drawn such as $\sum_{k=1}^{p} a_k < 1$, accept the proposed value of a_i, and update the value of a_i. Otherwise repeat step 1.
2. One component at a time, draw a proposed value for b_j from (5.24).
   (a) If b_j is drawn such as $\sum_{l=1}^{q} b_l < 1$, accept the proposed value of b_j, and update the value of b_j. Otherwise repeat step 2.
3. Draw λ from (5.25).
4. Use the following Metropolis-Hastings procedure to update $(\boldsymbol{v}, \boldsymbol{y}, \mathbf{z}_n)$. For $t = 1, \ldots, n$
   (a) Draw $y'_{t,i}$ from $Bin(x_{t-i}, a_i)$, for $i = 1, \ldots, p$;
   (b) Draw $v'_{t,j}$ from $Bin(z_{t-j}, b_j)$, for $j = 1, \ldots, q$;
   (c) If $x_t < \sum_{i=1}^{p} y'_{t,j} + \sum_{j=1}^{q} v'_{t,j}$, then repeat steps (a) and (b);
   (d) Set $z'_t = x_t - (\sum_{i=1}^{p} y'_{t,j} + \sum_{j=1}^{q} v'_{t,j})$. Thus, the proposal distribution for $(\boldsymbol{v}'_t, \boldsymbol{y}'_t, z'_t)$ is independent of $(\boldsymbol{v}_t, \boldsymbol{y}_t, z_t)$ and depends only upon $\boldsymbol{x}$ and $(\mathbf{z}_{t-1}, \mathbf{z}_{IN}, \mathbf{a}^{(p)}, \mathbf{b}^{(q)})$;
   (e) Calculate the acceptance probability, ω, for the move

$$\omega = \min\left\{1, \frac{f(\boldsymbol{v}', \boldsymbol{y}', \mathbf{z}'_n | \boldsymbol{x}, \mathbf{z}_{IN}, \mathbf{a}^{(p)}, \mathbf{b}^{(q)}, \lambda)}{f(\boldsymbol{v}, \boldsymbol{y}, \mathbf{z}_n | \boldsymbol{x}, \mathbf{z}_{IN}, \mathbf{a}^{(p)}, \mathbf{b}^{(q)}, \lambda)}\right.$$
$$\left.\times \frac{q_t(\boldsymbol{v}_t, \boldsymbol{y}_t, z_t | \boldsymbol{x}, \mathbf{z}_{IN}, \mathbf{z}_{t-1}, \mathbf{a}^{(p)}, \mathbf{b}^{(q)})}{q_t(\boldsymbol{v}'_t, \boldsymbol{y}'_t, z'_t | \boldsymbol{x}, \mathbf{z}_{IN}, \mathbf{z}_{t-1}, \mathbf{a}^{(p)}, \mathbf{b}^{(q)})}\right\},$$

   where $q_t(\boldsymbol{v}'_t, \boldsymbol{y}'_t, z'_t | \boldsymbol{x}, \mathbf{z}_{IN}, \mathbf{z}_{t-1}, \mathbf{a}^{(p)}, \mathbf{b}^{(q)})$ is the conditional proposal probability for $(\boldsymbol{v}'_t, \boldsymbol{y}'_t, z'_t)$ given $(\boldsymbol{x}, \mathbf{z}_{IN}, \mathbf{z}_{t-1}, \mathbf{a}^{(p)}, \mathbf{b}^{(q)})$.
   (f) If the move is accepted set $(\boldsymbol{v}_t, \boldsymbol{y}_t, z_t)$ equal to $(\boldsymbol{v}'_t, \boldsymbol{y}'_t, z'_t)$, otherwise leave $(\boldsymbol{v}_t, \boldsymbol{y}_t, z_t)$ unchanged.
5. Finally, in order to update $\mathbf{z}_{IN}$ we proceed as follows:
   (a) Draw z'_t from Po(λ);
   (b) If $z'_t > x_t$, then repeat step (a); otherwise the proposal probability is $q(z'_t|\lambda, \boldsymbol{x}) = g(z_t|\lambda, \boldsymbol{x})/F(x_t)$, where $F(\cdot)$ denotes the cumulative distribution of a Poisson random variable with mean λ.
   (c) Calculate the acceptance probability, ω, for the move

$$\omega = \min\left\{1, \frac{f(\boldsymbol{v}', \boldsymbol{y}', \mathbf{z}'_n | \boldsymbol{x}, \mathbf{z}_{IN}, \mathbf{a}^{(p)}, \mathbf{b}^{(q)}, \lambda) g(z'_t|\lambda, \boldsymbol{x})}{f(\boldsymbol{v}_t, \boldsymbol{y}_t, \mathbf{z}_n | \boldsymbol{x}, \mathbf{z}_{IN}, \mathbf{a}^{(p)}, \mathbf{b}^{(q)}, \lambda) g(z_t|\lambda, \boldsymbol{x})} \times \frac{q(z_t|\lambda, \boldsymbol{x})}{q(z'_t|\lambda, \boldsymbol{x})}\right\},$$

6. Repeat stages (1)–(5) say, N times, to generate a MCMC sequence of non-independent observation from the posterior distribution of both the model parameters $(\mathbf{a}^{(p)}, \mathbf{b}^{(q)}, \lambda)$ and the augmented data $(\boldsymbol{v}, \boldsymbol{y}, \mathbf{z}_n, \mathbf{z}_{IN})$. Moreover, the assessment of the convergence of the algorithm can be carried out through ad hoc methods such as plotting and inspecting traces of the observed MCMC sample. These ad hoc methods, however, tend to fail in the presence of a phenomenon called metastability, i.e. the Markov chain appears to have converged to the true equilibrium value, but after some period of stability around this value the Markov chain may suddenly move to another region of the parameter space. Along with the ad hoc techniques described above, a number of more formal methods exist which are prevalent in the literature. Two of the most popular were proposed by Geweke (1992) and by Raftery and Lewis (1992).

5.4 Model Selection

In this section an automatic criterion for identifying the most appropriate INARMA model is presented, with emphasis on the underlying ideas rather than on the mathematical details. The basic motivation behind model selection is to choose a model from a family of models so that the selected model describes the data best according with some criterion.

In analyzing INARMA processes, model-selection criteria based on the likelihood approach are in general intractable due to the complexity of the likelihood function. To tackle this issue, an alternative approach has been suggested by Enciso-Mora et al. (2009) who developed an efficient reversible jump Markov Chain Monte Carlo (MCMC) method for conducting inference for the orders p and q for moving between competing INARMA models. For completeness and reader's convenience basic results on the general reversible jump MCMC methodology are given below. We follow closely Brooks et al. (2003).

Suppose that we have a countable collection of candidate models, say, $M_1, \ldots, M_k, \ldots$, where model M_i has a continuous parameter space Θ_i containing elements $\boldsymbol{\theta}^{(i)}$ lying in $\mathbb{R}^{n_i}$, where the dimension n_i may vary from model to model. Let $\pi(M_i, \boldsymbol{\theta}^{(i)})$ denote the density part of the target distribution $\pi(\cdot)$ restricted to model M_i. Thus, for an arbitrary set B it follows that

$$\pi(B) = \sum_i \int_{B \cap \Theta_i} \pi(M_i, \boldsymbol{\theta}^{(i)}) d\boldsymbol{\theta}^{(i)}.$$

Attention is focused on moves between models M_i and M_j with $n_i < n_j$. Note that by reversibility this also characterizes the reverse move and moves between all collection of pairs of models can be dealt with similarly.

In a typical application, given that the chain is currently in state $(M_i, \boldsymbol{\theta}^{(i)})$, a new value for the chain $(M_j, \boldsymbol{\theta}^{(j)})$ is drawn from some proposal distribution $q(\boldsymbol{\theta}^{(i)}, d\boldsymbol{\theta}^{(j)})$, which is either accepted or rejected. Green (1995) proved that, if $\pi(d\boldsymbol{\theta}^{(j)})q(\boldsymbol{\theta}^{(i)}, d\boldsymbol{\theta}^{(j)})$ is dominated by a symmetric measure and has Radon-Nikodym derivative, say $r(\boldsymbol{\theta}^{(i)}, \boldsymbol{\theta}^{(j)})$ with respect to this symmetric measure, then detailed balance is preserved if the proposed new state is accepted with probability

$$\omega\{(M_i, \boldsymbol{\theta}^{(i)}), (M_j, \boldsymbol{\theta}^{(j)})\} := \min\{1, A_{i,j}(\boldsymbol{\theta}^{(i)}, \boldsymbol{\theta}^{(j)})\},$$

where

$$A_{i,j}([M_i, \boldsymbol{\theta}^{(i)}], [M_j, \boldsymbol{\theta}^{(j)}]) = \frac{r(\boldsymbol{\theta}^{(j)}, \boldsymbol{\theta}^{(i)})}{r(\boldsymbol{\theta}^{(i)}, \boldsymbol{\theta}^{(j)})}.$$

For the large majority of model selection problems, however, this general formulation can be simplified by restricting attention to certain jump constructions, as follows. To move from model M_i to M_j, first a random vector $\mathbf{V}$ of length $n_j - n_i$ consisting of variables drawn from some proposal density, say $\varphi(\cdot)$ is generated. For simplicity in notation, the joint density of $\mathbf{V}$ will be denoted by

$$\varphi_{n_j-n_i}(\mathbf{v}) := \prod_{i=1}^{n_j-n_i} \varphi(v_i).$$

Secondly, we propose to move from $\boldsymbol{\theta}^{(i)}$ to $\boldsymbol{\theta}^{(j)} = h_{i,j}(\boldsymbol{\theta}^{(i)}, \mathbf{V})$, where the so-called jump function $h_{i,j} : \Theta_i \times \mathbb{R}^{n_j-n_i}$ denotes an injection, mapping the current state of the chain together with the generated random vector $\mathbf{V}$ to a point in the higher dimensional space. Finally, this move is accepted with probability $\omega\{(M_i, \boldsymbol{\theta}^{(i)}), (M_j, \boldsymbol{\theta}^{(j)})\}$ being $A_{i,j}$ given by

$$A_{i,j}(\boldsymbol{\theta}^{(i)}, \boldsymbol{\theta}^{(j)}) = \frac{\pi(M_j, \boldsymbol{\theta}^{(j)}) s_{j,i}(\boldsymbol{\theta}^{(j)})}{\pi(M_i, \boldsymbol{\theta}^{(i)}) s_{i,j}(\boldsymbol{\theta}^{(i)}) \varphi_{n_j-n_i}(\mathbf{v})} \left| \frac{\partial h_{i,j}(\boldsymbol{\theta}^{(i)}, \mathbf{v})}{\partial(\boldsymbol{\theta}^{(i)}, \mathbf{v})} \right|, \tag{5.26}$$

with $s_{i,j}(\boldsymbol{\theta}^{(i)})$ denoting the probability that a proposed jump to model j is attempted at any particular iteration, starting from $\boldsymbol{\theta}^{(i)}$ in Θ_i. The case $n_i > n_j$ follows easily by taking $A_{i,j}(\boldsymbol{\theta}^{(i)}, \boldsymbol{\theta}^{(j)}) = A_{j,i}(\boldsymbol{\theta}^{(j)}, \boldsymbol{\theta}^{(i)})^{-1}$. Extensions, variations and modifications of reversible jump MCMC methodology can be found in Brooks et al. (2003).

For the general situation in which the order p and q of the INARMA process are unknown the reversible jump MCMC method is adopted to estimate model parameters and also the optimal order. Within this framework, the updating of the parameters and the data augmentation is carried out in a two-step procedure:

- The MCMC algorithm of Neal and Subba Rao (2007) introduced in Sect. 5.3.4 is used as a sub-algorithm for the within model moves (which corresponds to update both the parameters and the data augmentation when assuming p and q as fixed); and

- A new order for p and q is proposed by means of the efficient reversible jump MCMC introduced by Enciso-Mora et al. (2009) to perform moves between models.

The order determination algorithm, introduced by Enciso-Mora and co-authors, is as follows: let (X_t) be the INARMA(p, q) process in (5.20). For within-model moves the choice of prior distributions for p and q deserves special care since, as either p and q increases by one, the number of parameters within the model increases by $n+1$ and n new augmented data. For large sample sizes ($n \geq 400$) a good choice is $g(p) \propto n^{-p/2}$ and $g(q) \propto n^{-q/2}$. Finally, for $1 - \max\{p, q\} \leq t \leq 0$ we take $g(z_t|\lambda, \boldsymbol{x})$ as being Poisson-distributed with parameter λ subject to the constraint that $z_t \leq x_t$, since a priori not only Z_t is Poisson distributed with parameter λ, but also by definition $z_t \leq x_t$.

Thus, the full joint likelihood of $(\boldsymbol{V}, \boldsymbol{Y}, \boldsymbol{Z}_n, \mathbf{a}^{(p)}, \mathbf{b}^{(q)}, \lambda, p, q)$ given the data takes the form

$$f(\boldsymbol{v}, \boldsymbol{y}, \mathbf{z}_n, \mathbf{z}_{IN}, \mathbf{a}^{(p)}, \mathbf{b}^{(q)}, \lambda, p, q|\boldsymbol{x}) = g(p)g(q)p!q! \times$$
$$\times \prod_{t=1}^{n} \left\{ \frac{\lambda^{z_t}}{z_t!} e^{-\lambda} \prod_{i=1}^{p} \binom{x_{t-i}}{y_{t,i}} a_i^{y_{t,i}} (1-a_i)^{x_{t-i}-y_{t,i}} \times \right.$$
$$\left. \times \prod_{j=1}^{q} \binom{z_{t-j}}{v_{t,j}} b_j^{v_{t,j}} (1-b_i)^{z_{t-j}-v_{t,i}} \right\} \lambda^{A_\lambda} e^{-B_\lambda \lambda} \times$$
$$\times \prod_{i=1}^{\max\{p,q\}} \frac{\lambda^{z_{1-i}}}{z_{1-i}!} e^{-\lambda},$$

subject to the constrains in (5.22).

We now describe the order switching step proposed by Enciso-Mora et al. (2009). We restrict our attention to moves within either INAR model or INMA models. At each iteration alter the dimensionality of the INAR model (increasing or decreasing) by one with each move being proposed with probability $1/2$. This is subject to constraints at $p = 0$ and $p = p_{max}$, with p_{max} representing the maximum value allowed for p. Next adopt the same procedure for the INMA order with constraints at $q = 0$ and $q = q_{max}$. In order to develop an efficient reversible jump MCMC algorithm, a natural suggestion when constructing the order switching step is to keep μ_X fixed. This procedure remains valid as long as we are not proposing to move the INAR (INMA) order p (q) either to or from $p = 0$ $(q = 0)$. When dealing with INAR models the algorithm is as follows: consider extending an INAR(p) model to an INAR(p') with $p' = p + 1$. In order to retain μ_X fixed, $\mathbf{a}^{(p')}$ has to be choose such that

$$\sum_{i=1}^{p'} a_i' = \sum_{i=1}^{p} a_i, \tag{5.27}$$

with $\mathbf{b}^{(q')} = \mathbf{b}^{(q)}$ and $\lambda' = \lambda$. This can be accomplished as follows: let U be a uniformly distributed random variable in the interval $[0, 1]$ and let K be drawn uniformly at random from $\{1, \ldots, p\}$. Then for $i \in \{1, \ldots, p\}/K$ set a'_i. Let $a'_K = Ua_K$ and $a'_{p+1} = (1-U)a_K$. In addition, for $1 \leq t \leq n$, let N_t be a Binomial-distributed random variable with parameters $y_{t,K}$ and U respectively, and set $y'_{t,k} = N_t$, $y'_{t,p+1} = y_{t,K} - N_t$. The rest of the augmented data terms are kept fixed. Note that in this case, $\boldsymbol{\theta}^{(i)} = \boldsymbol{\theta}^{(p)} = (\mathbf{a}^{(p)}, \mathbf{w}^{(p)}, p)$ with $\mathbf{w}^{(p)} = (\mathbf{y}^{(p)}, \mathbf{v}^{(p)}, \mathbf{z}^{(p)})$. Thus the acceptance ratio in (5.26), can be expressed as

$$A_{p,p'}(\boldsymbol{\theta}^{(p)}, \boldsymbol{\theta}^{(p')}) = \frac{L(\mathbf{x}|\boldsymbol{\theta}^{(p')})}{L(\mathbf{x}|\boldsymbol{\theta}^{(p)})} \times \frac{g(\boldsymbol{\theta}^{(p')})g(M_{p'})}{g(\boldsymbol{\theta}^{(p)})g(M_p)} \times \frac{s_{p',p}(\boldsymbol{\theta}^{(p')})}{s_{p,p'}(\boldsymbol{\theta}^{(p)})\varphi(v)} \times |J|$$

$$= \text{likelihood ratio} \times \text{prior ratio} \times \text{proposal ratio} \times \text{Jacobian},$$

where $g(M_{p'})/g(M_p) = 1$, $s_{p',p}(\boldsymbol{\theta}^{(p')}) = s_{p,p'}(\boldsymbol{\theta}^{(p)}) = 1/(2p)$,

$$\frac{L(\mathbf{x}|\boldsymbol{\theta}^{(p')})}{L(\mathbf{x}|\boldsymbol{\theta}^{(p)})} = \prod_{i=1}^{n} \frac{\binom{x_{i-K}}{y'_{i,K}} (a'_K)^{y'_{i,K}} (1-a'_K)^{x_{i-K}-y'_{i,K'}}}{\binom{x_{i-K}}{y_{i,K}} a_K^{y_{i,K}} (1-a_K)^{x_{i-K}-y_{i,K}}}$$

$$\times \binom{x_{i-p'}}{y'_{i,p'}} (a'_{p'})^{y'_{i,p'}} (1-a'_{p'})^{x_{i-p'}-y'_{i,p'}},$$

$$\frac{g(\boldsymbol{\theta}^{(p')})}{g(\boldsymbol{\theta}^{(p)})} = \frac{g(\mathbf{a}^{(p)})g(\mathbf{w}^{(p')})}{g(\mathbf{a}^{(p)})g(\mathbf{w}^{(p)})} = \frac{g(p')}{g(p)} \times (p+1)$$

and

$$\varphi(v) = \varphi(U) = \prod_{i=1}^{n} \binom{y_{i,K}}{y'_{i,K}} U^{y_{i,K}} (1-U)^{y_{i,K}-y'_{i,K}}.$$

Finally, the Jacobian $|J| = a_K$. The reverse move from p to $p' = p - 1$ is simply the reverse of that given above.

The INMA order step procedure is basically identical to that given above with condition (5.27) replaced by

$$\sum_{i=1}^{q'} b'_i = \sum_{i=1}^{q} b_i,$$

with $\mathbf{a}^{(p')} = \mathbf{a}^{(p)}$ and $\lambda' = \lambda$. In particular, choose K uniformly from $\{1, \ldots, q\}$, splitting the Kth term with $b'_K = Ub_K$ and $b'_{q+1} = (1-U)b_K$. Afterwards split the corresponding augmented data terms and set $\nu'_{t,K} = S_t$ and $\nu'_{t,q+1} = \nu_{t,K} - S_t$ with S_t being a Binomial-distributed random variable with parameters $\nu_{t,K}$ and U respectively.

5.5 Extremal Behavior

Within the reasonably large spectrum of integer-valued models proposed in the literature, only a few have already been studied with regard to their tail behavior and extremal properties. In part this is due to the fact that many integer-valued distributions do not belong to the domain of attraction of a extreme-value distribution. Anderson (1970) gave a remarkable contribution to the study of the extremal properties of integer-valued i.i.d. sequences and as an example the author analyzed the behavior of the maximum queue length for $M/M/1$ system. Extensions of Anderson's results were proposed by Hooghiemstra et al. (1998) who provided bounds and approximations for the distribution of the maximum queue length for $M/M/s$ queues, based on an asymptotic analysis involving the extremal index. McCormick and Park (1992) were the first to study the extremal properties of some models obtained as discrete analogues of continuous models, replacing scalar multiplication by random thinning. Hall (1996) analyzed the asymptotic behavior of the maximum term of a particular Markovian model. Hall (2001) provided results regarding the limiting distribution of the maximum of sequences within a generalized class of INMA models driven by i.i.d. heavy-tailed innovations. Extensions in the INMA context were also proposed for systematic (Hall and Scotto 2003), periodic (Hall et al. 2004), and randomly sub-sampled (Hall and Scotto 2008) INMA sequences. Extremal properties of random coefficients INAR and INMA models have been considered by Roitershtein and Zhong (2013) and Hall et al. (2010), respectively. The extremal behavior of exponential type-tailed innovations has been analyzed by Hall (2003), Hall and Scotto (2006), and Hall and Temido (2009, 2012). Hall and Moreira (2006) derived the extremal properties of a particular moving average count data model introduced by McKenzie (1986).

The analysis of integer-valued sequences of i.i.d. non-degenerate r.v's $(Z_1, \ldots, Z_n)$ with common distribution F require extra care since, in many cases, there is no non-degenerate limiting distribution for the maximum term $M_n(Z) := \max(Z_1, \ldots, Z_n)$. This is due to the fact that one needs to impose certain continuity conditions on F at its right endpoint. The following result quoted in Leadbetter et al. (1983, Theorem 1.7.13) is crucial to understand which conditions of F ensure that the limit of $P(M_n(Z) \leq u_n)$ as $n \to \infty$, exists for an appropriate sequence of real numbers (u_n).

Theorem 5.5.1. *Let F be a distribution function with right endpoint $x_F = \sup\{x : F(x) < 1\}, (x_F \leq \infty)$, and let $\tau \in (0, \infty)$. There exist a sequence (u_n) satisfying $n(1 - F(u_n)) \to \tau$ iff*

$$\lim_{x \to x_F} \frac{1 - F(x)}{1 - F(x-)} = 1. \tag{5.28}$$

Unfortunately, condition (5.28) tends to fail when F is a discrete distribution such as the Binomial, Poisson or Negative Binomial. To tackle this problem Anderson (1970) defined a particular class of discrete distributions for which the maximum term (under an i.i.d. setting) possesses an almost stable behavior in the sense of the following theorem:

Theorem 5.5.2. *Let F be a distribution function whose support consists of all sufficiently large integers. Then, there exists a sequence of real numbers (d_n) so that*

$$\begin{cases} \limsup_{n\to\infty} F^n(x + d_n) \leq e^{-e^{-\alpha x}} \\ \liminf_{n\to\infty} F^n(x + d_n) \geq e^{-e^{-\alpha(x-1)}} \end{cases},$$

for some $\alpha > 0$ and for every $x \in \mathbb{R}$, iff

$$\lim_{n \to \infty} \frac{1 - F(n)}{1 - F(n+1)} = e^{\alpha}, \tag{5.29}$$

with $d_n = F_c^{-1}(1 - \frac{1}{n})$ where F_c is any continuous distribution in the domain of attraction of the Gumbel distribution with $F_c([n]) = F(n)$.

Whenever a distribution F satisfies the conditions of the above theorem, we shall denote it by $F \in D_\alpha(Anderson)$. The Geometric and the Negative Binomial distributions are well-known members of this class. An important family of distributions belonging to Anderson's class are those having exponential-type tails of the form

$$1 - F(n) \sim K n^{\xi} (1 + \lambda)^{-n}, \; n \to \infty, \tag{5.30}$$

with $\xi \in \mathbb{R}$ and $\lambda, K > 0$. Note that for this class of distributions the expression in (5.29) holds with $e^{\alpha} = 1 + \lambda$.

Extensions for stationary sequences exhibiting weak, long-range dependence at high levels (Leadbetter's $D(u_n)$ condition) along with the local dependence condition $D^{(k)}(u_n)$, $k \in \mathbb{N}$, were first proposed by McCormick and Park (1992) for the case $k = 1$ and by Hall (1996) who obtained the following result for the general case.

Theorem 5.5.3. *Suppose that for some $k \geq 1$, conditions $D(u_n)$ and $D^{(k)}(u_n)$ hold for the stationary sequence (Z_t) with marginal $F \in D_\alpha(Anderson)$, where (u_n) is a sequence of the form $u_n = x + d_n$. Then there exists a value $0 \leq \theta \leq 1$ such that*

$$\begin{cases} \limsup_{n\to\infty} P(M_n(Z) \le x + d_n) \le e^{-\theta e^{-\alpha x}} \\ \liminf_{n\to\infty} P(M_n(Z) \le x + d_n) \ge e^{-\theta e^{-\alpha(x-1)}} \end{cases},$$

iff

$$P(M_{2,k}(Z) \le u_n | Z_1 > u_n) \underset{n\to\infty}{\longrightarrow} \theta, \tag{5.31}$$

where $M_{2,k}(Z) = -\infty$, *for* $2 > k$, *and* $M_{2,k}(Z) = \max_{2\le r\le k}(Z_r)$, *for* $2 \le k$.

Remark 5.5.1. If $k = 1$ then $\theta = 1$.

Here, θ in (5.31) is the extremal index.

When attention is focused on the characterization of the extremal properties of models involving the binomial thinning operator, it is necessary to analyze the effect of this operator upon the tail distribution of the thinned random variable $Y := b \circ Z$. Hall (2003) proved that if the tail of F_Z satisfies (5.30) then the tail distribution of Y is of the same type.

Theorem 5.5.4. *Let* Z *be a non-negative integer-valued random variable with* $F_Z \in D_\alpha$*(Anderson) of the form (5.30). Then* $F_Y \in D_\alpha$*(Anderson) and*

$$1 - F_Y(n) \sim K^* n^\xi (1 + \lambda/b)^{-n},$$

with $K^* = Kb(\frac{1+\lambda}{b+\lambda})^{\xi+1}$.

It is also possible to assess the influence of the thinning operator on the random variable Y for distributions not belonging to Anderson's class. For example, assume that the random variable Z has distribution function belonging to the domain of attraction of the Fréchet distribution with parameter $\alpha > 0$, that is

$$1 - F_Z(n) = n^{-\alpha} L_Z(n), \; n > 0, \tag{5.32}$$

for some slowly varying function $L_Z(n)$ at $+\infty$, in short $F_Z \in D(\Phi_\alpha)$. This class of distributions includes some usual discrete distributions such as the Zeta distribution ($Z \sim Zeta(\rho)$, $\Leftrightarrow$ $P(Z = n) = n^{-(\rho+1)}/\zeta(\rho + 1)$, $n \in \{1, 2, \dots\}$, $\rho > 0$ where $\zeta(\cdot)$ is the Riemann zeta function). In this case, the thinning operator produces the same asymptotic effect as multiplying the random variable Z by the real constant b.

Lemma 5.5.1. *Let* Z *be a non-negative integer-valued random variable with* $F_Z \in D(\Phi_\alpha)$, $\alpha > 0$. *Then* $F_Y \in D(\Phi_\alpha)$ *and*

$$\lim_{n\to\infty} \frac{1 - F_Y(n)}{1 - F_Z(n)} = b^\alpha.$$

A general approach to look at the extremal properties of integer-valued ARMA models is through the analysis of INMA(∞) models of the form[7]

$$X_t = \sum_{i=0}^{\infty} b_i \circ_t Z_{t-i}, \tag{5.33}$$

with (Z_t) being an i.i.d. sequence of non-negative integer-valued r.v's, with common distribution F_Z satisfying either (5.32) or (5.30). Furthermore, in order to ensure the almost sure convergence of the sum on the right-hand side of (5.33), for every t, the coefficients need to satisfy $\sum_{i=0}^{\infty} b_i^{\delta} < \infty$, with $\delta < \min(\alpha, 1)$, for the heavy-tailed case and $\sum_{i=0}^{\infty} b_i < \infty$, for the exponential-type case. The following results characterize the tail behavior of the marginal distribution F_X of X_t whenever $F_Z \in D(\Phi_\alpha)$.

Lemma 5.5.2. *If $F_Z \in D(\Phi_\alpha)$ then $F_X \in D(\Phi_\alpha)$. Moreover for some sequence of real numbers (u_n), with $u_n = a_n^* x$, $a_n^*, x > 0$, it holds that*

$$\lim_{n\to\infty} n(1 - F_Z(u_n)) = \tau \Rightarrow \lim_{n\to\infty} n(1 - F_X(u_n)) = \tau \sum_{i=0}^{\infty} b_i^{\alpha},\ \tau > 0.$$

In contrast, however, the analysis of the tail behavior of the marginal distribution F_X of X_t with F_Z satisfying (5.30) is completely different since in this case only the summands corresponding to the largest coefficients have influence in the tail.

Lemma 5.5.3. *Assume that the sequence of coefficients (b_i) in (5.33) are such that $b_i = O(|i|^{-\delta})$, as $i \to \pm\infty$, for some $\delta > 2$. In addition, define $b_{max} := \max_{0\le i\le\infty}(b_i)$ and $\Upsilon = \{i_1, \ldots, i_k\}$, $i_1 < \cdots < i_k$, as the set of indices such that $b_i = b_{max}$. Then*

$$1 - F_X(n) \sim \breve{K} n^{\breve{\xi}} (1 + \breve{\lambda})^{-n},\ n \to \infty,$$

for $\xi \neq -1$, where $\breve{\lambda} = \lambda / b_{max}$

$$\breve{\xi} = \begin{cases} k\xi + k - 1 & \xi > -1 \\ \xi & \xi < -1 \end{cases},$$

and

$$K^* = \beta_{max} K \left(\frac{1+\lambda}{\lambda + \beta_{max}}\right)^{\xi+1}$$

with

---

[7] In what follows we will omit the index t below the thinning operator if there is no risk of misinterpretation.

$$\breve{K} = \begin{cases} \breve{\lambda}^{k-1} K^{*k} \frac{(\Gamma(\xi+1))^k}{\Gamma(k(\xi+1))} E\left[(1+\breve{\lambda})^{\sum_{i \notin \gamma} b_i \circ Z_{-i}}\right] & \xi > -1 \\ kK^*(E[(1+\breve{\lambda})])^{k-1} E\left[(1+\breve{\lambda})^{\sum_{i \notin \gamma} b_i \circ Z_{-i}}\right] & \xi < -1 \end{cases}.$$

We now proceed to obtain the limiting distribution of the maximum term of the INMA(∞) model in (5.33), where (Z_t) is an i.i.d. sequence of non-negative integer-valued r.v's, with common distribution F_Z satisfying either (5.32) or (5.30).

Theorem 5.5.5. *Let (X_t) be the integer-valued time series sequence defined in (5.33).*

1. *Assume that $F_Z \in D(\Phi_\alpha)$, $\alpha > 0$ and that*

$$\lim_{n\to\infty} n(1 - F_Z(u_n)) = \frac{\tau}{\sum_{i=0}^{\infty} b_i^\alpha}, \quad \tau > 0,$$

for some sequence of real numbers $(u_n)_{n\in\mathbb{N}}$ where $u_n = a_n^* x$ with

$$a_n^* \sim \left(\sum_{i=0}^{\infty} b_i^\alpha\right) F_Z^{-1}(1 - n^{-1}).$$

Then

$$\lim_{n\to\infty} P(M_n(X) \le u_n) = e^{-\theta x^{-\alpha}},$$

where the extremal index θ is given by

$$\theta = \frac{\max_{0\le i\le\infty}(b_i^\alpha)}{\sum_{i=0}^{\infty} b_i^\alpha}. \tag{5.34}$$

2. *Assume now that F_Z satisfies (5.30). Then under the conditions of Lemma 5.5.3 with $k = 1$, it holds that*

$$\begin{cases} \limsup_{n\to\infty} P(M_n(X) - d_n \le x) \le e^{-(1+\breve{\lambda})^{-x}} \\ \liminf_{n\to\infty} P(M_n(X) - d_n \le x) \ge e^{-(1+\breve{\lambda})^{-(x-1)}} \end{cases},$$

with $d_n = (\ln(1+\breve{\lambda}))^{-1}(\ln n + \breve{\xi} \ln\ln n + \ln \breve{K})$.

Note that, whereas for the heavy-tailed case the extremal index is in general less than one reflecting the influence of the dependence structure on the extremes, the integer-valued moving average driven by a sequence of innovations with exponential-type tail has similar extremal behavior as an i.i.d. sequence with the same marginal distribution.

It is important to mention that the second statement in Theorem 5.5.5 shows the well-known fact referred to above that for distributions belonging to Anderson's class, the maximum limiting distribution cannot be expressed in a closed form. However, in order to overcome the presence of these limiting bounds, Hall and Temido (2007) proved that if instead of looking at the maximum term of the n observations we consider only the maximum term of the first k_n observations, with (k_n) being a nondecreasing positive sequence satisfying[8]

$$\lim_{n\to\infty} \frac{k_{n+1}}{k_n} = e^{\alpha}, \ \alpha > 0. \tag{5.35}$$

we can then derive a well-defined limiting distribution for the maximum term. Hall and Temido (2007) obtained the following results for the maximum term for integer-valued moving average driven by a sequence of innovations with exponential-type tail.

Theorem 5.5.6. *Let (X_t) be the integer-valued time series sequence defined in (5.33). Assume that F_Z satisfies (5.30). Then under the conditions of Lemma 5.5.3 with $k = 1$, it holds that there exist a nondecreasing positive integer sequence (k_n) satisfying (5.35) and a real sequence (d_n) such that*

$$\lim_{n\to\infty} P(M_{k_n}(X) \leq x + d_n) = e^{-(1+\check{\lambda})^{-[x]}}.$$

5.5.1 Some Extensions

In order to make the integer-valued time series models more flexible, several extensions may be considered. For example, it may be of interest to include covariates in the model to account for the dependence of the thinning probabilities on several factors. Alternatively, it is possible to study integer time series models with random coefficients. Further extensions may include periodic integer-valued sequences. Potential applications of integer-valued time series models with periodic structure can be found in the analysis of demographic data sets. In this section some results related with the extremal behavior of integer-valued moving averages with random coefficients and periodic integer-valued time series are presented.

Assume first that X_t admits the representation

$$X_t = \sum_{i=0}^{\infty} B_i \circ Z_{t-i}, \ t \in \mathbb{Z}, \tag{5.36}$$

[8]Note that condition (5.29) is necessary and sufficient for the existence of such a sequence and of a real sequence (u_n) such that $k_n(1 - F(u_n)) \to \tau > 0$, as $n \to \infty$.

where (Z_t) constitutes an i.i.d. sequence of non-negative, integer-valued r.v's, with common distribution $F_Z \in D(\Phi_\alpha)$, and (B_i) consists in a sequence of independent r.v's, being (Z_t) and (B_i) mutually independent. The thinning operator $\circ$ is defined as follows:

$$B_i \circ Z_{t-i} := \sum_{j=1}^{Z_{t-i}} \xi_{j,t}(B_i),$$

where $\xi_{j,t}(p), j = 1, 2, \ldots,$ are i.i.d. Bernoulli r.v's with success probability $p = B_i \in [0, 1]$. All thinning operators involved in (5.36) are independent, for each t. Nevertheless, dependence is allowed to occur between the thinning operators $B_{i_1} \circ Z_t$ and $B_{i_2} \circ Z_t$, $i_1 \neq i_2$ (which belong to X_{t+i_1} and X_{t+i_2} respectively). Thus we consider a general class of models consisting of all the stationary sequences defined by (5.36) for which the vector of terms $(B_0 \circ Z_t, B_1 \circ Z_t, \ldots)$ has some fixed dependence structure for every n, and all other thinning relations are independent. Furthermore, the sequence of coefficients (B_i) will be taken to satisfy

$$\sum_{i=0}^{\infty} E(B_i^\delta) < \infty, \; \delta < \min(\alpha, 1),$$

in order to guarantee the a.s. convergence of the infinite series in (5.36).

Motivation for considering this class of models with random coefficients comes from the desire to focus on non-negative, integer-valued time series assuming low values with high probability but exhibiting at the same time sudden burst of large positive values. Examples of applications can be found in the analysis of time series of count data that are generated from stock transactions, where each transaction refers to a trade between a buyer and a seller in a volume of stocks for a given price; see Quoreshi (2006) for details. As a particular example of times series exhibiting this type of behavior, a simulated sample path for the model

$$X_t = B_0 \circ Z_t + B_1 \circ Z_{t-1} + B_2 \circ Z_{t-2} + B_3 \circ Z_{t-3}, \tag{5.37}$$

where $B_i \sim U[0, 1]$, $i = 0, 1, 2, 3$ and $F_Z(x) = e^{-\frac{1}{[x+1]^{0.5}}}$, $x > 0$, is displayed in Fig. 5.2 below. The study of the tail properties of the marginal distribution F_X of X_t, is carried out in two stages: first, the tail behavior of $Y := B \circ Z$ is obtained; later we will obtain the tail properties of F_X. Hall et al. (2010) proved the following result.

Lemma 5.5.4. *Let Z be a non-negative integer-valued random variable with $F_Z \in D(\Phi_\alpha)$, $\alpha > 0$. In addition, assume that B is a random variable distributed over the interval $[0, 1]$. Then it follows that $F_Y \in D(\Phi_\alpha)$ and*

$$\lim_{n \to \infty} \frac{1 - F_Y(n)}{1 - F_Z(n)} = E(B^\alpha).$$

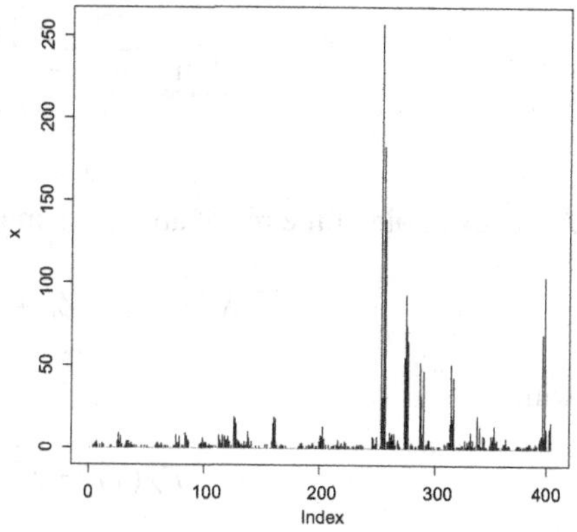

Fig. 5.2 Simulation results for model (5.37) with $n = 400$

The following results totally characterize the tail behavior of the marginal distribution F_X of X_t.

Lemma 5.5.5. *If $F_Z \in D(\Phi_\alpha)$ then $F_X \in D(\Phi_\alpha)$. Moreover for some sequence of real numbers (u_n), with $u_n = a_n^* x$, $a_n^*, x > 0$, it holds that*

$$\lim_{n\to\infty} n(1 - F_Z(u_n)) = \tau \;\Rightarrow\; \lim_{n\to\infty} n(1 - F_X(u_n)) = \tau \sum_{i=0}^{\infty} E(B_i^\alpha), \quad \tau > 0.$$

Our main task now is to derive the limiting distribution of the normalized maxima of the integer-valued moving average sequence (X_t). This is formalized through the next result.

Theorem 5.5.7. *Let (X_t) be the integer-valued time series sequence defined in (5.36). Assume that $F_Z \in D(\Phi_\alpha)$, $\alpha > 0$ and that*

$$\lim_{n\to\infty} n(1 - F_Z(u_n)) = \frac{\tau}{\sum_{i=0}^{\infty} E(B_i^\alpha)}, \quad \tau > 0,$$

for some sequence of real numbers (u_n) where $u_n = a_n^* x$ with

$$a_n^* \sim \left(\sum_{i=0}^{\infty} E(B_i^\alpha) \right) F_Z^{-1}(1 - n^{-1}).$$

Then

$$\lim_{n\to\infty} P(M_n(X) \le u_n) = e^{-\theta x^{-\alpha}},$$

where the extremal index θ is given by

$$\theta = \lim_{m\to\infty} \frac{\sum_{j=0}^{m} E\left(B_j^{\alpha} \prod_{i\neq j} F_{B_i}(B_j)\right)}{\sum_{i=0}^{\infty} E(B_i^{\alpha})}.$$

As an example of the result above assume that X_t admits the representation

$$X_t = B_0 \circ Z_t + B_1 \circ Z_{t-1}, \ t \in \mathbb{Z}$$

with

$$F_Z(x) = e^{-\frac{1}{[100x+1]^{\alpha}}}, \ x > 0$$

and $B_i \sim U[0,1]$, $i = 0, 1$. The choice of this distribution is justified by the fact that it is the natural integer analogue of the Fréchet distribution. From Theorem 5.5.7 the extremal index is given by

$$\theta = \frac{E[B_0^{\alpha} F_{B_1}(B_0)] + E[B_1^{\alpha} F_{B_0}(B_1)]}{E[B_0^{\alpha}] + E[B_1^{\alpha}]} = \frac{\alpha+1}{\alpha+2}.$$

Note that if $\alpha = 0.5$ then the extremal index $\theta = 3/5$ which is considerably smaller than one and hence reveals a relevant dependence of the high-threshold exceedances.

Next we turn our attention to the extremal properties of periodic integer-valued times series. We make use of the term periodic in a different sense than in the literature of periodic stochastic processes in which a sequence (Y_t) is said to be periodically stationary (in the wide sense) if its mean and autocovariance structure are periodic functions of time with the same period. This class of processes, however, does not appear to be sufficiently flexible to deal with data which exhibit non-standard features like nonlinearity and/or heavy tails, since in this case the autocovariance function is relatively uninformative and their empirical counterparts can behave in a very unpredictable way. Thus, by periodic sequence with period say T, we mean that for a sequence of r.v's (Y_t) there exists an integer $T \geq 1$ such that, for each choice of integers $1 \leq i_1 < i_2 < \cdots < i_n$, $(Y_{i_1}, \ldots, Y_{i_n})$ and $(Y_{i_1+T}, \ldots, Y_{i_n+T})$ are identically distributed. The period T will be considered the smallest integer satisfying the above definition.

In this case the setting is as follows[9]: let (Z_t) be a sequence of T-periodic independent integer-valued r.v's. We will assume that for some $\alpha > 0$

$$1 - F_{Z_r}(n) = P(Z_r > n) = n^{-\alpha} L_r(n), \ r = 1, \ldots, T, \ n \in \mathbb{N},$$

[9]Results for periodic sequences with exponential-type tails as in (5.30) can be found in Hall and Scotto (2006).

for some slowly varying functions $L_r : \mathbb{R}_+ \to \mathbb{R}_+$ $(r = 1, \ldots, T)$ at infinity, in short $F_{Z_r} \in D(\Phi_{\alpha,r})$, $r = 1, \ldots, T$. We further assume that the tails are equivalent in a sense that

$$\lim_{n\to\infty} \frac{1 - F_{Z_l}(n)}{1 - F_{Z_k}(n)} = \gamma_{l,k}, \quad (0 < \gamma_{l,k} < \infty) \ l, k \in \mathbb{Z}. \tag{5.38}$$

Note that $\gamma_{l,k} = \gamma_{l+T,k}$. Let (X_n) be a T-periodic non-negative integer-valued moving average sequence defined as in (5.33).

We start with the analysis of the tail behavior of X_r, $r = 1, \ldots, T$. In doing so, the following representation for X_r is very useful

$$X_r = \sum_{s=0}^{T-1} \sum_{j=0}^{\infty} b_{jT+s} \circ Z_{r-jT-s}.$$

From the representation above Hall et al. (2010) proved the following result which completely characterizes the tail behavior of X_r.

Lemma 5.5.6. *For the process defined in (5.33), it holds that for $r = 1, \ldots, T$*

$$\lim_{n\to\infty} \frac{P(X_r > n)}{P(Z_r > n)} = \sum_{s=0}^{T-1} \gamma_{r-s,r} \sum_{j=0}^{\infty} b_{jT+s}^{\alpha}, \ n \in \mathbb{N}.$$

Now we can obtain the limiting distribution of the normalized maxima of the periodic sequence (X_t). The result is as follows:

Theorem 5.5.8. *Let $(X_1, \ldots, X_n)$ be the T-periodic non-negative integer-valued moving average sequence defined in (5.33). Assume that $F_{Z_r} \in D(\Phi_{\alpha,r})$, $r = 1, \ldots, T$ satisfying*

$$\lim_{n\to\infty} n(1 - F_{Z_r}(u_n)) = \frac{\tau_r}{\sum_{s=0}^{T-1} \gamma_{r-s,r} \sum_{j=0}^{\infty} b_{jT+s}^{\alpha}}, \quad \tau_r > 0,$$

for some sequence of real numbers (u_n) and

$$\tau_r = \frac{\sum_{s=0}^{T-1} \gamma_{r-s,r} \sum_{j=0}^{\infty} b_{jT+s}^{\alpha}}{\sum_{s=0}^{T-1} \gamma_{1-s,1} \sum_{j=0}^{\infty} b_{jT+s}^{\alpha}} \gamma_{r,1} \tau_1,$$

with $\tau_1 = x^{-\alpha}$. Then, the distribution of X_r satisfies

$$\lim_{n\to\infty} n(1 - F_{X_r}(u_n)) = \tau_r, \ r = 1, \ldots, T.$$

If $u_n = a_n^* x$ *with*

$$a_n^* \sim \frac{1}{T}\left(\sum_{r=1}^{T} \gamma_{r,1}\right)\left(\sum_{j=0}^{\infty} b_j^{\alpha}\right) F_{Z_1}^{-1}(1 - n^{-1}),$$

then

$$\lim_{n\to\infty} P(M_n(X) \leq u_n) = e^{-\theta x^{-\alpha}}$$

and the extremal index θ *is given by*

$$\theta = \frac{\max_{0\leq j\leq\infty}(b_j^{\alpha})}{\sum_{j=0}^{\infty} b_j^{\alpha}}. \tag{5.39}$$

Note that the expression for the extremal index in (5.39) coincides with the one obtained by Hall (2001) in (5.34) for the stationary case. This means that the extremal index is not affected when considering the tail equivalence condition in (5.38). In other words, for these models the clustering tendency of high-threshold exceedances is completely determined by the coefficient values and the tail index of the marginal distributions in the same way as in the stationary case.

References

Ahn S, Gyemin L, Jongwoo J (2000) Analysis of the M/D/1-type queue based on an integer-valued autoregressive process. Oper Res Lett 27:235–241

Al-Osh MA, Aly E-EAA (1992) First order autoregressive time series with negative binomial and geometric marginals. Commun Stat Theory Methods 21:2483–2492

Al-Osh MA, Alzaid AA (1987) First order integer-valued autoregressive INAR(1) process. J Time Ser Anal 8:261–275

Al-Osh MA, Alzaid AA (1988) Integer-valued moving average (INMA) process. Stat Pap 29:281–300

Alosh M (2009) The impact of missing data in a generalized integer-valued autoregression model for count data. J Biopharm Statist 19:1039–1054

Aly E-EAA, Bouzar N (1994) Explicit stationary distributions for some Galton-Watson processes with immigration. Commun Stat Stoch Models 10:499–517

Aly E-EAA, Bouzar N (2005) Stationary solutions for integer-valued autoregressive processes. Int J Math Math Sci 1:1–18

Alzaid AA, Al-Osh MA (1988) First-order integer-valued autoregressive process: distributional and regression properties. Stat Neerl 42:53–61

Alzaid AA, Al-Osh MA (1990) An integer-valued pth-order autoregressive structure (INAR(p)) process. J Appl Probab 27:314–324

Anderson CW (1970) Extreme value theory for a class of discrete distributions with applications to some stochastic processes J Appl Probab 7:99–113

Andersson J, Karlis D (2010) Treating missing values in INAR(1) models: An application to syndromic surveillance data. J Time Ser Anal 31:12–19

Bakouch HS, Ristić MM (2010) Zero truncated Poisson integer-valued AR(1) model. Metrika 72:265–280

Blundell R, Griffith R, Windmeijer F (2002) Individual effects and dynamics in count data models. J Econom 108:113–131

Brännäs K (1995) Explanatory variables in the AR(1) count data model. Umeå Econ Stud 381:1–22

Brännäs K, Hall A (2001) Estimation in integer-valued moving average models. Appl Stoch Models Bus Ind 17:277–291

Brännäs K, Hellström J (2001) Generalized integer-valued autoregression. Econom Rev 20:425–443

Brännäs K, Nordström J (2006) Tourist accommodation effects of festivals. Tour Econ 12:291–302

Brännäs K, Hellström J, Nordström J (2002) A new approach to modelling and forecasting monthly guest nights in hotels. Int J Forecast 18:19–30

Brooks SP, Giudici P, Roberts GO (2003) Efficient construction of reversible jump Markov chain Monte Carlo proposal distribution. J R Stat Soc B 65:3–55. (With discussion)

Bu R, McCabe BPM, Hadri K (2008) Maximum likelihood estimation of higher-order integer-valued autoregressive processes. J Time Ser Anal 29:973–994

Cui Y, Lund R (2009) A new look at time series of counts. Biometrika 96:781–792

Du J-G, Li Y (1991) The integer valued autoregressive (INAR(p)) model. J Time Ser Anal 12:129–142

Enciso-Mora V, Neal P, Subba Rao T (2009) Efficient order selection algorithms for integer-valued ARMA processes. J Time Ser Anal 30:1–18

Fokianos K (2011) Some recent progress in count time series. Stat Pap 45:49–58

Fokianos K, Rahbek A, Tjøstheim D (2009) Poisson autoregression. J Am Stat Assoc 104:1430–1439

Freeland RK, McCabe B (2005) Asymptotic properties of CLS estimators in the Poisson AR(1) model. Stat Probab Lett 73:147–153

Garcia-Ferrer A, Queralt RA (1997) A note on forecasting international tourism deman in Spain. Int J Forecast 13:539–549

Geweke J (1992) Evaluating the accuracy of sampling-based approaches to the calculation of posterior moments. In: Bernardo JM, Berger JO, Dawid AP, Smith AFM (eds) Bayesian statistics 4. Oxford University Press, New York, pp 169–194. (With discussion)

Gomes D, Canto e Castro L (2009) Generalized integer-valued random coefficient for a first order structure autoregressive (RCINAR) process. J Stat Plann Inference 139:4088–4097

Green PJ (1995) Reversible jump Markov chain Monte Carlo computation and Bayesian model determination. Biometrika 82:711–732

Hall A (1996) Maximum term of a particular autoregressive sequence with discrete margins. Commun Stat Theory Methods 25:721–736

Hall A (2001) Extremes of integer-valued moving averages models with regularly varying tails. Extremes 4:219–239

Hall A (2003) Extremes of integer-valued moving averages models with exponential type-tails. Extremes 6:361–379

Hall A, Moreira O (2006) A note on the extremes of a particular moving average count data model. Stat Probab Lett 76:135–141

Hall A, Scotto MG (2003) Extremes of sub-sampled integer-valued moving average models with heavy-tailed innovations. Stat Probab Lett 63:97–105

Hall A, Scotto MG (2006) Extremes of periodic integer-valued sequences with exponential type tails. REVSTAT 4:249–273

Hall A, Scotto MG (2008) On the extremes of randomly sub-sampled time series. REVSTAT 6:151–164

Hall A, Temido MG (2007) On the maximum term of MA and max-AR models with margins in Anderson's class. Theory Probab Appl 51:291–304

Hall A, Temido MG (2009) On the max-semistable limit of maxima of stationary sequences with missing values. J Stat Plan Inference 139:875–890

Hall A, Temido MG (2012) On the maximum of periodic integer-valued sequences with exponential type tails via max-semistable laws. J Stat Plan Inference 142:1824–1836

Hall A, Scotto MG, Ferreira H (2004) On the extremal behaviour of generalised periodic sub-sampled moving average models with regularly varying tails. Extremes 7:149–160

Hall A, Scotto MG, Cruz JP (2010) Extremes of integer-valued moving average sequences. Test 19:359–374

Hall P, Heyde CC (1980) Martingale limit theory and its application. Academic, New-York/London

Hooghiemstra G, Meester LE, Hüsler J (1998) On the extremal index for the $M/M/s$ queue. Commun Stat Stoch Models 14:611–621

Ispány M, Pap G, van Zuijlen MCA (2003) Asymptotic inference for nearly unstable INAR(1) models. J Appl Probab 40:750–765

Joe H (1996) Time series models with univariate margins in the convolution-closed infinitely divisible class. J Appl Probab 33:664–677

Jung RC, Tremayne AR (2006) Binomial thinning models for integer time series. Stat Model 6:81–96

Jung RC, Tremayne AR (2011) Useful models for time series of counts or simply wrong ones? Adv Stat Anal 95:59–91

Jung RC, Ronning G, Tremayne AR (2005) Estimation in conditional first order autoregresion with discrete support. Stat Pap 46:195–224

Kedem B, Fokianos K (2002) Regression Models for Time Series Analysis. John Wiley & Sons, New York

Kim HY, Park Y (2008) A non-stationary integer-valued autoregressive model. Stat Pap 49:485–502

Lambert D, Liu C (2006) Adaptive thresholds: monitoring streams of network counts. J Am Stat Assoc 101:78–88

Latour A (1998) Existence and stochastic structure of a non-negative integer-valued autoregressive processes. J Time Ser Anal 4:439–455

Leadbetter MR, Lindgren G, Rootzén H (1983) Extremes and related properties of random sequences and processes. Springer, New York

Leonenko NN, Savani V, Zhigljavsky AA (2007) Autoregressive negative binomial processes. Ann de l'I.S.U.P LI:25–47

McCabe BPM, Martin GM (2005) Bayesian prediction of low count time series. Int J Forecast 21:315–330

McCormick WP, Park YS (1992) Asymptotic analysis of extremes from autoregressive negative binomial processes. J Appl Probab 29:904–920

McKenzie E (1985) Some simple models for discrete variate time series. Water Res Bull 21:645–650

McKenzie E (1986) Autoregressive analysis of extremes from autoregressive negative binomial processes. J Appl Probab 29:904–920

McKenzie E (1988) Some ARMA models for dependent sequences of Poisson counts. Adv Appl Probab 20:822–835

McKenzie E (2003) Discrete variate time series. In: Rao CR, Shanbhag DN (eds) Handbook of statistics. Elsevier, Amsterdam, pp 573–606

Monteiro M, Pereira I, Scotto MG (2008) Optimal alarm systems for count processes. Commun Stat Theory Methods 37:3054–3076

Monteiro M, Scotto MG, Pereira I (2010) Integer-valued autoregressive processes with periodic structure. J Stat Plan Inference 140:1529–1541

Monteiro M, Scotto MG, Pereira I (2012) Integer-valued self-exciting threshold autoregressive processes. Commun Stat Theory Methods 41:2717–2737

Moriña D, Puig P, Ríos J, Vilella A, Trilla A (2011) A statistical model for hospital admissions caused by seasonal diseases. Stat Med 30:3125–3136

Neal P, Subba Rao T (2007) MCMC for integer-valued ARMA processes. J Time Ser Anal 28:92–110

Nordström J (1996) Tourism satellite account for Sweden 1992–93. Tour Econ 2:13–42
Quoreshi AMMS (2006) Bivariate time series modelling of financial count data. Commun Stat Theory Methods 35:1343–1358
Raftery AE, Lewis S (1992) How many interactions in the Gibbs sampler? In: Bernardo JM, Berger JO, Dawin AP, Smith AFM (eds) Bayesian statistics 4. Oxford University Press, New York, pp 763–773
Ristić MM, Bakouch HS, Nastić AS (2009) A new geometric first-order integer-valued autoregressive (NGINAR(1)) process. J Stat Plan Inference 139:2218–2226
Ristić MM, Nastić AS, Miletić Ilić AV (2013) A geometric time series model with dependent Bernoulli counting series. J Time Ser Anal 34:466–476
Roitershtein A, Zhong Z (2013) On random coefficient INAR(1) processes. Sci China Math 56:177–200
Rudholm N (2001) Entry and the number of firms in the Swedish pharmaceutical market. Rev Ind Organ 19:351–364
Scotto MG, Weiß CH, Silva ME, Pereira I (2014) Bivariate binomial autoregressive models. J Multivariate Anal 125:233–251
Silva I, Silva ME, Pereira I, Silva N (2005) Replicated INAR(1) processes. Methodol Comput Appl Probab 7:517–542
Steutel FW, van Harn K (1979) Discrete analogues of self-decomposability and stability. Ann Probab 7:893–899
Thyregod P, Carstensen J, Madsen H, Arnbjerg-Nielsen K (1999) Integer valued autoregressive models for tipping bucket rainfall measurements. Environmetrics 10:395–411
Tjøstheim D (2012) Some recent theory for autoregressive count time series. Test 21:413–438. (With discussion)
Villarini G, Vecchi, GA, Smith JA (2010) Modeling of the dependence of tropical storm counts in the North Atlantic basin on climate indices. Mon Wea Rev 137:2681–2705
Weiß CH (2007) Controlling correlated processes of Poisson counts. Qual Reliab Eng Int 23:741–754
Weiß CH (2008a) The combined INAR(p) models for time series of counts. Stat Probab Lett 78:1817–1822
Weiß CH (2008b) Thinning operations for modelling time series of counts–a survey. Adv Stat Anal 92:319–341
Weiß CH (2008c) Serial dependence and regression of Poisson INARMA models. J Stat Plan Inference 138:2975–2990
Weiß CH (2009) Modelling time series of counts with overdispersion. Stat Methods Appl 18:507–519
Weiß CH (2013) Integer-valued autoregressive models for counts showing underdispersion. J Appl Stat 40:1931–1948
Weiß CH, Kim HY (2013) Binomial AR(1) processes: moments, cumulants, and estimation. Statistics 47:494–510
Ye N, Giordano J, Feldman J (2001) A process control approach to cyber attack detection. Commun ACM 44:76–82
Yu X, Baron M, Choudhary PK (2013) Change-point detection in binomial thinning processes, with applications in epidemiology. Sequential Anal 32:350–367
Zhang H, Wang D, Zhu F (2010) Inference for INAR(p) processes with signed generalized power series thinning operator. J Stat Plan Inference 140:667–683
Zheng HT, Basawa IV, Datta S (2006) Inference for pth-order random coefficient integer-valued autoregressive processes. J Time Ser Anal 27:411–440
Zheng HT, Basawa IV, Datta S (2007) First-order random coefficient integer-valued autoregressive processes. J Stat Plan Inference 173:212–229
Zhou J, Basawa IV (2005) Least-squared estimation for bifurcation autoregressive processes. Stat Probab Lett 74:77–88

Zhu R, Joe H (2003) A new type of discrete self-decomposability and its applications to continuous-time Markov processes for modeling count data time series. Stoch Models 19:235–254

Zhu R, Joe H (2006) Modelling count data time series with Markov processes based on binomial thinning. J Time Ser Anal 27:725–738

Zhu R, Joe H (2010) Negative binomial time series models based on expectation thinning operators. J Stat Plan Inference 140:1874–1888

Data Sets

Data set	Author, site or entity that provided the data
Monthly numbers of cases of Brucellosis, Typhoid, Hepatitis C and Leptospirosis	Report from the Division of Epidemiology Health General Bureau, Lisbon http://www.dgs.pt/
Canadian Linx	See Tong (1990)
Number of deceased observed in the district of Évora (Portugal)	INE – Statistics Portugal http://www.ine.pt/
PSI 20	http://www.bolsapt.com/historico/
Wolf's sunspot numbers	See Tong (1990)
Tagus river flow	www.snirh.pt and Macedo (2006)
Daily temperatures recorded in Lisbon	IPMA (Instituto Português do Mar e da Atmosfera)

Reference

Macedo ME (2006) Caracterização de Caudais Rio Tejo. Direção de Serviços de Monitorização Ambiental

Tong H (1990) Non-linear time series. Oxford Science Publications, Oxford

K.F. Turkman et al., *Non-Linear Time Series*, DOI 10.1007/978-3-319-07028-5,
© Springer International Publishing Switzerland 2014

Printed in the United States
By Bookmasters